PENGUIN BOOKS

PLANT AND PLANET

Anthony Huxley was born in 1920, the elder son of the late Sir Julian Huxley, FRS. He began his career as a gardening editor and writer in 1949, after working during the war in Operational Research with the RAF, and subsequently with BOAC. He worked on the long-established magazine *Amateur Gardening* until 1971, eventually becoming its editor – a position he relinquished to devote himself to full-time horticultural and botanical writing. Among the thirty-three books he has written or edited are *An Illustrated History of Gardening* (a study of gardening methods through the centuries), *The Penguin Encyclopedia of Gardening, The Macmillan World Guide to House Plants* and *Green Inheritance* (which supports the World Wildlife Fund's recent plant conservation campaign). Apart from gardening his great interests are photography and wild flowers, which – combined with extensive travel abroad – have resulted in notable field guides, including *Mountain Flowers, Flowers of the Mediterranean* (with Oleg Polunin), *Flowers of Greece and the Aegean* (with Dr William Taylor) and *Wild Orchids of Britain and Europe* (with P. and J. Davies). He is a member of the Royal Horticultural Society's Council and has been awarded the Society's Veitch Memorial Medal and their highest award, the Victoria Medal of Honour, for his services to horticulture and his contributions to horticultural literature. *Plant and Planet* is the culmination of all Anthony Huxley's experience, reading and first-hand study of plants in the wild.

ANTHONY HUXLEY

Plant and Planet

REVISED AND ENLARGED

PENGUIN BOOKS

Penguin Books Ltd, Harmondsworth, Middlesex, England
Viking Penguin Inc., 40 West 23rd Street, New York, New York 10010, U.S.A.
Penguin Books Australia Ltd, Ringwood, Victoria, Australia
Penguin Books Canada Limited, 2801 John Street, Markham, Ontario, Canada L3R 1B4
Penguin Books (N.Z.) Ltd, 182–190 Wairau Road, Auckland 10, New Zealand

First published in Great Britain by Allen Lane 1974
First published in the United States of America by The Viking Press 1975
Revised edition published in Pelican Books 1978
Revised and enlarged edition published in Penguin Books 1987

Filmset in 10/12 Aldus by
Rowland Phototypesetting Ltd
Bury St Edmunds, Suffolk
Reproduced, printed and bound in Great Britain by
Hazell Watson & Viney Limited,
Member of the BPCC Group,
Aylesbury, Bucks

Contents

Preface

This book has been written to stimulate interest in the world of plants and our perception of it. I have throughout assumed that the reader will have more than a passing interest in the plant world – he or she will be gardeners, amateur botanists or wild flower lovers, whose day-to-day contact with plants, together with echoes of school biology courses, gives them an almost subliminal understanding of plant arrangements. I have therefore summarized much basic detail of how plants function while concentrating on the more extraordinary aspects of that functioning. But at the same time the subject is virtually inexhaustible, and only some of an infinity of possible examples are included; and while there are inevitable technicalities much that is really complex, such as modern views of the internal workings of the cell, has been omitted. With the proviso of an intrinsic interest in the subject, I hope the book can be read and enjoyed by anyone with or without much technical knowledge.

The book begins with a brief exposition of the process of evolution and its vast time-scale, and of how plants evolved. In this section all the main plant groups mentioned in the text are briefly described. Then chapters on photosynthesis, structure, growth and its regulation – always with the emphasis on the ingenuity displayed in carrying out these necessities – lead on to other aspects of plant life. These include all-important reproduction with or without sex, and the dispersal of the consequent spores or seeds, so often in conjunction with animals; ageing, illness and death; the society of plants, their associations, how they live together in peace or in enmity.

The last five chapters form a distinct section and are in an altogether different key: they are about the impact of man – his exploitation, manipulation and destruction of the natural plant kingdom, the artificial plants he has created and the primitive ones he is trying to domesticate. The final chapter looks at the possibilities ahead, the options available to man and those being forced upon the plant world.

Where possible, colloquial plant names have been used; the Latin equivalents are in the index. Many of my examples, however, inevitably have no familiar name, and I have had to use the botanical Latin nomenclature, without stressing the synonymy which bedevils so much botanical literature. But since my aim has been to describe the structures, devices, mechanisms and adaptations that exist, the names of the plants concerned are not really important. Their absence would doubtless irritate the more botanical reader, but they are there mainly to help anyone who wishes to pursue any point further.

In brief, I have tried to make this a book which will awaken interest, create wonder and stimulate consideration about a form of life which is all about us, sharing our planet, and yet is taken far too much for granted.

I must acknowledge a debt of inspiration to *The Natural History of Plants*, by Kerner von Marilaun of Vienna, published in an English translation by F. W. Oliver in 1894. 'Kerner and Oliver', as it is usually called, has for many years been one of my favourite books. Likewise much inspiration has come from the writings of Dr E. J. H. Corner, notably *The Life of Plants* published in 1964; if I have quoted him a number of times it is because his vivid writing cannot be bettered, especially when he is describing some personally observed occurrence.

My grateful thanks are due to Mr Richard Gorer and Miss Janet Mortimer, who both read the typescript and made many pertinent comments; to Mr Desmond Meikle of the Royal Botanic Gardens, Kew, who most kindly checked the nomenclature; and to the late Miss Thalia Bence, also of Kew, who assisted me in indexing the colloquial names and in checking many points of detail. The index itself has been prepared by Mrs Dorothy Frame, for whose careful work I and the publishers are indebted.

*

When this book first appeared an American reviewer, a professor of botany, wrote that he 'had not previously believed that a book could (to paraphrase James Thurber) tell me more about plants than I wanted to know'. After rereading the text for this third revision, I remain quite unashamed at the professor's reaction, which no one else has echoed. I continue to be amazed and delighted by the innumerable variations in plant life forms and behaviour.

This reviewer also chided me for 'frequent anthropomorphisms'. This – the attribution of human characteristics or purposes to organisms like plants – is a matter, I think, of phrasing and approach. To suggest that plants have in certain areas 'experimented' with different mechanisms and outcomes may suggest mind and purpose to some, but it is not meant to. It is just a more direct way, in a non-academic work for the layman, of explaining plant life and bringing it more sharply into focus. But most of the time I have let the extraordinary facts speak for themselves.

*

In this third edition I have amplified a number of passages, mainly to bring the text up to date with scientific discoveries and deductions, in particular in dealing with the complex details of plants' internal communications systems, which are beginning to appear to depend on electrical impulses and potentials as much as on chemical ones, and to exhibit fascinating parallels with the methods occurring in animals.

Most of the necessary alteration has been made to the final two chapters, where I have attempted to deal with the interaction of plants and mankind, and the possibilities for both in the future. A great deal has happened in the eight years since my last revision, most of it, alas, very much for the worse. It makes depressing reading, but I have always considered that without looking at the latest and perhaps final era of plant evolution at man's hands, any account of the plant world would be incomplete.

I have tried not to overdo the tale of pollution and destruction, for it has been very widely written about in the last few years, and the conservation message has begun to be repetitive. But it has to be said that the ever-accelerating obliteration of plant habitats seems at present to be unstoppable. There is, unfortunately, all too little the individual can do about this devastation. But there are conservation bodies to be supported, notably the World Wildlife Fund and the International Union for the Conservation of Nature, and politicians to be incessantly badgered.

Part One

1 · *The Planet Sharers*

All around us on this planet Earth are plants. They were there before us: and today, everywhere we allow it, is the green of the plant world – in Amazonia, in Siberia, on mountains and in oceans. Even in the middle of towns there are trees we have planted and weeds which take advantage of the least opportunity.

What is this world of plants, this world which shares our planet, forming the one life-group whose mode of existence differs so much from all others, without bone, shell, muscle, blood or nerves?

This book takes what one might call the 'plant's-eye-view'. But the eye is also squinting at the human race. Plants do many things that we can and many that we cannot. Man is as the blink of an eye in evolutionary terms, while plants are the earliest form of life; it is a great pity that in this blink man has turned into such a scourge of nature. If we let wild plants exist they will survive, and thrive. But can, or will we?

It is easy to be both anthropocentric and anthropomorphic about any living thing, and all the more so when the things seem so different from ourselves, like that 'other kingdom' of plants. It is easy to say that a plant is vicious because it has thorns, or is a pest because it has good reproductive powers; but these judgements are in man's mind only. While I have tried to avoid being too anthropomorphic the occasional close resemblances between plants and animals (including man) have proved too strong a lure to avoid entirely.

There is more to it in any case than the plants seen by the casual eye. Water and soil teem with microscopic plants, some unicellular, some forming relatively uncoordinated groupings of cells gradually defining themselves into structures as they become more elaborate and visible. Even the one-celled plants are intricate and as beautifully designed for their purpose as the most complex, and they proliferate unceasingly and innumerably. Larger plants exist from the shallower seas to the shore, from the coast to high up on mountains, one kind or other finding it possible to live somehow in almost every natural situation that occurs on the land we also inhabit. Only about 30 per cent of the earth's surface is land, and of this land some 40 per cent is counted uninhabitable by ourselves, yet plants cover much of this terrain.

The really notable and unique characteristic of plants is their capacity to manufacture energy from light. In so doing the plants provide for every other living organism. Christ was in fact mistaken when he said that 'they toil not, neither do they spin'; for, as E. J. H. Corner has written, 'plants *do* toil – they spin the fabric of living matter'. Plankton in the sea makes food for fish and whales; land-living plants of every kind are food for herbivorous animals; and the marine or terrestrial carnivores eat the herbivores. 'All flesh is as grass' indeed, and this is something the human race cannot afford to forget – the quotation (from the First Epistle of Peter) continuing 'and all the glory of man as the flower of grass. The grass withereth, and the flower thereof falleth away.' The evolution of animals could not have occurred without that of plants before them, while in the aeons since the initial creation of life the two worlds have continually affected and conditioned each other.

The plants upon which we depend for food, and a great deal else besides, have been exploited and in some cases have been developed by man. A few 'camp followers' have arisen which grow only with cultivated crops. In general, however, the green kingdom in its proper wild sense is entirely separate from mankind, indifferent and sometimes even inimical to us. Plants arise, live, mate, fruit and die in their many different ways without needing any assistance from mankind at all, although many have developed liaisons with animal life – insects in particular – to assist in their mating, and many others take advantage of animals to disperse themselves.

It is indeed in taking advantage of other life-forms and of other circumstances that plants excel. They are superb opportunists, making the most of different combinations of water, air, soil and climate. Their grip on the planet, their capacities for colonization and their integration with the environment are due to an astounding diversification and variety.

There are some twenty different *classes* of plant alone. Whereas, going up the evolutionary ladder, the numbers of animals decrease (in a total estimated at 1,125,000 species, of which almost a million are insects, there are only 4,237 species of mammal, 193 species of monkey and ape) the species of 'higher' or flowering plants are relatively very numerous; a recent considered estimate is 250,000, but the number may be as high as 300,000.

Within each class of plants, and to a certain degree between groups of classes, activities such as respiration, feeding and creation of rigidity have a basic similarity, and sometimes in this book I have had to generalize about the ways in which plants meet life's problems. But as Theophrastus wrote in the third century BC, 'In fact your plant is a thing various and manifold, and it is difficult to describe in general terms.' Throughout, therefore, I have brought in examples, and if these sometimes appear to be exceptional or odd, deviations from the norm, it is the better to display the fantastic capabilities, adaptations and resilience of the plant world. The appearances of a plane tree or a fig, a petunia or a begonia, a cabbage or a maize plant, are presumably familiar enough; but there are plants which most of us have no chance of seeing and which therefore deserve to be singled out.

Although I have written much about the more recent and highly developed plants we cannot properly understand the evolution, growth habits and structures of these without some examination of the earlier, more primitive and less complicated, although even among these there are most ingenious mechanisms. Many of these simpler plants are in any case vital in the world context and some of the most elementary are indeed likely to be part of our human salvation. For many reasons they deserve consideration.

As I have already suggested plants resemble man in certain ways, or at least their 'inventions' are comparable to our own. Charles Darwin once wrote in his diary, 'prove animals like plants'. Plants seldom move from place to place but they have many powers of movement; in their reactions to external stimuli they do exhibit percipience of a kind. Like animals, too, they feed and breathe, mate and produce offspring. Though, as in most lower animals and birds, these activities are entirely 'programmed' in response to stimuli it is seldom that a plant is beaten by an unusual combination or sequence of circumstances.

It can be shown that interaction between zoology and botany in the past stimulated research in both subjects. Thus the discovery of the circulation of the blood led to that of sap movement. Sexuality in plants was demonstrated by 'castration' of tulips by removing their stamens, which resulted in a failure to set seed. Mobility and sensitivity to irritation were found in both kingdoms. Especially in their chemistry and physics, there is less difference between the kingdoms than early students and also philosophers believed, although analogies can be taken too far.

Certainly in their development and adaptability plants have never been outdone by animals. Indeed, if plants could think, one would marvel at their inventiveness. They display sophisticated examples of aerodynamics, hydrodynamics, structural engineering, insulation, plumbing and internal communications systems.

They bait, lure and delude animals to aid in their mating, often employing complex mechanical devices to ensure that, once in place, the animal will, willy-nilly, carry out the required function. Some flowers act exactly like prostitutes to this end; others might be termed illusionists.

Equally they have evolved means, ranging from the crude to the clever, of making certain that animals and the elements disperse their seeds or spores. In other cases dispersal occurs by various ingenious mechanical devices working on explosion, tension, jet reaction and so on.

As if that were not enough, many plants have means for spreading not associated with reproduction, separating or rooting themselves to make new individuals in such a way that, genetically speaking, some species may be considered virtually immortal.

Plants are not without their vices, both in our biased view and to their own kind. They can sting and scorch, scratch and cut; some contain deadly poisons, and a handful can trap and digest small animals. Many are parasites on their own relations.

In all forms of life nature is preoccupied with *how* to do things successfully or at least adequately; she does not explain *why* things are done, why structures and habits exist. Nature is always experimenting and changing; it often appears that she abhors perfect mechanisms and absolute principles. This is very much the case in the world of plants where change seems sometimes to occur virtually for its own sake. But we can speculate upon the elaboration of plant structures and devices or how they have achieved their fitness for purposes, even if in some cases the mystery is beyond man's imagination. One aspect on which I shall constantly harp is how nature 'hedges her bets' with alternatives.

I have already mentioned man's dependence on plants, which has led, during the tiny fraction of the total time it has taken man to become 'civilized', to a huge measure of control over them. He increases them artificially, cultivates them for maximum growth, and in particular selects and breeds for high yields and other desirable qualities, often producing strains which cannot survive in the wild. Besides this he is able, in an increasing number of cases, to control the growth of plants, from germination onwards; he dwarfs and enlarges them, makes them produce flower and fruit at specific times.

Although in nature animals are only occasionally markedly destructive to plants – for an equilibrium is usually established – under man's cultivation plants become more susceptible to attack by pests and diseases. Therefore man devises methods not only of growing plants more luxuriantly, but of protecting them as best he can from their enemies, a battle which he has by no means won.

At the same time man is infinitely careless of natural resources. He builds over plants or digs under them; he cuts them down or burns them when they are in the way, notably to make room for the artificial plants which his increasing population increasingly requires. Then he over-exploits soil and natural grazing, so that even weeds, those spivs of the vegetable world, may eventually succumb to extreme situations, with deserts as the end result.

But, given any kind of a chance, the resilience of the plant kingdom as a whole is remarkable. It can conceal man's monuments in a remarkably short time; it repopulates an atom-bombed atoll; it will attempt to regenerate in one way or another on an over-grazed steppe or a semi-desert.

All these aspects form the basic themes of this book, in which I have tried to explain and describe the ubiquitous world of plants that so many of us take for granted, yet with which we must co-operate if we are to survive, let alone retain pleasure in our surroundings. The study of plants under natural conditions is an aspect of ecology and environment, to use these now fashionable words. We are beginning to realize that every change made in our environment brings a series of further changes in its wake, and that these are frequently unfortunate to ourselves, to natural life and to the thin surface of the earth itself. Our resources are indeed surprisingly slender. The 'biosphere' in which life exists in our planet, even with the extremes of mountains and oceans, is the thinnest of layers which has been described as equivalent to a lick of paint on a football.

Seldom, alas, does our impact on the green world result in change favourable to it; man finds it hard to improve on nature. But, to put it bluntly, we *must* try to understand this world and to appreciate the effect of our activities upon it. We have to live within the earth's income, as it were, if we are in the long term to survive at all. There are ways of co-operating with the wild world which would provide not only for our own physical needs and commodities in part, but for our well-being. A garden may be a lovesome thing, and a green plant on a windowsill some refreshment to the office worker or housewife; but it is the big tracts of unspoilt natural landscape, plants, animals and all, that are most likely to refresh our tiring spirits and keep us sane.

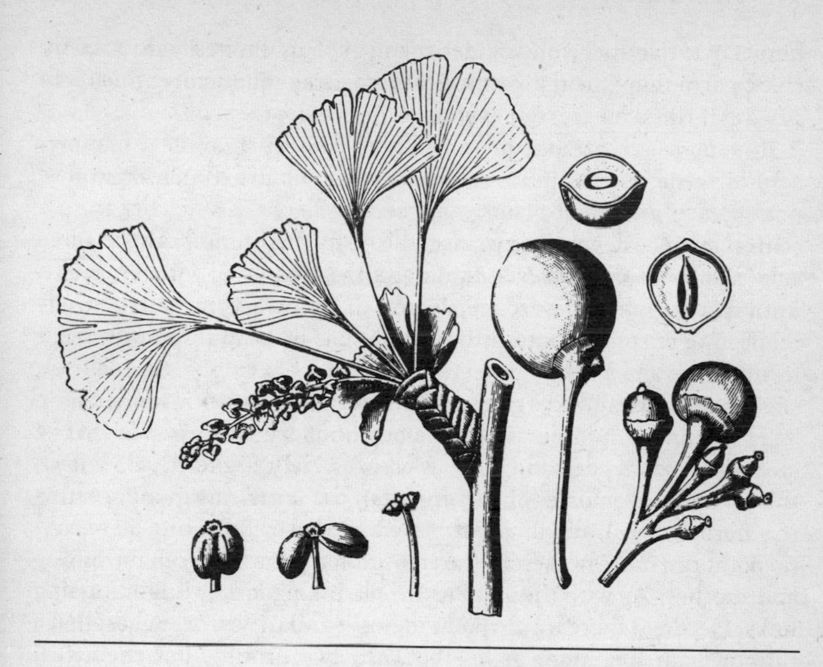

2 · *The Ways of Change*

'Many species . . . were simultaneously called into existence on the third day of creation each distinct from the other and destined to remain so.' The eminent botanists Hooker and Arnott wrote this in their *Flora* published in 1860 (a year after publication of *The Origin of Species*) – the very year that Charles Darwin and Thomas Henry Huxley exploded the myth that the existing order of life was the result of simultaneous creation in 4004 B C.

The plant kingdom as we see it today consists of many very different kinds of organism. Their connecting link is chlorophyll, that great invention of creation and the mainspring of this book, a pigment which makes use of solar energy to transmute carbon dioxide and water into food materials in the process called photosynthesis. In other ways plants may appear as different from each other as are barnacles and horses in the animal world. Yet they can be roughly fitted into a family tree by arranging them in an order of increasing complexity which is in general terms a measure of evolution, or development.

It is unwise to arrange present-day organisms in a linear manner, assuming that each simpler group gave rise to a more complex one. The present-day groups of plants must be considered as branches from a central stem of development. At the base of this stem there are three side branches: the bacteria, fungi and the elemental viruses, which until fairly recently were considered to be barely more than self-replicating chemical units, although there is no doubt that they are a form of life, and a very aggressive one at that – possibly reduced from slightly more complex organisms to undertake a parasitic life.

The main stem of our vegetable evolutionary tree ascends via the algae (e.g. seaweeds), and off it at progressively higher levels branch the bryophytes (mosses and liverworts), the ferns, the gymnosperms (coniferous trees), and finally the angiosperms or flowering plants. At no point can one say, here is the transitional point between one group and another. As with the animals the plant kingdom is full of missing links. Here and there we can point to close similarities, or suggest that a certain plant lies more or less between two groups; but the actual bridging from one group to the next remains conjectural.

Some of the groups that we see today may closely resemble their ancestors, and we might assume that these are lines which have remained virtually unchanged for many millions of years. Though this may be partly true of their external features the simpler plants are by no means living fossils; at least they have become adapted to changing external circumstances over millennia, and this plasticity is a very important part of evolution.

How do we know anything about the development of organisms over millions of years? Quite simply, from the record of the rocks which show fossils at various levels. Geologists have painstakingly charted and named these levels. As the diagram on page 34 shows, the oldest level, the Pre-Cambrian, ended about 570 million years ago, and extends backwards in time to rocks very roughly 3½ billion years old and the formation of the earth about a billion years before that. Developments in microscope techniques and highly specialized chemical analyses of rocks have made it possible to detect that organisms capable of photosynthesis existed more than three billion years ago.

As plants became larger leaves, shoots, cones and seeds were frequently preserved, more or less complete, by becoming carbonized in films of mud which eventually became rock. One must always remember what a haphazard business this is, and that leaves, stems and woody structures are much more likely to be preserved than flimsy flowers and similar readily decomposing fragments. Curiously enough, minute pollen grains, with their indestructible coats, persist indefinitely and form an infallible way of establishing the presence at a given time and place of any pollen-producing plant. The difference between the results obtainable from the pollen record and from the large fossils is shown by work on several thousand metres of Tertiary and Cretaceous sediments in Venezuela and Borneo, which yielded a detailed pollen record but virtually no macrofossils.

It is also necessary to remember that bogs preserve plant remains much more effectively than upland areas, where wind and erosion remove them, so that the fossil record may well favour mud- or bog-loving plants rather than those of mountain and desert. Therefore palaeobotany consists partly of observation and analysis of records, partly of inference and the drawing of parallels from what is clearly known. It is also necessary to bear in mind that even the earliest fossil record is likely to occur when the plant was already widely spread, and not when it first appeared.

The diversity of plants and the continuous change implied by this diversity must raise the question of how evolution operates. It does so first of all by the introduction of sexual reproduction which ensures regular recombination of the genetic characters.

The characters of any organism are derived from programming units known as genes which are carried on bar-like structures called chromosomes within the cells. The chromosomes are collected within a nucleus in all but very primitive organisms. The basic chemical structure of a chromosome is very similar throughout the plant and animal kingdoms. The evolution of any organism – its development into more complexity, or merely its capacity to withstand changing circumstances – depends initially on recombination of this genetic material. To ensure this the material has to be 'shuffled' as often as possible, and this is achieved in sexual reproduction. No shuffling can occur in simple division. In sexual reproduction, gametes from either parent, each with one set of chromosomes (haploid) fuse to produce a 'generation', such as the familiar seed, with two sets of chromosomes (diploid).

The other basic mechanism of evolution is the capacity of the genes – the units of heredity, determining the characters of the organism concerned – to mutate, or 'sport' as the gardener calls it. Mutation involves a change, most often minute but sometimes quite considerable, in the characters controlled by the affected gene. By occasional mutation the shuffling of characters which is provided by normal interchange of genes from the two parents in sexual reproduction is given the potentiality of variation, and in the course of time the permutations become astronomical in number. This is accentuated by the fact that, although an individual gene is associated with one or several specific characters, its actual influence over these characters is affected by the other genes involved. Changes are thus controlled by a gene-complex: a kind of collective decision. The gene-complexes help to smooth out the effects of mutations, and to make character changes gradual rather than sudden.

Mutation can also occur in non-sexual reproduction, especially where the organism, for instance a bacterium, is one which multiplies itself extremely rapidly. Certain modern scientists believe that sexual reproduction is in fact unnecessary and that evolution could have taken place by mutation alone.

The cause of mutations is usually obscure; it appears typically to be an internal fault which can be likened to the occasional slipping of a cog in a clock. It seems possible that external stimuli such as cosmic rays can also effect genes; certainly artificial radiation will produce mutations. Some of the most remarkable examples of this are the tortuous beeches of Europe. In these trees the branches writhe like serpents, often rejoining the trunk or other branches. All are about the same age, and they exist in Denmark, Germany, northern France and Brittany, mainly in small numbers within an assemblage of normal trees. The remarkable thing is that the localities are in a straight line, and one implication is that a radioactive meteorite passing over these places about a century ago affected young trees immediately below it.

Interestingly enough, the tortuous form is reproduced by seed from the affected trees.

Remarkable mutations are also reported from the Tunguska crater in the USSR, where in 1908 a meteorite, comet or other heavenly body – some even postulate a nuclear-powered spaceship from outer space – exploded and devastated an area of around 5,000 sq km of forest. Something – the radiation from the explosion, the fires or even the magnetic storm which followed it – has caused remarkable alterations in the flora, transforming an area of stunted trees into a tall forest, and the natural rate of genetic change has been estimated to have multiplied twelve-fold.

Mutations are basically random. The majority are either inconsequential or positively harmful to the organism; only a few are beneficial. The harmful ones become swamped under the control of the gene-complexes, or if they are really destructive the individual organism concerned is likely to perish. Beneficial mutations become assimilated. This is the essence of the process of natural selection oversimplified by Herbert Spencer as 'the survival of the fitter' (so often misquoted as 'fittest').

Any given gene can mutate as frequently as 1 in 2,000 individuals, whether these be plants, insects or mammals, but the average rate of mutation is of the order between 1 in 10,000 and 1 in 100,000. This is an adequate rate for changes to be assimilated in the evolutionary time-scale, and for the ever-increasing variability of the organism to become useful.

Periods of consistent conditions tend to favour the status quo; but when conditions change it is the plants which can mutate most quickly to adapt themselves that will survive, while those with a slow mutation-rate will lose out. Conversely, one can say that it is in 'difficult' conditions or habitats that new development goes on most quickly; new conditions create new opportunities for adaptable plants, such as those which endure the desert and mountain habitats created by geological and climatic upheavals. New conditions do not just involve such upheavals. The rise of the grazing animal, for instance, had a profound influence on the plant kingdom, notably in the reduction of forests and the increase of prairies and savannas composed largely of grasses.

When environments change, plants must either evolve or become static, which implies the road towards extinction taken by the venerable American bristle-cone pines or the maidenhair tree.

In animals a linear succession of development is often clear, as in that of the horse, and it is possible to say, like Gavin de Beer, that 'it is selection, not mutation, that determines the direction of evolution'. In animal terms mutation is change and, quoting again, 'no mutant gene has the slightest chance of maintaining itself against even the faintest degree of adverse selection'. Selection is in effect purposeful and, although it has been defined as 'a mechanism for generating an exceedingly high degree of improbability', it is *not* a game of roulette. Indeed, the more complex and detailed an adaption, the more improbable it becomes that it *could* arise by mere chance. H. J. Muller made an estimate of the number of chance beneficial mutations needed for an amoeba to evolve into a horse. The figure was of the order of 1,000 raised to the power of 1 million – which even the millions of years that have elapsed since amoebae began would not begin to encompass. But we have horses, so selection and purpose seem to have overridden mere chance.

But with plants one is forced to wonder whether a rather different process has taken place. There are two levels of evolution to be considered, the overall and the specific. In the former there are certainly general adaptive trends, such as those towards ultra-succulence in arid conditions, and yet one sees the ultra-succulents alongside grasses adapted only by having limited surface area and precautions against excess transpiration. And what is the point of all the variation in leaf shape? When one tries to analyse the differences between the flowers of different families one realizes what a variety of solutions has been found for a limited number of problems, what a kaleidoscopic reshuffling of a few repeated parts there are, reminding one of a football pools permutation. It is often difficult to imagine how the process of selection can have evolved characteristics which have a function today, as, for instance, with the flower of a bee orchid which deludes a male insect into believing that the flower is a female fit to mate with. The changes in the 'Ur-orchid' towards such mimicry seem almost incapable of having had functional value as they occurred.

However, most evolutionary biologists would agree with C. K.

Sprengel (except for his reference to the Creator) when he wrote in 1787 that he was 'convinced that the wise Creator of nature has brought forth not even a single hair without some particular design'.

But I must leave this perhaps contentious scene. Whatever their cause there is no doubt that mutations tend to 'add up' and direct a development once begun. It can be shown that such developments tend to move along lines of least resistance and, quoting W. F. Ganong,

> this means that when, through a change in some condition of the environment, the necessity arises for the performance of a new function, it will be assumed by that part which happens at the moment to be most available for that purpose, regardless of its morphological nature, either because that part happens to have already a structure most nearly answering to the demands of the new function, or because it happens to be set free from its former function by change of habit or for some other non-morphological reason.

Evolutionary trends seem often to be towards simplification of organs by reduction. Thus non-petalled flowers probably evolve from multi-petalled ones, unisexual from bisexual. At the same time, more of the reduced organs may be produced per plant. After such reductions, too, several reduced organs may combine to form a new, complex structure, as in the mechanisms which aid the distribution of grass seeds. One set of organs may develop while others remain static or may even degenerate.

As evolution in a group of related plants proceeds hybridization may occur between more primitive and more advanced members of a group, and this gives rise to complex combinations and innumerable further opportunities for variation and improvement. In any case originally similar populations may spread geographically, when they will tend to become physically separated. Barriers such as oceans, mountain ranges and deserts can divide a population and create new geographical races, the raw material from which new species or entities are formed (we need look no further than Darwin's original observations in the Galapagos Islands on finches and giant tortoises to see how a very small geographical separation produced new circumstances and hence distinct species). In a long-isolated land mass like Australia massive divergence occurs and unique families of plants – as of animals – appear.

One may also note how plants of quite different ancestry can, in similar habitats, evolve structurally in the same way, like the American cacti and the African euphorbias, both groups succulent to withstand arid conditions. This is called parallel evolution. Storage organs such as tubers have been repeatedly evolved as need arose, from primitive groups like hornworts and horsetails to many flowering plants. The parasitic habit has likewise been evolved on several distinct occasions.

As already mentioned static conditions tend to slow down the mutation rate, while in some cases an equilibrium is reached in which there is apparently a selective elimination of mutations. Thus the conifers, after developing actively for many millions of years, have remained virtually unchanged for the last 70 million. In fact, the longer a group of plants has existed the lower its mutation rate; as with animals, the rate of change, and that of producing new species, is highest in a group near the start of its development. Herbaceous plants are the most recent type and they are evolving much faster than other groups of higher plants. There are various reasons for this: the herbaceous plants can produce seeds in a year from 'birth' (annuals) or little longer (perennials); and they often live in unstable habitats as opposed to the relatively stable forests and woodlands inhabited by slow-generation trees and shrubs.

Sometimes, especially when new 'lines' are radiating from an unspecialized ancestor, mutation rates will be very high; it has been estimated that in certain circumstances a new plant species could evolve within fifty to a hundred generations.

The word species needs to be defined. It refers to a population of individual organisms which are true-breeding – that is, their offspring resemble them closely – and which have resemblances in structure, habit, flower shape and colour, while in close relations some or all of these characters will differ. Also, this population does not normally interbreed with any other. The extent to which one species differs from its relations varies considerably among families, and it is not completely immutable – the concept may permit some variations, as in flower colour and size. Further, as this chapter will have suggested, a species is only a stage in a never-ending process. Early stages of separation may lead to the botanist calling the different populations varieties of the same species.

We must always remember in any case that a species is a man-made conception, and the criteria for separating one population from another differ also in the minds of the botanists who do taxonomical research. The species is, however, an essential concept in handling and thinking about plants at any level.

It has also been realized quite recently that plants which can look the same, and may even have the same number of chromosomes (a useful diagnostic feature for the scientist) can behave quite differently from each other. Members of one apparent species may have high- and low-altitude forms, which cannot live in each other's habitats, or live in different latitudes so that the time of flowering and the length of day needed to ensure flowering are different.

Such entities tend to have intermediate forms, which makes it all the more difficult to define them except at the ends of their ranges, and they are normally spoken of as geographical races.

Other kinds of superficially similar race can also develop, and sometimes demonstrate very rapid evolution. Thus metal-tolerant plant races have evolved on smelting wastes in the Swansea valley within two hundred years; plants growing under galvanized wire fences in the Breckland have become tolerant of the zinc washed on to them within thirty-five years.

One problem arises when a plant can be shown to vary slowly but surely across a geographical range. Here the extremes may be given different specific names, but the centre of the range is composed of intermediates. Such a range as, for instance, found in rhododendrons and birches along the Himalayas is technically known as a cline, a word coined by my father, Julian Huxley.

Evolution, then, appears to carry out experiment after experiment, making the occasional breakthrough, and eventually solving environmental problems so that plants can occupy almost every kind of habitat and combination of external conditions. On another level the flowering plants put up a virtuoso display of variations on a theme – as to their leaf form, habit, and in particular flower design. But we see also how end results are often surprisingly similar in quite different groups and in very different areas – not only in function but in terms of their place in the environment or ecosystem. At the same time the reasons for character and functions in the animal world often seem to have different standards in the world of flowers. One is tempted to postulate a dynamic natural force – what Henri Bergson called *'l'élan vital'* – which cannot stop itself from infinitely producing novelties – a kind of cosmic doodling, an innate urge to change and modify. But in principle everything can be adequately explained, without any need to personify evolution, in terms of mutation and recombination of characters, the two essential factors in the ways of change of all living organisms.

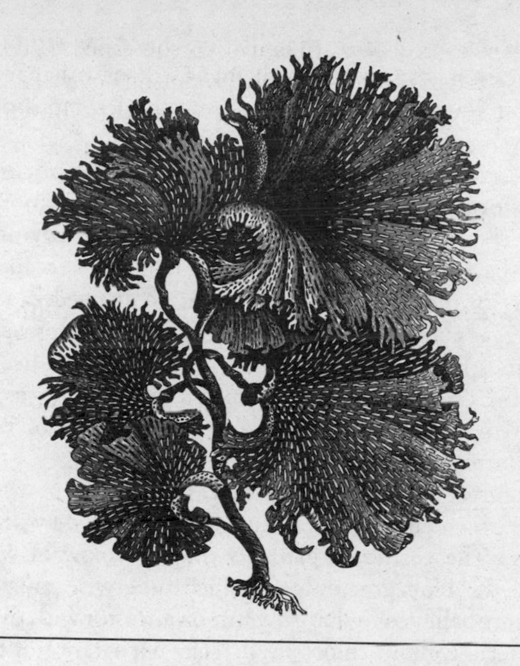

3 · *From Cell to Tree*

It is usually considered that life began in water, in the seas that formed as water vapour condensed around the cooling, recently formed earth. One can imagine these seas as a mineral-rich 'soup' in which, aided by the sun's energy, complex organic molecules could develop, notably amino-acids, essential to the formation of proteins. The presence of ammonia, derived from the atmospheric nitrogen, was probably necessary for the creation of amino-acids. Another theory is that lightning discharges through the nitrogen-dominated atmosphere on to water containing calcium carbonate could yield complex molecules from which chemical evolution and the earliest spark of life could have followed. Other theories locate the origin of life in clay crystals, or even in cosmic dust.

At any rate at some point the physical and chemical conditions and the energy levels were such that life – defined by the capacity for self-multiplication – sparked into being. Perhaps it happened many times, at first unsuccessfully because of the decomposition of the molecules.

The final successful reaction led, one assumes, to an organism which could divide for self-replication, and a surrounding membrane which enclosed its chemistry but allowed food materials to enter by permeation. A bacterium is just such a simple cell. To make use of minerals for its metabolism and duplication it needed a source of energy; whereas plants today use oxygen for energy, this was absent in the early atmosphere. Other elements must have been used, sulphur probably being one of the most important energy sources; there are still sulphur-oxidizing bacteria today. Their source of carbon was chemically formed organic matter.

The next great invention of creation was that of the pigment chlorophyll, which enabled organisms to produce food with the aid of solar energy. The earliest organisms that we know of with such a pigment are the blue-green algae, which otherwise much resemble bacteria and are believed to share a common ancestor with them. (Some modern bacteria contain chlorophyll.) The appearance of these algae suggests an oxygenic environment. The oxygen may have been generated biologically or by photodissociation of water vapour. Whatever its origin, an oxygenic atmosphere provides an ozone layer: this absorbs lethal ultra-violet radiation and thus allows the surface water to be colonized. Oxygen utilization in transpiration also increases available energy twenty-fold.

The blue-green algae formed felted mats in conjunction with bacteria, and successive layers of sediment upon these and new algal mats above created what are called stromatolites – quite large dome-shaped, columnar or conical structures the oldest of which, among rocks 3,000 million years old, are known from the western Australian coasts. Bacteria and blue-green algae have genetical or character-carrying material like all other organisms, but in such elementary organisms it is spread about in the cells: there is no well-defined nucleus. Such organisms are called prokaryotes.

In the next evolutionary jump – which took an estimated 2,000 million years – this genetical material became mainly confined and compacted with a nucleus. Cells with a nucleus are termed eukaryotes.

In the same way the chlorophyll is diffused in the blue-green algae, while in more developed algae, as in all other plants, it is confined in disc-shaped bodies called chloroplasts. It has been suggested that chloroplasts are the descendants of blue-green algae which penetrated the bodies of elementary animals and existed there symbiotically, each organism benefiting from the other, a situation which certainly exists today, as elaborated in Chapter 22. Eventually the algae lost any semblance of independence. In the same way it has been postulated that the mitochondria – other bodies within cells, both in plants and animals, which are involved in respiration and provide cells with their energy – are the descendants of certain bacteria which inhabited primitive organisms symbiotically.

Once users of solar energy – autotrophs, as they are called – had come into being the way was open for organisms which preyed upon them, so that they had body-building and energy-producing materials ready made. The words 'all flesh is as grass' thus took on meaning from the earliest days. It seems certain that such a second level of activity in the 'food web' increases the diversity of species in the original level, and this 'cropping principle' increased autotroph species in these early days as it did with much more advanced species in later epochs.

Spontaneous mobility was another experiment for the single-celled creations, and today we can still see both green and colourless single-celled animalcules, superficially very similar, swimming with whip-like flagella or cilia for propulsion. These are major components of the sea's plankton, existing together in the surface layers and forming the chief diet of many fish and even some whales. One wonders if any ancestor had the capacity both to photosynthesize and to ingest prey; the sole possible descendant today, the unicellular *Euglena*, can exist without its usual chlorophyll. Once these early plants and animals existed, bacteria presumably took to feeding upon their decomposing bodies after death, and at some stage to attacking them while alive, as they continue to do now. Some bacteria, like the purple *Rhodospirillium* of tropical coasts, can either photosynthesize or, if in the dark, gain energy from decomposing organic matter.

At this stage, then, the division between plant and animal is blurred;

an observer would have found it impossible to predict the development of plants, but might have supposed the most likely to be an increasingly complex mobile organism. Blue-green algae seem to make communities almost by accident; there is little if any communication between cells. Over 2,500 million years back, however, filamentous algae – strands one cell wide – and other structures began to appear. The mobile plants *did* develop into multi-cellular colonies such as the modern *Volvox*, self-propelled and combining many functions, but it seems unlikely that they ever grew much beyond this small size, about a millimetre across; it has been argued that the metabolic system of larger mobile conglomerations would have become unstable. At any rate the development of plants turned to the non-mobile, at this stage still floating, in which the energy once expended on movement went into more growth and better reproductive capacities. To develop the latter, a multi-celled structure with cells of different functions is desirable.

Sexual reproduction then began; it was an agent for improvement as compared with division or budding. I have described how sexual gametes from each parent fuse together; the gametes each have one set of chromosomes (haploid) and the result has two sets (diploid). Whereas in higher plants the sexual process is a relatively brief one carried out by transient haploid organs (as in flowers), borne on a relatively long-lived diploid plant, in most 'lower' plants there are two distinct generations.

This fact alone makes one wary of equating 'lower' with 'simpler', for the dual life-cycle of these plants can be complex. They involve a diploid non-sexual generation, the spore-bearer (sporophyte) in distinction to the haploid sexual generation, the gamete-bearer (gametophyte). The sporophyte produces spores non-sexually. These tiny spores, composed of a single cell, have the capacity to germinate, and when they do they give rise to the sexual generation.

Why two generations? Spores seem to be a logical extension of the budding of very simple plants such as yeasts, allowing the multi-cellular plants to continue producing single-celled reproductive 'buds'. Spores are designed for easy dispersal and are often resistant to hard external conditions, whereas sexual reproduction is often hazardous, especially in earlier plants such as seaweeds whose gametes have to swim in moving water.

Sex was at first very much an experiment. Nature as usual hedged her bets by providing an alternative means of dispersal and continuity. It is notable that the sporophyte is most frequently larger and longer lived than the gametophyte; in higher plants the sporophyte and the diploid seeds resulting from mating provide the entire life continuity.

A slow rate of evolution can often be correlated with poorly developed sexual reproductive methods. Mating also avoids the relatively unsatisfactory methods of many filamentous algae, in which every cell produces spores, leaving the original plant an empty skeleton. The persistence of the parent, with only some parts of it involved in reproduction, ensures continuity if conditions such as water turbulence arise to make it difficult for spores to survive. Both sporophytes and gametophytes may be capable of increasing also by completely non-sexual means.

At this point a simple figure may be helpful to show the time-scale of evolution, and the different epochs with their names and approximate ages. When one remembers that the whole evolution of man in anything like his modern form has taken less than a million years – represented by barely the thickness of a line in the figure – we can gain some idea of the immensely long periods of time involved, in which the most minute developments slowly add up to vast changes.

Beside the time-scale the initial appearances of major plant and animal groups are shown, as far as the fossil record can tell us. (See p. 34.)

The scale of the past is such that the largest part of the figure shows only the last sixth of the earth's biological history, some 500 million years in which most of the evolutionary development has occurred after an incredibly long period of very little change. The Pre-Cambrian all-plant era had few different niches, so there was no need for many species. Once plant-eaters appeared, as I have outlined, the balance was disturbed ever after, and new species appeared in ever-increasing numbers.

Fossils and calcareous algae are found in the late Pre-Cambrian period and algae have continued to develop for many of the intervening millennia. What we see today are very disparate groups, in hardly any

Million Years Ago	Geological Periods		Seas Widespread + or Restricted –	Plants		Other Organisms
↑ NOT TO SCALE ↓	QUATERNARY	TERTIARY (CAENOZOIC)	(Ice Ages)	Forests dwindle, herbaceous plants increase		Man
2	PLIOCENE					
7	MIOCENE					
26	OLIGOCENE					Whales, horses, early primates
38	EOCENE			Modern angiosperms well established		
54	PALAEOCENE					
65	CRETACEOUS	MESOZOIC	–	Diversification of angiosperms Earliest angiosperms	← Age of dinosaurs →	Large-scale extinction of animals due to climatic change
135	JURASSIC		+	*Ginkgo*		Birds
195	TRIASSIC		–	Gymnosperms (first conifers); cycads, lycopods and horsetails dwindle		Mammals Many early groups in decline
225	PERMIAN Pangea formed	PALAEOZOIC				Reptiles, insects, spiders
280	CARBONIFEROUS (COAL MEASURES)		+	Lycopods First seed-bearing plants (seed-ferns)		Amphibians
350	DEVONIAN		–	Ferns, horsetails Earliest forests First vascular land plants		Fishes
400	SILURIAN				← Marine algae prominent →	Ammonites
440	ORDOVICIAN		+			First vertebrates (fish-like) Corals
500	CAMBRIAN					Trilobites dominant Trilobite fossils
600			+			Multi-cellular invertebrates Plant-eating animals
1000	PRE-CAMBRIAN			Algae evolve First multi-cellular organisms First cells with nuclei (eukaryotes) – unicellular green algae; fungi – sexual reproduction starts		
2000	Free oxygen in atmosphere					
3000				Filamentous blue-green algae First unicellular organisms – prokaryotes – bacteria; blue-green algae; strombatolites Earliest supposed stromatolites		
4000	Earliest known rocks		+	Start of photosynthesis Earliest life forms 'Primeval soup'		

ORIGIN OF EARTH (4600 m.y.)

THE TIME-SCALE OF EVOLUTION

The lowest part of this chart, the Pre-Cambrian era, is on a compressed scale compared with the upper part.

way similar or interrelated, some of the minor groups being very intricate considering their small size, and with many different versions of sexual plus asexual methods of reproduction. Some have developed external skeletons of lime and resemble corals, building reefs in the same way. There is, indeed, among the present-day algae more diversity than in any comparable plant group, and in some cases, as with the familiar brown algae of the seaside, there are no obvious fossil or living ancestors for the well-developed forms that now exist. It is a group which has had an exceedingly long time in which to experiment, and has evolved very slowly and steadily in an environment where major change did not take place as it has on land.

The most advanced algae – the kelps and wracks of our shores, leathery and rubbery, which include one of the largest plants in the world, the giant kelp *Macrocystis pyrifera,* which is reputed to grow to 200 m long (perhaps not more than 60 m normally) – have a structure and a reproductive system comparable with those of land plants. They have a 'holdfast', root-like but with no absorbing function, and sometimes a distinct stem carrying the fronds. Indeed one can say that the jump from kelp to land plant is relatively small compared with the development of a single-celled alga into a kelp. But all the same the change from sea to land was a very considerable one: it was by no means just a case of a seaweed crawling its way up a shore-line until it no longer needed daily immersion. Indeed, it was the lack of an absorbing root which could absorb moisture from the soil which probably prevented the seaweeds from taking this jump.

First of all, remembering that much of the light striking water is reflected, and that what does enter the water is quite rapidly absorbed, the first thing a land plant has to face is a vastly greater amount of light. While this would improve the possibilities of growth by increasing the amount of photosynthesis, it also brings up problems of scorching and evaporation.

It has been shown that algae lose into the water that surrounds them a good deal of the food materials they produce by photosynthesis; losses of up to 30 per cent of the carbon fixed by the plants have been measured in fresh water. A land plant, most of whose structure is based upon carbon, cannot afford to lose it. Carbon retention and high photosynthesizing lead in turn to potential excesses of carbohydrates, which were used to produce thick cell walls, and to manufacture wood. This eventually enabled plants to raise themselves above damp habitats, stand up to dry conditions in air and wind, and once more attain considerable size.

Another problem for land plants was their immobility. An anchor in the sea is a good thing, for it prevents the plant being moved about at the mercy of tide or current out of its chosen set of conditions, being thrown up on a coast, or broken up by a gale. It also keeps it at the most appropriate depth. But once the plant is on land, immobility is potentially a hazard. Not only has the seedling no chance of changing its habitat if it proves unsuitable, but the adult cannot avoid natural catastrophes like fire or eruptions. Yet it must have roots capable of finding out and drawing up very large amounts of water. We can speculate on the mechanics of a walking tree or even invent mobile plants, as John Wyndham did with triffids. But leaving science fiction aside, we can place against the sometimes disadvantageous fixedness of the plant the often enormous mobility and dispersal capacity of its spores or seeds.

Despite their complexity, and their often remarkable powers of withstanding desiccation, it was probably not the seaweeds that gave rise to land plants but, from their similar photosynthetic pigments and photosynthetic mechanism, the green algae (*Chlorophyta*). There is one terrestrial green alga (*Cladophorella*), a branching filamentous plant of damp rock and mud, which makes an exterior skin of cutin, resembling that of higher plants. But one can only speculate about the actual succession.

The earliest land plants doubtless colonized shallow water that might dry out seasonally, wet places and bogs; over the course of yet more millennia, they spread away from these to drier habitats. One of their major problems was that they retained the reproductive system of mobile male gametes and stationary female gametes or eggs, which required moisture for the males to reach the females. Shallow pools might initially provide this, but eventually only a thin film of moisture on the plant's surface was required to allow the male gametes – by now equipped with powerful flagella – to swim to the eggs which eventually became enclosed in chambers, called archegonia, sunk into the plant surface. These flask-shaped chambers safeguard the egg and provide a rendezvous for the male gametes.

Another improvement in reproduction is that the gamete-producing generation becomes small and fast-maturing, so that gametes are available quickly. This speeding up means more generations in a given time and thus accelerates the possibility of further evolution.

In the Devonian period in the most obvious terrestrial successors to the marine plants, the mosses and liverworts, the gamete-producing generation is the dominant one. As mentioned above there are considerable similarities between mosses and liverworts on land and green algae, for instance, the photosynthetic pigments and mechanism, the materials of the cell walls, the type of food reserve and also the type and number of flagella on the gametes. But the bryophytes, as mosses and liverworts are termed, have little similarity with other land plants, notably in their vascular or conducting structure, and we have to consider them as a long-established branch of the main evolutionary stem. They have rather more cell differentiation than algae, and simple breathing pores, but they are small plants – the biggest known moss, *Dawsonia*, exceptionally has stems up to a metre long. They have root-like structures but these are primarily anchors on the damp soil, wet rocks, tree boles or decaying twigs on which they grow.

Liverworts make flat structures on the ground, often without much definite form, although one group has overlapping 'leaves' which look like those of flowering plants, and some branch in a symmetrical way. One exceptional liverwort is subterranean and without chlorophyll. They bear their reproductive apparatus aloft like miniature umbrellas; the spore-bearing generation is essentially a parasitic structure on the gamete-bearing plant.

Mosses live almost everywhere capable of supporting life. In the tropics mosses may actually live on the leaves of higher plants. They are also to be found in water. Their leaves begin to resemble those of higher plants, sometimes having midribs, and there is a primitive conduction system for water and nutrients which also gives a little mechanical strength. An exception to this is in the sphagnum mosses, in which the leaves contain large empty spaces, perforated in such a way that they can hold an amazing amount of water; you can squeeze and squeeze a handful of sphagnum and it will produce water more plentifully than a soaked sponge. Their appearance is such that one submerged species is popularly called 'drowned cats'.

Here we are faced with another great gap in the evolutionary story, for neither bryophytes nor algae offer an obvious ancestor for the vascular plants with well-defined conducting tissues, dating from late Silurian times, of which the most primitive present-day examples are the horsetails and club-mosses. All these vascular plants – which end up with the flowering plants – are characterized by the dominance of the spore-bearing generation and the eventual total submergence of the gamete-producing one as a separate entity. Xylem – woody conducting tissue – is the distinguishing common factor of all future developments in terrestrial plants, including many large groups known only from fossils.

At this point I must mention the fungi, the misfits in the order of things; they are certainly not animals but neither, since they do not contain chlorophyll, are they real plants. Like many bacteria, they either attack live hosts for food, when they are called parasites, or absorb food from dead matter, when they are called saprophytes. They have the power of digesting cellulose and lignin. In short they are the scavengers in the green plant's main world of light-fixers. Structurally fungi are simple, including single-celled spore-like yeasts, perpetually budding, but most roughly resembling some filamentous algae in being composed of a mass of cell-strands which form threads or spreading mats like fine silk, and are virtually undifferentiated most of the time. These strands do, however, give rise to the fruiting bodies in which the strands are formed into structures. These are fairly simple in microscopic fungi but can develop into elaborate mushrooms and 'toadstools' in many bizarre shapes, as well as the bracket fungi on trees and the plate-like 'fruits' of the dry-rot fungus, which can reach 250 cm in diameter. The structures are formed by the 'false block' method, in which filaments of cells, with the protoplasm passing through pores along the length of a strand, are pressed very tightly together, branching and interlacing so that rigidity is created.

Fungi have existed since Pre-Cambrian times, living in the sea as some still do today; they exist in fresh water and others live in and on the earth, plants, and decaying substances. Some can grow without oxygen; others can dissolve inorganic elements such as silica, iron and magnesium from rocks by releasing citric acid. They include diseases like rusts, smuts and moulds. Wherever there is something organic to attack, fungi will be found. The tissues of fossil vascular land plants are often impregnated with fungus threads (recognizable by their spores and reproductive organs), themselves sometimes suffering from secondary invasions of threads. Whether these early land fungi – many of which closely resemble present-day kinds – were parasites or saprophytes living on decaying tissue is not always clear, but some of their modern counterparts are certainly parasitic diseases, and some attack insects and other animals. Equally we can find quite early evidence of symbiotic root-fungi and also of lichens (as described fully in Chapter 22).

What are we to make of the fungi in evolutionary terms? They do not share with green plants the unique photosynthetic chemistry, and, although their cell walls usually contain cellulose as in most plants, they may also contain chitin, the substance used by animals such as insects and crustaceans to form their external skeletons. Can we possibly suggest that the ancestors of fungi were single-celled animalcules which became filamentous? It seems more likely that they originated from the filamentous algae, becoming adapted to living in the dark and absorbing food in distinct ways, and hence losing their chlorophyll as the need and indeed the possibility of photosynthesis disappeared. E. J. H. Corner has suggested that it should not be beyond the capabilities of scientists to introduce chloroplasts into fungus cells and so turn the clock back.

Had one been writing this book 200 million years ago, one would unhesitatingly have said that the club-moss group was one of the most successful. Superficially like mosses, they had a fairly complex conducting and strengthening structure, an efficient reproductive mechanism, and a capacity to produce forests of tree-like plants suitable to the lush growing conditions in many parts of the world. It would have seemed a climax without any possible alternative.

At the same time there were tree-like horsetails up to 30 m high, some with massive trunks rather like those of a modern palm, and numerous erect branches. These are quite different from club-mosses in structure, with a ring of vertical air canals and, between these and the hollow centre, another ring of stiff conducting bundles; the strength of the stem lying largely in the rigid exterior, which is reinforced with silica. The leaves are reduced to scales.

The antique club-mosses, horsetails and another, now extinct, group, the calamites, dominated the earth's vegetation. But, as later with the dinosaurs, climatic change found these giants unable to adapt themselves sufficiently rapidly. They could not withstand the increasingly dry conditions, and soon their massive corpses were making rich fossil deposits in the shales we now call the Coal Measures.

Though many club-mosses of today are very tough plants, found for example in the Arctic, they are quite unimportant in relation to the whole plant world, both in numbers and in size. Horsetails equally have no present dominance, but they have not shrunk so much: there is one (*Equisetum giganteum*) reputed to reach 10 m, though most are smaller. They are successful, invasive plants, which many gardeners know and curse because of their fearful persistence as weeds.

These two groups made a major evolutionary step in that they were the first in which (in a few species) the gamete-producing generation became entirely subordinate, being carried on the spores. The young sporophyte grows out of the spore – which contains food reserves – very much like a germinating seedling in a higher plant. This allows the spores to be far more independent of conditions of moisture or dryness. However, this step seems to have occurred too late for these groups to have taken full advantage of it in evolutionary terms.

Fern ancestors are as ancient as the bryophytes, club-mosses and horsetails. First known from the later Devonian period, they carried on successfully in the climatically varied epochs which followed; today they are found in most parts of the world, often as forest undergrowth and also growing on mossy trees or in their branch-forks. They have fairly well-developed roots, although these are used mainly for anchorage, and a rigid, elaborate conduction system. Some grow into palm-like trees up to 18 m tall; a few grow in water. Their sexual generation is a small 'prothallus': the dominant generation is the spore-bearing one. Most ferns carry spores on their leaf-like fronds; in a few, like royal fern and moonwort, a separate stem carries clusters of spore capsules, almost a premonition of the flowering plant.

All the plant groups so far described rely on spores for their dissemination. The next evolutionary jump is the production of seeds.

Although some seeds are little more than improved spores, they typically contain a ready-made embryo plant and a food reserve to start the infant on its way; also, they are usually enclosed in at least one protective layer.

Quite early in the fossil record are found fern-like plants with nut-like seeds, which probably had a common ancestor with the spore-bearing ferns. However, these faded out without any clear line of descent to the modern seed-bearing plants. The fossil record is full of such tantalizing morsels, in which a plant may resemble an earlier one in one feature but not in another, just as the algae diverged greatly in the sea, land plants did so in the much more varied conditions on dry land, and one must imagine different groups of plant carrying one primitive feature or another onwards and evolving different ones around it – a hotbed of trial and error.

In the seed-bearing plants – the paramount members of the plant kingdom today – the seed-carrying stage, which we can still call the sporophyte, is totally dominant. There is no independent other generation, however insignificant, and the male and female gametes are minute, the female always being carried on or within the sporophyte and having no individual existence. The male gametes are pollen, dispersed in a variety of ways.

Seed-bearing plants differ further from most spore-bearers in having a well-elaborated conducting system which also strengthens them, so that they include really big trees which can withstand gales, and in which water and nutrients can move from root to tops and vice versa.

The present-day seed-bearers are divided into two main groups, the gymnosperms or 'naked-seeded' and the angiosperms or 'enclosed-seeded'. In general the gymnosperms carry their ovules, which develop into seeds, on leaf-life cone-scales; in fact they are usually as well protected as in angiosperms, but in a different way, and when the seeds are ripe the cone-scales simply open out to release them, in contrast to the often elaborate devices resorted to by the angiosperms. In the latter the ovules are almost always carried within closed structures (carpels).

The earliest seed-bearing plants still with us are the conifers which originated in the Carboniferous era; the palm-like cycads and the maidenhair tree, *Ginkgo*, virtually unchanged since its earliest fossils, date from the late Palaeozoic. All these retain 'memories' of the ancient seas in which their algal ancestors mated, in the shape of a 'pollination drop' of fluid at the neck of the egg-chamber. In cycads and *Ginkgo* it may take three or four months for pollination to occur, and the germinating pollen grains give rise to mobile male gametes like sperms – sometimes large enough to be visible to the naked eye – which swim for weeks before achieving fertilization of the female nucleus. The cycad gametes (antherozoids) look like the whipping tops children used to play with, activated by flagella in a spiral groove.

In conifers a long period elapses – in pines a whole year – while the pollen tube grows towards the female nucleus to liberate its sperm nuclei for final fusion. This intermediate device enabled all seed-bearing plants to escape from the tyranny of external moisture for reproduction.

Cycads, incidentally, have a sexual apparatus much larger than in any other living plant. Individual ovules can reach 6 cm long (*Macrozamia*), while the female cones which house them, often bright yellow or orange, can weigh up to 45 kg (*Encephalartos*). The egg cells themselves are up to 6 mm across and their nuclei may reach ½ mm. *Ginkgo* produces large egg-like seeds about 2 cm long, with a thin rather soft 'shell' and a distinctly soft interior.

Some of the fossil cycads differed considerably from those we know today, and indeed one group, the *Bennetitales*, almost appear to be prototypes of flowering plants. Their sexual parts were enclosed in leaf-like organs, while their ovules had scales making them comparable with carpels. One group resembled a buttercup in shape and pattern. However, there are too many technical differences to say that there is a direct link with flowering plants, and the *Bennetitales* and early cycads apparently became extinct as the first of the angiosperms developed.

Cycads, like the three very anomalous, distinct plants classified as *Gnetales*, probably descended from seed ferns; conifers (which hardly need description) and *Ginkgo* derive from different branches of an extinct family, the *Cordaitales*, whose own beginnings are obscure. *Gnetum*, one of the *Gnetales*, is a genus of lianas, trees and shrubs, and although its method of reproduction is clearly gymnospermous it has flat leaves with netted veining – the only gymnosperm with this character, although *Ginkgo* has flat, fan-shaped leaves with branching but not netted veins. These cannot be called missing links, but they are clearly intermediates of a kind.

Although long static in evolutionary terms, the conifers are as highly successful today as they were in the times of the dinosaurs, and have been given a new lease of life, if one were needed, by their present utility to man. Their strong, flexible woody structure was a great improvement on the rigid, brittle ramifications of tree-like club-mosses and horsetails, and enabled them to maintain their size during the great geological upheavals and corresponding climatic changes that occurred at the end of the Coal Measures.

Such were the antecedents to the flowering plants, latest of the great plant groups to develop while most of these earlier lines continued with little change.

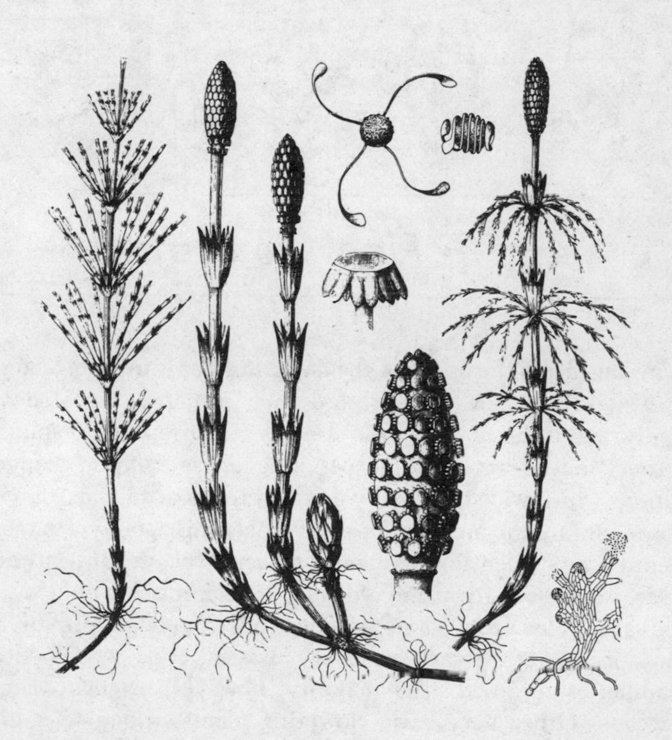

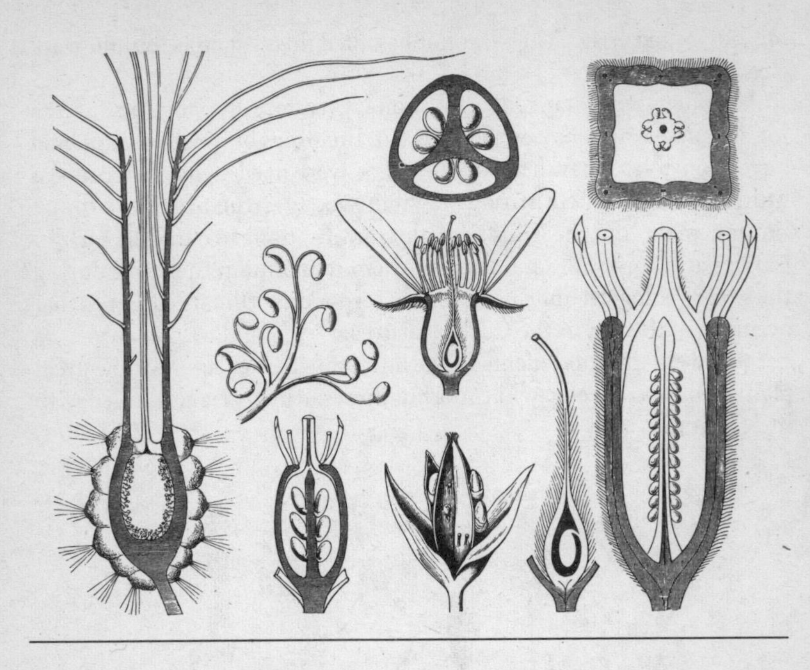

4 · *The Final Flowering*

Darwin considered the origin of the flowering plants to be an 'abominable mystery'. In the fossil record they certainly appeared very suddenly, and at a time when little else was occurring on the evolutionary front; indeed, in the same period as an apocalyptic change of conditions which wiped out many diverse groups of animals, of which the dinosaurs are the most notorious. Perhaps this apocalypse actually accelerated the sudden development of flowering plants. But once again there are no obvious fossil antecedents, no clear 'missing links'.

The earliest flowering plants, as we can tell from the records of fossil pollen, appeared at the very end of the Jurassic period, when pollen of water-lilies is recorded. The majority, however, originated in the Cretaceous. Other very early flowering plants belong to a group represented today by magnolias and buttercups. These flowering plants, or angiosperms, had a good spring-board. They inherited from the gymnosperms their already highly developed seeds, conducting and strengthening tissues, leaves and branch systems, which they proceeded to refine. At the same time they elaborated the reproductive structures, namely the flowers. I mentioned in the last chapter that the seeds of angiosperms develop within closed structures called carpels, initially containing embryo sacs; one part of these carpels developed a receptive surface, called the stigma, for pollen grains, and the pollination drop of the gymnosperms disappeared. The pollen itself is never mobile.

The angiosperm flower is usually much more elaborate than that of gymnosperms; it is a condensed structure, built up at least in part from modified leaves, which protects and deploys pollen and ovules, and is typically designed to secure fertilization, although in cases like grasses and spurges it may be reduced to the bare essentials of male pollen-carrying stamens and female pollen-receptive stigmas.

Some angiosperms are wind-pollinated, but the majority rely on animals. Indeed, the evolution of the flower was so rapid because of this. Insects in particular, which first appeared in the Carboniferous, evolved fast: they sought food, and at first went after pollen. To distract them from this plants developed nectar as a substitute, while their flowers became increasingly insect-attractive with scents, colours and guide-patterns. Some plants became dependent on single insect species, so that complex mechanisms for that insect alone would develop. The vast number of insects, and later other animals such as birds and bats, made possible, or rather necessary, a much greater number of new adaptations than had the relatively few combinations of climate and topography for earlier plants.

The angiosperms equally refined the relatively clumsy microscopic mechanisms of fertilization of the gymnosperms; to quote Bell and Woodcock, 'Compared with the cytological elegance of fertilisation in an angiosperm, the clumsy antherozoid sperm of *Cycas* is not only barbarous, but also primitive.' With their effective reproductive systems the angiosperms could reproduce virtually anywhere on dry land. Finally, the group evolved innumerable and often very ingenious ways of ensuring seed dispersal, as well as developing the possibilities of non-sexual reproduction by vegetative means.

The flowering plants evolved fast, and by the middle of the long Cretaceous period they were dominant. Speedier reproduction and development of new individuals means continually greater potential for change. The group has a far greater range of life-forms than any other, as to size, habit of growth, and duration of life, all of which have a vital bearing on adaptability to almost every climatic and topographical niche.

A few of the earliest plants are still recognizable today, like magnolias and water-lilies, but the main ancestors of present-day angiosperms seem to have been another group of plants with no close resemblances to their ultimate descendants. Many of the earliest flowering plants have completely died out: the early Cretaceous flora must have been notably different in appearance from that of today.

There are two major, distinct groups of flowering plants, the monocotyledons and the dicotyledons. These words categorize the seedling according to whether it has one or two seed leaves or cotyledons; these seed leaves are incidentally usually quite different in shape from those of the adults in dicots (to use the habitual abbreviation). It is possible that the shapes of the seed leaves indicate those of the plant's ancestors, just as gills in mammalian foetuses point to the ancestral fish. Leaves in monocots are typically narrow and elongated, with parallel veins, as opposed to netted veins and leaves of most other shapes in dicots. The stem anatomy is distinct, as described in Chapter 6, while the monocots almost always have flowers with parts in threes or multiples of three, the dicots more often having four, five or indefinite numbers of parts.

None of these differences provides an infallible guide to assigning a plant to one category or another but such a distinct impression is given by a plant – which Ronald Good has expressed as a king of 'streamlining' in monocots – that anyone who handles plants regularly is rarely outside the tropical forest and steadily colonizing less easy habitats; all available into chemical energy is on average less than 1 per cent – much more varied than the monocots. It seems likely that the separation of the flowering plants into these two distinct groups took place quite early in their evolution, and equally likely that the monocots descended from the original dicots.

It is by examining the changing structure of the wood, or xylem, that we can in fact work out most accurately the evolutionary status of any family. Relative primitiveness can also be deduced from the seed, whose enclosure, in a box-like carpel, segregates the flowering plants from the naked-seeded gymnosperms and the spore-bearers. The simplest carpels resemble the enclosure of an object in a piece of pie-crust, which is simply folded over so that the edges of the crust meet on one side. It is along this join that the stigmatic surface (receptive to pollen) appears in more primitive flowering plants. Later, the stigmas are carried on top of the carpel, or in specialized structures above it.

Flower characters may also point to antiquity of origin. The flower contains various distinct sets of parts: sepals, petals, stamens, stigmas and carpels. In more primitive flowers, as indeed in the early water-lily, one set may merge gradually with the next and not be sharply differentiated (inner petals may merge with stamens, outer petals with sepals); there are likely to be many of each part, and each individual part will be separated from its neighbours. Flowers with apparently fused parts (like the trumpets which the petals of, say, a bindweed compose) are likely to be more advanced.

Another distinct evolutionary tendency, at least in certain groups of families, is to produce numerous flowers in a compact head, eventually producing a 'capitulum' as in the daisy family, or in sea hollies of the cow-parsley family.

When an organ has become modified for a completely new function, as the leaves of pitcher plants and sundews have for trapping insects, or when a plant changes its entire metabolism, as in saprophytes (plants without chlorophyll) and parasites, or chooses a novel habitat, as in epiphytes (plants growing upon trees), these are equally likely to be relatively recent.

It seems probable from the available evidence that the earliest angiosperms were trees, which probably arose in the warmer parts of the globe, forming tropical forests somewhat similar to those of today. From these there developed liana-type climbers, shrubs, and eventually herbaceous plants – those without woody stems, dying down to a crown each season, with shorter life cycles and thus more capacity for change. Annuals, which are also herbaceous, probably came later still, and have of course the fastest potential for change of all. The trees themselves diverged from those carrying large fleshy fruits appetizing to animals, to those with smaller dry seeds, capable of being carried outside the tropical forest and steadily colonizing less easy habitats; all the trees of sub-arctic areas today have small dry seeds. Further changes were initiated by the appearance of grazing animals, which discouraged trees and encouraged plant structures constantly renewed from the base as in the grass family. Flowering water plants probably did not, as might be surmised, develop from more primitive aquatics; they are almost certainly herbaceous plants which re-migrated into both fresh water and, exceptionally, the sea.

The development and spread of the angiosperms took place concurrently with the last important phase of geological change on earth, involving mountain formation including the Alps and Himalayas in the Miocene, and the establishment of the continents and oceans roughly in their present outlines in the Pliocene. Such conditions bear witness to the adaptability and colonizing powers of the flowering plants.

As has been suggested, the earliest fossils are those of what we now regard as the most primitive flowering plants, the *Ranales*, of which the buttercup and water-lily families, magnolias and tulip trees are examples, although very different in many ways. Trees like poplars and planes are also known from the earlier Cretaceous. It seems likely that the highly organized composites (daisy family), orchids and grasses began to emerge towards the end of the epoch.

The Eocene saw many basically modern types of plant well established. Leaves of the small tree *Cercidiphyllum* from this time, for instance, are almost unchanged in present-day species. Deciduous trees began to dominate. If anything, these eras are periods of geographical movement: the Eocene clays of London contain a remarkable number of fruits and seeds and show that the flora of southern Britain then resembled that of present-day India and Malaya. What is today the Arctic Circle had a flora like that of Europe now; north of the present circle vast forests of the dawn redwood, a deciduous conifer only discovered in 1944 after being considered extinct, flourished together with deciduous trees.

The forests began to diminish and the grasslands to spread in the Miocene, as the grazing mammals became more important. In this period many entirely new plant families make their appearance.

Among the plants were many modern prairie grasses, bulrushes, and a host of familiar deciduous trees: beeches, maples, walnuts, oaks, sweet gum, sour gum, Judas tree, mock orange and so on. Conifers persisted, pines, redwoods and cedars, for instance, being plentiful.

In the Pliocene Europe still had plants which are now confined to China and North America. During this period the plant species we know today began to settle down. It was during the second part of the Quaternary that the icing up of the polar regions narrowed the tropics and established the northern and southern temperate zones, driving tropical and sub-tropical plants into the band around the equator. Successive ice ages forced temperate plants into much smaller areas than previously, and eventually, after a rapid reafforestation about 8,000 years ago, produced the distribution we see today, sometimes remarkably discontinuous. The distributions of certain related plants living in southern South America, South Africa and Australia, and also in central Africa and central South America, have helped to confirm the concept of continental drift.

This concept is that the present-day continents were formed from two larger landmasses, Laurasia in the north, Gondwanaland in the south, about 280 million years ago. These themselves began to split apart about 190 million years back, at the beginning of the Jurassic, from a single 'supercontinent', Pangea or 'all-earth', situated in a very extensive ocean.

As the latter part of this book will show, man is today the most important agent furthering evolution and distribution. Apart from his activities, which speed up the processes to an enormous degree for some plants, there is no reason to doubt that some natural evolution continues in the sense that new wild species are emerging.

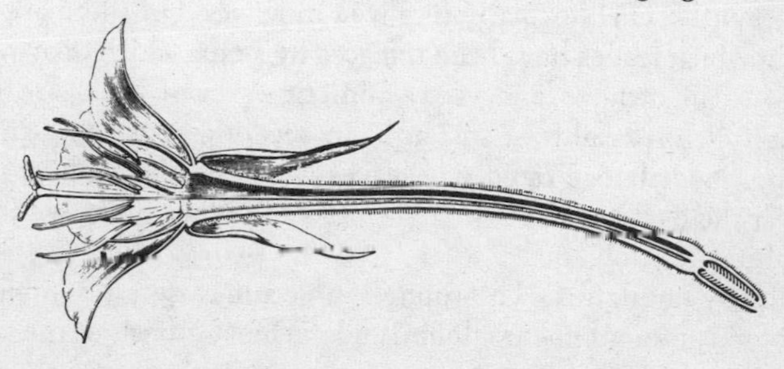

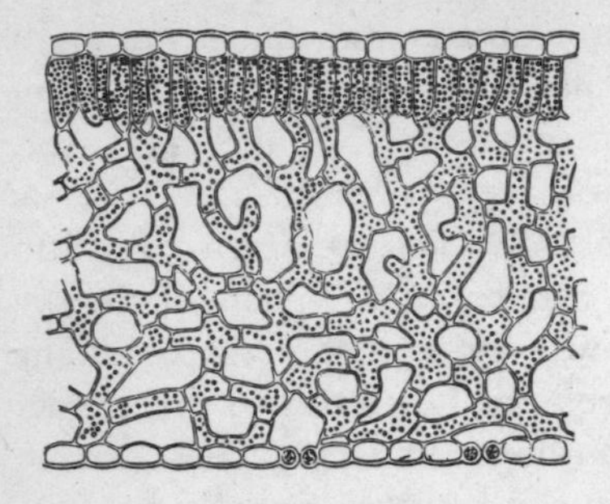
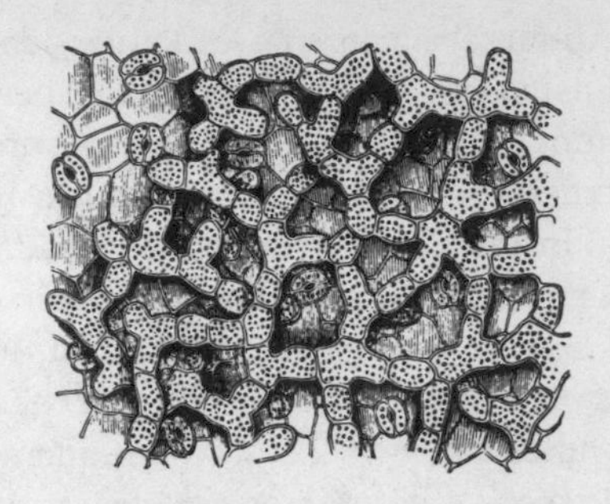

5 · *The Great Invention*

The great invention of the plant kingdom, the factor which differentiates it totally from all animals of whatever complexity, apart from a few borderline cases among single-celled organisms, is chlorophyll. Whether it is a one-celled alga or a giant forest tree, a plant will contain this substance.

Chlorophyll is a pigment which absorbs some, but not all, of the rays of sunlight – the red, orange and blue parts. Chlorophyll uses this radiant energy to combine water and carbon dioxide and to produce sugar in the form of glucose. At the same time oxygen is given off. Production is governed by the amount of light available, although some plants are able to get more out of poor light than others. This process is called photosynthesis, and it is an attribute unique to plants which makes them independent of external sources of carbohydrates.

Photosynthesis is normally localized in leaves, but there are many plants in which leaves have been replaced by stems which take over the process, as in brooms and cacti. Sometimes even roots are green. Chlorophyll forms almost instantly in seedlings emerging into the light, and indeed very rapidly in leaves which have been kept unnaturally in the dark, such as the hyacinths and other bulbs we force in winter for early flowers.

Curiously enough the chlorophyll molecule closely resembles that of the blood pigment haemoglobin, and the biosynthesis of the two has remarkable similarity, though the iron in haemoglobin is replaced by magnesium in plants. There are two ways in which animals can manufacture the basis of haem, and one of them – accounting for about a quarter of an individual's blood – is the same way as used by plants to manufacture chlorophyll. This mechanism points very clearly to a common ancestry of plants and animals.

Though chlorophyll is basically green, it can be masked by pigments of different colours, and is always associated with the yellow pigment xanthophyll and the orange-yellow carotin, whose functions remain rather mysterious. In seaweeds other pigments appear actively to assist the chlorophyll, and give these algae their various colours. In higher plants the chlorophyll is always in distinct, variously shaped bodies called chloroplasts in the appropriate cells; in seaweeds the equivalent bodies are called chromoplasts, because they are usually coloured other than green.

These chloroplasts have been described as 'the little green slaves' which enable a plant to pursue a calm, passive existence while a lion, say, has to expend energy in hunting and killing for meat. In fact the plant, as I hope to show throughout this book, is not really calm and passive at all.

The chloroplasts have striking similarities to the rod cells (visual receptors) in vertebrate eyes, and the parts of sunlight they absorb correspond very clearly to the range of the human eye. Both these pigmented bodies, if greatly magnified (chloroplasts are tiny discs less than a micron across) are seen to have a structure which strongly suggests a radiator, condenser or electric accumulator; one writer has likened them to a stack of several hundred hollow balls squashed flat, piled up and wrapped in a single membranous layer.

An eye rod, in technical terms, is a combination of transducer and amplifier, converting a light stimulus into a nerve signal. The chloroplast is an energy converter, changing light energy into chemical energy. In both cases very rapid photochemical reactions are involved. The chloroplast acts as a semiconductor, comparable to those used in spacecraft to convert radiant into electrical energy. Its processes have been shown to be reversible: plants actually emit light, if very feebly, for a short time at the moment of change from light to darkness.

In a few plants chloroplasts move within the cells in response to light conditions, controlled by filaments of a protein called actin, which have a surprising resemblance to animal muscle fibres. In the single-celled alga *Meugeotia*, they turn edge-on to the source of light if it is excessive, and full-face in poor light. In the thin-leafed duckweeds, the chloroplasts arrange themselves horizontally along the cell walls in poor light, vertically in strong. Usually, however, there are passive filtering devices to prevent excessive light from damaging the vital chloroplasts.

The glucose made in photosynthesis is the starting point for all the plant's metabolism – the production of energy for growth. Photosynthesis takes six molecules of carbon dioxide and six of water and, with the energy provided by light, combines these into one molecule of glucose, while six molecules of oxygen are discarded. The energy in the glucose is available at once, and much of it is swiftly converted, by a chain of complex reactions, into thirty-eight molecules of ATP (adenosine triphosphate). This universal 'energy currency' is the basic activator for all cell actions and growth processes.

This may suggest that the process is purely chemical but in fact light energy is initially converted into chemical potential energy by exciting electrons within the cells. These electrons travel away from the chlorophyll across a membrane to be seized by molecules called quinones; and are replaced by new electrons from water molecules. As is further discussed in Chapter 10, this positive/negative electron process drives many metabolic processes. Attempts to convert solar energy by artificial photosynthesis are based upon this basic electron movement.

Photosynthesis stops in the dark, of course, and at night the plant gives out carbon dioxide and consumes oxygen, exactly like an animal. The consumption of oxygen (oxidization) is the equivalent of burning and it enables the plant to make use of the stored energy produced by day. The energy released goes to building the structure by cell multiplication, so that roots extend, branches carry more leaves for more photosynthesis, stems and trunks thicken. The liquids and minerals also needed for this growth are transported through the conduits, the communications network is kept active, the reproductive organs come to maturity.

Surpluses of food go into store, the sugars being converted into more complex carbohydrates such as starch in which energy can be locked up. Other 'capital assets' produced are fats, which help build up cells, and proteins, the main component of protoplasm. In strong light and best nutritional conditions a cell produces much more protein than carbohydrate; in weak light protein and carbohydrate are roughly balanced; while in good light but poor nutritional conditions protein gives way to fats.

Plants such as seaweeds absorb all the sugars they produce; any surplus goes into thickening the cellulose cell walls and adding strength. They can photosynthesize less than land plants because of restricted light, but do not need to transport food materials about, for the minerals, like the carbon dioxide, are all around them in the water. Land plants, with far more light, sometimes make more sugar than they need for growth, and use it to produce sugary nectar or glandular secretions for other purposes, and sometimes have so much that it is actively excreted through the leaves or other parts.

Water plants have a particular problem if they are largely submerged, because although oxygen and carbon dioxide are dissolved in water they are present there in lower concentrations than in air. Many water plants thus have very spongy tissue, or large open canals as in water-lily stems, in which air diffuses all the way through the plant. In some cases this also gives them additional buoyancy.

Consuming energy is in effect respiration. This breathing sometimes goes on in the day too, but at a much lower rate, and the carbon dioxide it produces is immediately used up by the photosynthetic process, which can be said to mask daytime respiration. However, this 'photorespiration', as it is called, may waste half the carbon assimilated through photosynthesis.

The conversion rate by an average crop of the total radiant energy available into chemical energy is on average less than 1 per cent – much less in the case of plants inactive for part of the year. Even this low figure takes into account times when the crop is not deploying its leaves fully, or some of its adult leaves are partly shaded, and also climatic conditions in which cloud may reduce light intensity to as little as 5 per cent of that of full sun.

One of the limiting factors is the low concentration of carbon dioxide in air – only three parts per 10,000. Some plants, such as sugar cane and other tropical grasses, can photosynthesize much more efficiently than others, and can do so in lower CO_2 concentrations. They have a very efficient mechanism for fixing CO_2, and they waste no carbon in daytime respiration, like most plants. Apart from differences in the chemical processes involved, these plants (sometimes called 4C, C_4 or hatch-slack plants) have two types of chloroplast, one kind containing much more starch and possibly acting as temporary energy stores. The 4C plants are invariably tropical or sub-tropical, frequently come from arid regions and are resistant to drought conditions. There are several genera which include 4C tropical species and ordinary-cycle temperate ones. It seems likely that this is an evolutionary advance, for today's atmosphere is less rich in CO_2 than that of the Coal Measures.

This is clearly demonstrated by comparing the production rates of different plants measured in carbon grain per unit area. Plant plankton – virtually elemental cells – has the lowest productivity, only slightly bettered by temperate grasses. Birch forest is about six and a half times more efficient; tropical rain forest is over twenty times better, while maize and sugar cane are respectively twenty-five and thirty-five times more productive.

Once again we have to make an exception of the fungi. Having no chlorophyll they do not use the sun's light to produce energy; they have devised their own means of obtaining food from decaying or living matter. One cannot doubt that they are really plants; but without chlorophyll these link-men of the vegetable kingdom, so often the main agents in renewing simple food materials from the complexities of decaying animal and vegetable bodies, are certainly anomalous.

To grow, a plant needs not only the foods produced by radiant energy from photosynthesis; it must also have certain minerals and of course water. The results of this activity worldwide are staggering. It has been calculated that the annual growth, or yield, of plants – including the microscopic algal plankton of the oceans – is around 14×10^{10} tonnes of organic matter. In terms of energy this is very roughly equivalent to the output of 2,000 million large power stations. Land plants alone create 150 billion tonnes of new matter a year.

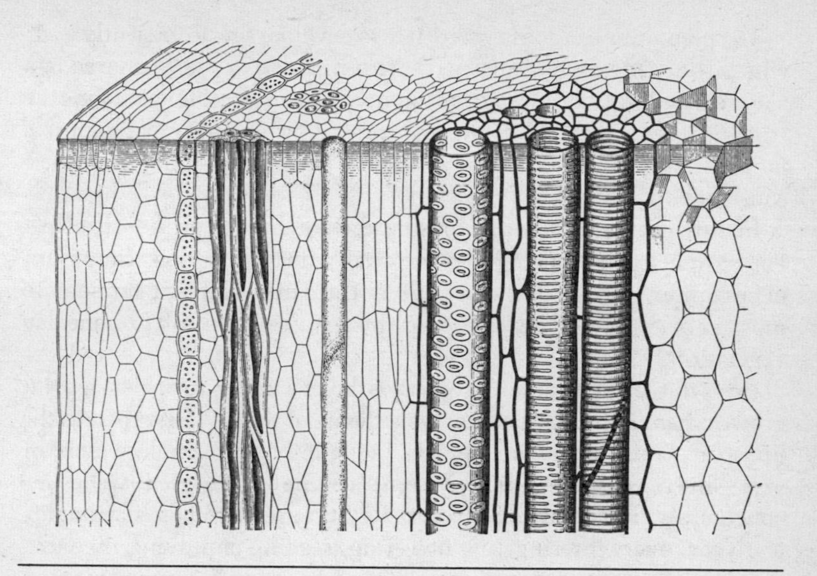

6 · *Nature the Engineer*

An animal has blood, a plant sap; the one haemoglobin, the other the partly similar chlorophyll. Animals have eyes, plants chloroplasts: these organs react to light and operate in similar ways. Animals have a nervous system, together with glandular secretions which form a second line of command; in plants, growth, movement, and most details of their organization seem to be primarily controlled by chemicals which act as messengers, but the action of electrical charges, only beginning to be understood, may show that these are as important as chemicals. But a plant has solid cell walls which prevent its taking in food in the least animal-like manner; even the carnivorous plants are only superficially animal-like. Plants and animals both have skeletons; a plant has no muscles, but it has pumping capacity. Both have circulation systems, for blood and for sap. Sexual processes in plants are comparable in their complexity to those of animals. A wounded plant is likely to regenerate parts; few animals can do this.

The plant cannot choose where it is to develop, settle down and feed; where the seed has germinated, there it must do or die, anchored in a way resembling, say, a polyp but in no way comparable to any motile animal. It is perhaps the polyps and corals which plants most resemble in their way of life, their replication of similar units and their construction.

In this chapter I want to outline the totality of an 'ideal' plant. For simplicity if nothing else we shall consider a flowering plant or a conifer, for this at least reflects the kinds of plant familiar to most people, and which compose the greatest bulk of present-day vegetation.

Even this simplification leaves us with very many variables in plant growth, habits and place in the total scheme of things. First there is the life span. Some plants are annuals, living only for a single season, or very much within it; indeed some of the short-lived are called ephemerals, living only a few weeks. Others are biennials, starting life one year, overwintering, and flowering, seeding and dying the next. Plants which live for several years, flowering every year when mature, are called perennials. Some put up fleshy stems each spring which, after flowering, wither before winter. The plant spends this time, or any resting period that the climate may decree, as a crown or clump more or less under the soil, its roots barely active. These are herbaceous perennials. Many aquatic plants are similar, spending cold winters resting as a sort of fleshy knob without leaves or stems.

Other plants are woody and more permanent, although some sub-shrubs, woody only near the base, have herbaceous upper parts which may die in the winter, like sage and lavender. Woody plants are roughly divided into shrubs and trees, although these categories inevitably overlap; shrubs having a quantity of relatively short stems or trunks, trees typically a single one, usually with a crown or head of branches.

There are certain plants which live several years but are able to flower and seed only once – these are termed monocarpic. Most of these are herbaceous plants, like many bromeliads, but there are a number of curious in-betweens, half herbaceous and half shrub, such as the great rosette-forming agaves and puyas of the Americas. Some of these may take a decade to mature, others very much longer, which is why agaves are often called century plants. A few palms are monocarpic, and so is one other tree group, the spathelias, which is related to the orange. One of them, the mountain-pride, is widely planted in Jamaica for ornament: it can grow 15 m tall although its trunk is only around 8 cm in diameter. After eight or ten years it produces an enormous cluster of red flowers, which may be 1·5 m high and 2·4 m across; and once the seeds ripen the entire plant dies.

Among herbaceous and woody plants, both annual and perennial, there are many which climb, making long fast-growing stems with devices or growth habits designed to allow them to clutch, hook or sprawl over surfaces or other plant structures.

There are also the very varied parasites, which depend to a greater or lesser degree upon other plants for their food; these are specialized adaptations which are irrelevant to the present general discussion. They are covered in detail in later chapters, for they are indeed fascinating exceptions to the general rule. So are the less vicious saprophytes, which live on decaying matter like some fungi, and equally have no chlorophyll.

To understand the functions of the parts of a plant one must consider its metabolism and the needs this creates. It is possible for a plant to exist for a considerable time with only water and the minerals dissolved in water, as long as the leaves are carrying out photosynthesis; but normal healthy growth needs inorganic elements in greater quantity. The main trio are nitrogen, phosphorus and potassium; others needed, though in far smaller quantities, include sulphur, calcium, magnesium, sodium, iron and certain other metals, and boron. Their absence can cause a plant to die although the amounts needed are infinitesimal. Farmers, foresters and gardeners go to considerable lengths to balance these minerals in the 'right' quantities, although a great deal obviously depends on the make-up of the soil being used, while the plant is distinctly selective in what it actually takes up. The proportions of minerals within plant cells are usually very different from those in the soil outside. How a plant takes up and absorbs minerals is a complex electro-chemical process known as ion transport; what matters to us here is that they *are* taken up.

Water is taken up by osmosis. This means that where a porous membrane exists, such as the wall of a root hair, a solution with a high concentration of salts on one side (i.e. within the root hair) will extract water from a solution with a lower concentration (i.e. typical soil water). This simple law of nature operates in many places including internal cells.

The main reason for water uptake is to assist in photosynthesis, but a great deal more water than is needed for this process is taken up and lost by the plant. This loss of water is called transpiration, and enormous quantities of water are involved. Although estimates of transpiration vary considerably, small experiments have shown an equivalent of over 500 tonnes of water per acre transpired by grass between May and July, the resultant hay-crop being 1½ tonnes per acre; over a somewhat longer period potatoes transpired 570 tonnes per acre, the crop being 2 tonnes. Besides this active transpiration there is also water loss from evaporation, the amount depending on the insulating efficiency of the leaf structure.

Water is also needed to keep the plant cells turgid; this is not just a matter of keeping the plant rigid, but of maintaining the health and efficiency of the cells. Below a level which varies from 70 to 95 per cent water in plant tissue the cells cannot grow satisfactorily, and at a certain point loss of water is fatal.

The roots have therefore to take up and transmit very large quantities of water, despite the fact that less than 1 per cent of the water absorbed is retained in the plant. But water does not come to the roots, which in any case are sucking it out of the soil as hard as they can: they must continually grow out in search of fresh moisture. This means that root systems are often enormous compared with growth above ground. Individual maize roots have been shown to extend 5 or 6 cm a day, and the average total daily increase in the length of a root system of a mature rye plant has been estimated at 5 *kilometres* a day. This excludes root hairs, the average daily growth of which on a single rye plant is nearly 90 km per day! The four-month-old rye plant in the experiment concerned was grown in a box of roughly 1,300 cu cm of soil. Its total root length was 622 km and the length of the root hairs a staggering 10,620 km. (These figures are *not* misprints!)

The basic plant, then, has three distinct zones of different function. Apart from absorbing from the soil the essential water and soluble mineral foods, roots anchor the plant in the soil; even the smallest annual needs its anchor. The roots lead up to the stem or trunk, sometimes via a 'crown' or gathering point if many stems are involved. On the stems or trunks, which may be more or less branched, the leaves, the photosynthesizing factories, are carried. Sometimes a stem is virtually absent, as in some alpines, and the leaves emerge from the crown. From this, or on the branches, are also carried the sexual organs which are usually responsible for perpetuating the race, but these are not essential for the metabolism of the plant, and are dealt with in later chapters. In some cases, indeed, for instance various trees, it may be many years before the structure is large and mature enough for flowers or sexual organs to be produced at all.

These three zones are all part of a system which combines liquid conduction with rigidity, like an architect's dream of a plumbing circuit which supports a whole house. In principle water is passed up to the leaves along with minerals, while a contrary flow takes solutions of sugars and other organic materials from the leaves to the roots and other parts of the plant needing sustenance. In the flow chemical hormones and other messengers are also transmitted. These pipelines can be likened to a combination of animal blood stream and digestive system, but substances can also pass readily from one flow to the other horizontally.

The basic conduit for the upward flow is known as xylem and for the downward, phloem. The first signs of a central conduit, acting also as a strengthener, come in various mosses. The two types of conduit are clearly differentiated in ferns and more advanced plants. In seaweeds, where conduction is usually unnecessary because the plant lives in water from which minerals are obtained, there can be a strong central core of linked cells to provide the necessary strength, and whatever conduction, for instance of hormones, is needed.

The two great classes of flowering plant – the dicotyledons and monocotyledons – differ somewhat in the positioning of their conduits. In the roots the phloem and xylem begin, at least, as a central bundle; in dicots the xylem forms a cross pattern with phloem in between the arms, while in the monocots the two sets of conduit cells form alternate groups in a circle around a central pith. In the former, older roots develop a central core of xylem and an outer ring of phloem.

Initially roots grow downwards, since they are geotropic – following the pull of the earth's gravity. Soon, however, they must grow sideways in order to seek moisture, doing this with a spiralling movement, as they continue growing outwards and downwards. Mostly they are below ground, although tree roots may emerge above the ground surface as the trees mature. Roots need oxygen in principle and perish in waterlogged soil; where plants do grow in such soils special structures may exist to ensure that oxygen is taken into the root system.

At their tips the roots are tender and delicate. They protect themselves from bruising and damage by a protective cap rather like a thimble, made of expendable cells constantly being sloughed off at the tip; the walls of these cells become mucilaginous, providing a lubricant to the root tip's steady probing between the particles of soil. The actual absorbing mechanism of the root usually lies just behind the tip in the form of a mass of small hairs, which are always in production near the root tip and die off behind as the root extends. These root hairs have the capacity of actually gripping soil particles, so that a root tip usually appears to be coated with soil. The root thickens behind the tip, helping to force apart a crack in the soil in which the root tip can probe.

Herbaceous plants continue their conduits into the stems and branches in a system of vascular bundles in which xylem and phloem are combined. In dicots these bundles are typically arranged in an outer ring surrounding the central pith, but in monocots they are scattered through the tissues, and this happens even in large species like palms.

A diagram of the bundles in a dicot looks rather like one of the blood circulation in a human being. In the stem the bundles, arranged in a circle, form a kind of open sheath with branches going off to side stems, forking continuously until they enter the leaves, and at the tip of the shoot entering the embryonic tissues making more growth or, as the case may be, flowers. In a monocot the apparent scatter of the bundles shown in a cross-section is revealed, in a dissection diagram, to have a very definite pattern; they grow obliquely, so that in very simple terms the stem is based on an intricate network, elongated vertically, the bundles dividing at various points to make contact with those in the leaves. It is a pattern resembling a geodetic structure, which makes for a great deal of strength, and this is why it is effective even in large plants; indeed it is almost impossible to saw a palm stem, let alone section it neatly for microscopic examination, because of the great mass of vascular bundles which are each sheathed in very hard, thick fibres, rather like individual electric cables.

In woody dicots, however, including the gymnosperms, a process called secondary thickening produces an arrangement in which there is an ever-widening central core of xylem. Cut across, this core is seen to have rings, each composed of larger and smaller cells, the larger produced in spring when growth is at a peak and overall expansion takes place, the smaller and denser in summer and winter as growth slows down. Every year a new ring is formed, and its width can tell us how good a season it was, while the number of rings tells us the exact age of a tree.

Outside the central xylem core is a sheath of cambium, a tissue responsible for the immediate growth of both the xylem within and the sheath of phloem without – a miraculous one-celled layer, constantly dividing, which in effect embodies the whole growing principle of even the largest tree.

A section of tree trunk will also show narrow radial lines crossing the annual rings. These are called rays and are built up of non-woody cells which allow food materials to pass across the radius of the trunk.

Woody trunks can dry up and decay from the centre outwards, sometimes leaving a very narrow exterior sheath which can, however, still support a head of branches and foliage, and connect these with the roots. As a tree ages, only the outermost annual rings are active in conduction. In any case the xylem in the centre does not operate actively, and ends up as a kind of internal refuse dump for unwanted by-products, although fortunately for us these often enhance the quality of the wood.

The xylem cells involved in conduction are usually woody, and without living content: they are really a kind of pipe, rot-proof and waterproof, with walls strengthened with rings, spirals, netted or pitted thickenings. They are called vessels and tracheids, and range from relatively wide to very long and narrow. Some are open at each end, others have pits and holes, not only at their ends but on their long sides, so that fluids can be passed sideways through adjacent cells – very important in the case of blockage, air-locks or other damage. In diameter these pipe-like vessels range from 20 to an exceptional

600 microns, in length from a few centimetres to many metres – sometimes, as in ash and oak, a tree has some vessels that are continuous from roots to branching crown. (A micron is 1/1000th mm.)

How is the water conducted or forced up the stem of a plant? One might imagine that capillary action was responsible. This is a phenomenon, dependent on surface tension, in which liquids will rise in extremely fine tubes; but its limit is a metre. Roots undoubtedly exert a pumping pressure in some cases, caused by osmosis, which under experimental conditions has been measured as about 8 atmospheres. Certain plants, notably trees and climbers of tropical forests, pump so hard that when there is plenty of water at the roots they exude moisture at leaf tips and edges (this is known as guttation), and many are provided with special pores at the ends of the veins for this purpose.

But this root pumping is largely dependent on good growing conditions, especially at the roots. Starvation, dryness, lack of oxygen and low temperatures will all curtail it. In any case, root pumping does not occur in all species, and in, for instance, deciduous temperate trees, it disappears after the initial burst of growth in spring. Further, as already mentioned, the quantities of water pumped are a fraction of the amount lost by a plant through transpiration; perhaps only 1 per cent of the water taken in is retained by a plant.

The final answer seems to lie in water tension combined with cohesion. A column of water in a narrow tube can be maintained by production of suitable tension at the upper end, such as a vast number of transpiring leaves exert, through the vascular bundles, on the main xylem flow of a tree. To maintain such tension beyond ordinary atmospheric pressure, under which a height of 10 m could be achieved, one has to involve the cohesive attraction of water molecules, by which a continuous column of liquid resists breaking. It is very difficult to measure tensions on living trees, because any insertion of instruments causes the water column being examined to collapse. It has, however, been possible to estimate tensions on cut branches by a special technique, and this has resulted in measurements of up to 80 atmospheres – the maximum being from desert and salt-marsh plants which have the greatest problems of extracting water from their surroundings. It is also significant that the upper branches of trees have a higher tension than lower ones. A tension of 30 atmospheres is sufficient to raise water to 150 m, and since the tallest trees ever recorded are less than this, plant capacities in this respect are adequate. (Tallest trees recorded include *Eucalyptus regnans* from Australia of 114 m and coast redwoods from California of 112 m. The existing record, another redwood, is 111·60 m tall.)

It must be said that there is a snag to this theory as it stands. If one creates an artificial conduction system with water under tension in capillary tubes, the water columns can very readily be disrupted and collapse if the tubes are shaken or tapped. But this does not happen in a plant stem, which can be knocked about, or sways in the wind. Additionally it is possible to make overlapping cuts into a tree trunk from opposite sides, without the xylem flow being disturbed; while air bubbles in the sap resulting from freezing in winter similarly do not interrupt the flow.

However, there is no doubt that sap can traverse the walls of the xylem cells, which become saturated if there is a blockage or interruption. This can most clearly be shown by examining the relation of roots and stems. Normally a given sector of the roots supplies only the part of the stem and whatever organs it may carry which are directly above it. But if a sector of the roots is removed, the corresponding upper parts do not suffer: they are undoubtedly supplied by lateral movement from the remainder of the root system. In any case, although the tension–cohesion theory may not be perfect, transpiration and vertical water conduction are facts.

It should finally be mentioned that the suggestion has been made that very tall trees may absorb moisture from the air high up, thus relieving the need for a total upward conduction system. Though this may well occur with the conifers in the foggy western United States, for instance, it can hardly do so when it comes to eucalyptus trees of similar height in dry Australia.

Movement of food materials in the phloem is if anything more extraordinary, because phloem cells have living contents. Among these cells those called sieve-tubes are prominent. These are long narrow cells at each end of which there is a wall pitted with small holes like a sieve; these walls are quite often at an oblique angle to the long sides of the cell, and abut on to those of the adjoining cells. Controversy has long existed over whether the sieve-holes are open or not. If open, the sugar solutions could be moved by hydrostatic pressure, stimulated by high osmotic pressures in the leaves. However, recent work strongly suggests that they are not open, but are passages for longitudinal strands containing protein filaments. One suggestion arising from this is that the bundles of filaments in the strands act like muscles, contracting rhythmically in order to propel the sugar solution along the inter-connected tube cells. Another theory, based on very recent observations, is that the sugars are moved by the streaming of the cell contents (cytoplasm) between the cells. Not the least peculiar thing about sieve-tubes is that materials can apparently travel through them in both directions simultaneously.

The amount of fluid moved in the phloem at some seasons can be gauged from the century plant. The Mexicans make a potent liquor called *pulque* from this monocarpic plant by removing the flower bud as soon as it appears in the centre of the rosette and collecting all the sap that subsequently exudes, which they later ferment. During the first few days about 375 g are produced daily, and the total amount produced over the four or five months the flower shoot would have taken to mature is around 50 kg.

Palms up to 10 m tall are treated in the same manner to produce wine from the sugar-rich sap; tapping sugar maples again takes advantage of the phloem flow which mainly originates in the plant's leaves.

The growth and functioning of the roots is largely activated by the sugars passed down from the leaves. The soil minerals combine with the organic materials in various ways. This is the essence of a very complex process, and in fact the sap, or fluid, passing upwards may contain high proportions of sugars, notably in spring when trees are leafing, while the leaf sap contains many minerals, especially phosphorus and potassium, although nine-tenths of the solid matter in it are sugars. So rich is the spring sugar flow that it is possible to tap certain trees for it. The sugar maple is tapped commercially and up to four holes 5 cm deep are drilled into each tree, from which sap flows for an average of thirty-seven days. I know an Irish canon who taps his local birch trees each spring in the same way to make wine from the sugary sap.

Aquatic plants are surrounded by nutrients and one would expect their conduction needs to be absent. However, they do have a rather slow upwards movement in their xylem, which is probably an evolutionary relic.

The simplest way a plant achieves rigidity is by filling itself up with water. The cells become fully distended, or turgid, like a blown-up balloon or rubber tyre. If an inner tube is really tightly blown up it is difficult to bend or indent; the same principle applies to the distended cells. However, this is only applicable to small plants, or to small parts of larger ones like leaf stems. A plant, like any other structure, is up against certain laws of nature: the strength of a stem is directly proportional to its cross-section, but its volume and weight increase as the cube of its exterior size. Thus a stem can be doubled in size and become four times stronger; but this incurs eight times the increase in weight.

In herbaceous plants generally, rigidity is largely due, as already suggested, to the strength, number and pattern of the vascular bundles, which typically have a group of fibres at their outer edges. These resemble in function the unseen, relatively slender but very strong steel girders in a ferro-concrete building. There is an unending variety in the number and arrangement of these girders, which can be separated or fused together, and are often combined with flanges or ribs of mechanical tissues. They are always symmetrically arranged. The girders in palms and their relations are reminiscent of geodetic engineering structures, often having an outermost sheath of tough fibres just below the skin. Girders arranged in a ring near the outside of the stem make the structure tube-like. Given a certain minimum strength, a tube is as rigid as a solid bar, and much less wasteful of material; the wider the tube the more rigid it is. Most herbaceous plants either have very soft pith in the centre of their stems, or they may indeed be hollow and are thus real tubes.

One of the most effective tubular structures is that of the bamboo. Like the palms it is a monocot, and so its vascular bundles are not strictly in a ring; but they are concentrated towards the outside of a relatively narrow tube of tissue surrounding a central hollow. At intervals the stem forms nodes, or joints, where a solid disc shuts off the tube. Further, the outer cells have walls containing so much silica that it is sometimes possible to strike a match on a bamboo stem. In this they are similar to other tubular plants, like grasses, notably cereals, and horsetails. The horsetails are indeed so full of silica that they have been used as abrasive pot-scourers.

Another strengthening method is the production of stone-cells or sclereids, which are woody, thick-walled cells of various shapes, normally without living contents, which are often massed together close to the surface skin or cuticle, forming a relatively stiff, unbending layer to leaves as well as stems. An associated form of dead cells forms fibrous bundles; the separate cells are elongated and have pointed ends so that they can mesh with their neighbours. The strength of this second form of sclerenchyma, as these dead tissues are called, is evident in that we take advantage of it for fibres, as in flax, jute and hemp, the latter two having woody fibres. Many of the cells are relatively very long; in flax they can be 20 to 40 mm, while in ramie or china grass they may be up to 30 cm long. This latter produces superb fibres in our terms but is extremely difficult to process because the fibre bundles are additionally strengthened by a cement-like sheath of gummy cells.

Such combinations of external sclereids and internal fibres can produce quite astonishing resilience, which a plant depending simply on its vascular bundles cannot count on. It is easy enough to break the soft stem of a small annual plant, although the 3 to 4 m stems of sunflowers have a good deal of strength. Some perennial stems are also relatively very strong; indeed in the garden I always make it a practice to keep the stems of plants such as some michaelmas daisies and the coneflower *Rudbeckia newmannii* when cutting them down in autumn, since they are quite stiff enough, once dry, to be used for light plant stakes the following season.

Once a plant starts to form a central woody core the whole of the stem virtually combines conduction and rigidity. Almost all plants have remarkable strength in relation to rigidity and lightness, as indeed do animal materials like bone and chitin, and the way in which these biological materials score over many man-made ones is in their non-brittle nature, their resistance to fracture. Plastics and metals can be moulded to requirements more easily than wood, but the specification of timber takes a great deal of beating. Engineers at the Royal Aircraft Establishment, Farnborough, used to quote the couplet, 'Plastics are made by fools like me, but only God can make a tree.'

In herbaceous plants the stems are covered by a skin or cuticle whose main function is to insulate and prevent the loss of water. The cuticle is usually, though not always, composed of varying proportions of wax and of cutin, a substance similar to cork also containing cellulose and pectin. Insulation can be increased by hairs. This skin seldom takes part in providing stiffness, but it is often backed by a layer of large cells with much-thickened walls of cellulose, called the collenchyma, which does provide strength and resilience. These cells are alive and quite frequently contain chloroplasts to carry out photosynthesis. Sometimes the collenchyma forms ridges on the outside of a stem, exactly in the way that man gives a strip of metal more rigidity. The square stems of the dead-nettle family are good examples of this reinforcement technique.

In shrubs and trees the skin of the stems becomes woody and eventually bark is formed. This protective layer is immediately outside the phloem sheath, and is usually composed of dead, corky cells constantly added to from within. In this way the steady expansion of the trunk does not cause a splitting of the innermost layer, although the exterior does split, or break into plates or scales which may eventually flake off, as one sees clearly on pines and planes. Other barks are fibrous or membranous. The corky cells may contain air or other substances, such as calcium oxalate, tannin in oaks, or the quinine for which *Cinchona* is prized. Sometimes the bark cells are mucilaginous, as in slippery elm. It is possible that these substances are waste products excreted to a dead-end position, but it is clear that in some cases at least they may act as deterrents to attacking animals, both the larger biting ones and the small boring ones such as beetles, and to fungi.

Bark can vary enormously in thickness. In some trees like beech or cherry it measures only a millimetre or two. In a pine the irregular scales can be an inch or more thick, and in the giant redwood the bark is so deep – 30 cm or more – and full of air that it can be punched with the fist with impunity.

Barks contain air pores or lenticels which allow gases to diffuse through the bark, which is otherwise impermeable; these air pores are groups of loose corky cells within a rather ragged aperture, and they usually form in the same place as the breathing pores of the original young green stems. It may seem rather curious that the internal cells of a conducting trunk need extra oxygen, but this is the case; likewise carbon dioxide and often excess water vapour are given off. Bark will usually regenerate if destroyed, although the process is slow. Some trees, such as the cork oak and those tropical species with fibrous bark used for clothing, regenerate bark quite rapidly.

Monocotyledons usually have a simple stem but in some cases these develop a cambium and a corky layer. In many groups, the old leaf bases create a protective layer; one only has to look at palms to see what a barrier these can make. In many palms the decay of the leaf bases results in a fearsome array of spiny fibres sticking out in all directions.

Stems and trunks carry leaves and flowers. To do this they are very often branched. This branching may be the minimal amount needed to deploy a number of flowers on an annual for it is seldom that a stem carries but one bloom. Herbaceous perennials will often make a number of stems from a root-crown, the number depending largely on the health and vigour of the crown; the stems typically make an inverted cone of growth so that each head of flowers receives its share of space and each flower is deployed to best advantage to attract its pollinators. In these herbaceous plants and in many annuals the leaves are often carried on the stems. There is also a big group of plants which make leaves on individual stems rising from a rootstock, the flower stem being produced independently, as for instance with rhubarb and docks.

Shrubs likewise tend to have leaves on the stems and carry flowers either at the stem ends or grouped along them in various ways. In trees, however, it is typical for the trunk to branch fairly high, the branches forking repeatedly till they end in twigs bearing leaves and flowers.

The angles at which branches are produced do not alter materially as the tree ages; if as often happens with cultivated fruit trees there is an excessive crop the branches may actually break because of the rigidity of these angles. But as a tree grows the branches become steadily thicker and heavier, and one might expect this to cause breaking. Like a cantilevered structure in engineering, the upper part of a branch is in tension – under a pulling strain – the lower part in compression from weight. If we study a branch from a deciduous tree the wood in the upper part differs from the rest in having narrower vessels and fibres with thick inner cell walls composed of cellulose rather than the normal wood-substance lignin. In conifers, it is the wood on the branch undersides which is modified, being denser and containing more lignin.

It is easy to assume that these differences in wood structure are responsible for the branches being able to endure ever-increasing stresses on their cantilever, but it is possibly an effect rather than a cause. The problem still needs much investigation.

Branches are initiated in a precise pattern, almost invariably spiral; the more primitive trees such as conifers exhibit this clearly, though some are whorled. There are some comparable examples among flowering trees, but in most of these the regularity of the spiral becomes obscured, the development of each branch depending greatly on that of its neighbours as well as on external factors such as light, so that the arrangement appears much more informal while creating an overall balance. In such trees the arrangement may appear more and more irregular as the branches thin out towards their tips, dividing and subdividing into branches and twigs.

Trees vary vastly in the architectural style – what the gardener and botanist call their habit – which the frequency, angling and size of branching creates. But whatever the arrangement it is designed to bring every leaf into a position of maximum light reception, and equally to present the flowers and fruits to best advantage according to the ways they are pollinated and distributed respectively.

The satisfactory deployment of these flowers, fruits and leaves is the ultimate purpose of all this complex of vegetable engineering, as Ruskin perceived. In *The Elements of Drawing*, Letter III, he wrote:

> Liberty of each bough to seek its own livelihood and happiness according to its needs, by irregularities of action both in its play and its work, either stretching out to get its required nourishment from light and rain, by finding some sufficient breathing-place among the other branches, or knotting and gathering itself up to get strength for any load which its fruitful blossoms may lay upon it, and for any stress of its storm-tossed luxuriance of leaves; or playing hither and thither as the fitful sunshine may tempt its young shoots, in their undecided states of mind about their future life.
>
> Imperative requirements of each bough to stop within certain limits, expressive of its kindly fellowship and fraternity with the boughs of its neighbourhood, and to work with them according to its power, magnitude and state of health, to bring out the general perfectness of the great curve, and circumferent stateliness of the whole tree.

7 · *The Power Station*

Without its chlorophyll a plant would not be a plant; and, in general terms, without leaves there would be nowhere for the chlorophyll to be housed. There are other places where photosynthesis can be carried on, mainly stems under special circumstances, but the leaf is a very specialized creation in its own right, each leaf a miniature power station providing energy for the rest of the plant.

Leaves vary enormously in size and shape, from thread-like to spherical and they provide a range of form, lobing, indentation, toothing which would make an interminable catalogue. Most leaves are symmetrical along their centre line but some are irregular and a few, like almost all those in the large begonia tribe, are lop-sided, sometimes in a spiral formation. Leaves vary enormously in their rigidity; most are relatively soft and pliable, but some, like those of yuccas, agaves and bromeliads in general, are extremely hard, and if they end in a sharp point could almost make a lethal weapon. Leaves vary in thickness too, being typically rather thin, but, especially in succulents, sometimes thick and fleshy and even with a circular section.

Leaves are joined to the main shoot or branch, sometimes via an intermediate stalk, usually at one end but sometimes more or less in the centre. The joining point is relatively narrow, and from it the leaf expands. At this joining point are concentrated the vascular bundles which continue into the leaf as veins. There is frequently a major central vein, the midrib, or in regularly lobed leaves a series of ribs extending towards each lobe-tip. From the midribs there may extend smaller raised ribs, and in any case the remaining area of the leaf is criss-crossed by minor veins, or veinlets. The word is apt, for these are very similar in appearance to the veins in animals with blood. They can remain as a bleached skeleton when the leaf tissues have decayed. These plant veins are the leaf's irrigation system. In dicots they form an irregular network, but in most monocots they are parallel.

In each vein the xylem and phloem fit together, so that the sugars from the leaf cells can be passed into the remainder of the plant, while the water essential for photosynthesis and transpiration, as well as the minerals which assist in building up cells, are passed into these cells.

The basic leaf is flat. This gives it the most surface and least volume, so that it is best designed to receive maximum light and carbon dioxide without any waste of cells in the centre. Succulents need to photosynthesize and transpire less, and water storage is essential, so they have fleshy leaves. The leaf has a transparent, waterproof skin or cuticle on both sides, an epidermal layer of colourless and sometimes dead cells, usually containing water, under the skin, and on the upper surface a dense layer of cells containing numbers of chloroplasts, the minute bodies containing chlorophyll. The centre and lower parts of the leaf contain spongy tissue whose main feature is wide irregular air spaces between the cells.

On the underside of the leaf are the breathing pores or stomata (the word is Greek for mouths). Seen vertically, these have a lip-like cell, known as a guard-cell, on either side of a gap. They are designed so that when they are turgid with water absorbed from the neighbouring epidermal cells the slit opens; if they lose water to the surrounding cells the slit is closed. Only by such a mechanism could apertures in the rigid skin be opened and shut without making it buckle. Through the stomata gases and water vapour diffuse in and out according to external circumstances and the leaf's internal operations.

As already noted, the skin is waterproof, and indeed in many cases it may be thick and waxy to ensure this. The round leaf of that most beautiful of plants, the Indian lotus, carried above the water on a slender central stem, is a fine example. Shining drops of water run about like quicksilver if the leaf is moved; the surface, covered with microscopic waxy outgrowths, is quite unwettable. The outer wax layer controls the permeability of the skin to water vapour; for example, when it was removed from the leaf of a cultivated apple tree permeability increased from 15 mg per sq cm to 107 mg. Without its wax a leaf would rapidly shrivel up as its internal moisture passed into the air.

Leaves may also have insulation in the shape of hairs in order to keep the leaf cool, often reflecting the sunlight in hot climates, or the ultra-violet light at high altitudes. With the transparent epidermal cells the skin forms a layer of sufficient rigidity to protect the tender tissues within. At the leaf edges further strength is given by several layers of epidermal cells which hold the transparent 'bag' firmly and resist damage to this vulnerable area.

The number of stomata is normally so great that if they were always open the leaf would evaporate almost as much water as it would without its skin. The capacity to open and shut, however, can render the surface almost entirely impermeable to movement of water vapour. The efficiency of stomata varies, but in principle they control the evaporation from the leaf when external conditions are dry or windy so that water vapour cannot collect round the openings.

Plants which grow in damp conditions, or have easily wetted leaves, may have stomata raised on stalks composed of epidermal cells, as in the cucumber; any gardener will know that this plant prefers a very moist atmosphere under glass because the leaves have to encourage loss of water in order to create an adequate exchange of gases. Conversely, plants which live in dry conditions very often have pores sunk well into the leaf surface, so that basic evaporation is as low as possible, a little pocket of water vapour tending to remain in the sunken pit above the pore. Such sunken pores may additionally have a mass of hairs above them to reduce vapour movement still more.

It is always necessary to remember that these pores are not there solely to control evaporation loss; they also control the intake of essential carbon dioxide. If they are closed in dry conditions, photosynthesis slows down and stops. Thus in a drought plants lose out not only from lack of water but also from starvation. The 4C plants mentioned in Chapter 5, which live in arid conditions where the stomata are frequently almost closed, owe their efficiency to a specialized process which makes it possible for the plant to operate adequately at very low levels of carbon dioxide, and extremely well in damper conditions when the pores are open.

Lastly, the stomata normally close up at night when the plant as it were recharges itself with water, and gaseous exchange is unnecessary since photosynthesis does not occur. The need for recharging with water is great, because during the day the plant passes out a very large amount. It has been suggested that the cooling effect of evaporation is necessary to protect leaves from scorching on hot days, but this is a minor feature even if true. The overriding object is to absorb carbon dioxide in order to make food, which is hard work considering that the air contains only 0·03 per cent of this gas, so that a great deal of air has to move around inside the leaf tissues. To absorb carbon dioxide adequately the pores must be as open as possible, and so water vapour escapes and more must be dragged up through the trunk or stem and sucked in by the roots. This relatively inefficient system is a necessary evil which works under normal conditions and has developed because the photosynthesizing operation is of paramount importance – one of the major problems plants had to solve on emerging from the sea. Highly adapted plants such as succulents have partly overcome the problem and reduced their transpiration rates enormously.

Stomata, incidentally, occur on green stems, especially in plants such as succulents and switch-plants which have lost all their leaves, and also occasionally on roots. Simple breathing pores occur in lower land plants like liverworts because in the moist conditions in which such plants live there is no need for lips to close the mouth. In liverworts the pore is at the top of an air chamber, on the floor of which are special photosynthetic cells to take advantage of the entering gases.

I have already mentioned how the twigs terminating the branches on a tree carry the leaves in such a way that each receives the maximum light and air. Sometimes this is mainly a matter of holding all leaves at right angles to the light source; in others a 'leaf mosaic' is produced.

This may not be very evident in trees such as a beech where a great mass of leaves is produced on slender, whippy twigs, but in some trees it is quite striking. If the leaves of a shoot of the ordinary sycamore or field maple are viewed from the end, a remarkable interlocking of forms can be seen, which gives both older and younger leaves the maximum light reception. Palm trees and others such as tree ferns and cycads with a crown of spreading leaves or fronds, especially if they are divided pinnately like feathers, radiate to give every leaflet light. A similar effect is achieved by the pinnate leaves of the tree of heaven. Many tropical trees have very long-stalked leaves radiating from a crown, each leaf making radiating deeply cut lobes. One such is known as the parasol tree because it absorbs so much of the sunlight with its foliage that it is used as a shade tree in cocoa plantations.

It is of course not only trees that produce leaf mosaics. An ivy makes a good example of a tight jigsaw; the leaves of begonias are lop-sided to fit a rather looser pattern. Some plants have two sizes of leaf so that every available space is used, such as the deadly nightshade. The elm combines a small unevenness of form with difference in size. The most striking examples of leaf mosaics are usually found in plants of shady places, where every part of every leaf needs to receive all the available light.

Leaves are able to put themselves into a good light-receiving position, especially if they have long stalks, because frequently at the base of each stalk there is a swollen area called the pulvinus which can twist the whole leaf according to its needs: a little knot of cells operating like a servo-motor. The elongation of the stalk ensures that older leaves are thrust further and further away from new ones; the resulting leaf-pattern is therefore composed of leaves with different stalk-lengths. Such arrangements are very strikingly demonstrated by aquatic plants with floating leaves, which must of course deploy their foliage in one plane only. Water plants which live in running water frequently have thread-like or much-cut foliage which lessens the danger of tearing.

The rosette is a classic example of providing maximum light-reception, and is best seen in virtually stemless plants such as house-leeks and saxifrages, although it is adopted by many others. The leaves radiate from a central stem, the oldest underneath ones being the longest, and the newest the shortest. This ensures that each leaf has roughly the same amount of light.

Certain trees do not seem to obey these rules, notably the conifers, which tend to cover their twigs in 'needles' or to have leaves reduced to scale-like proportions covering the actual surface of the shoots like a snakeskin. These are mainly trees of difficult climates, and their object may be to produce as many photosynthesizing positions as possible to take advantage of whatever light and air there may be.

Some conifers are notably symmetrical, carrying their branches in elegant flat tiers, like the Norfolk Island pine and other monkey puzzle relations. This arrangement is not restricted to conifers; the Panama tree looks like a candelabrum a hundred feet tall. The leaves of the lowest tier may be about 110 by 55 cm, and as the tiers ascend the leaves become smaller and smaller. The deeply lobed leaves of the pawpaw radiate from its crown like the spokes of a wheel. *Cornus controversa* is called the wedding-cake tree because of its tiered arrangement; *C. alternifolia* the pagoda tree. Although most palms adopt the spreading crown, one of the most remarkable growth-forms is that of the travellers' tree, whose 8-m banana-like leaves spread out from the root like a gargantuan fan.

In very hot areas it may not be an advantage for leaves to receive the full intensity of sunlight. There is one group, mostly relations of the lettuce, which twist their flat, erect leaves so that they are edge-on to the midday sun. They are known as compass plants, and their directional accuracy was used by the early hunters on the American prairies. Some plants move their leaves to the vertical when in the sun and the horizontal in the shade, like the silver lime which, with its white leaf-undersides, presents an intriguing two-tone effect on a sunny day. Where the sun is strong, as in Australia, the eucalyptus trees hang their leaves vertically by day. Many monocots, the gladiolus is an example, have long narrow leaves folded down the centre so that the upper surface is shielded from the sun most of the time. Leaves whose lower ends run wing-like down the stems are a similar sun-avoiding device.

Leaves may be arranged not only in relation to the reception of light but also of rain. In trees the pattern of the leaves is often such that rainwater is constantly conducted outwards, so that eventually most of it falls to the ground at the outer edge of the leaf canopy, and it is at that point that most of the tree's absorptive roots will be found. This direction of water is usually associated with grooves in the leaves, as in the horse chestnut. Sometimes hairs in the grooves help to direct the flow. In herbaceous plants a smaller array of leaves may act in the same way when the roots spread outwards, but when the roots spread mainly downwards, or the plant has a vertical taproot, channels created by the leaves and their stems may direct water inwards so that it runs down the main stem. Hyacinths and tulips demonstrate this characteristic as well as plants such as the wild lettuces whose leaves end in flanges which clasp the stem. Other tall perennials, for example mulleins, have series of leaves decreasing in size as they ascend the stem, whose outer ends turn downwards so that there is both an inner flow to the stem and an outer one eventually channelled to the ground around the perimeter of the bottom leaves.

It has already been noted that the first leaves of seedlings, the cotyledons, may be quite different from the adult ones: this is almost always so in dicots, where the seed leaves are often fleshy and form a food store while the seedling's roots become established. In monocots such as palms they are usually similar to an adult leaf. In gymnosperms such as conifers there may be several narrow cotyledons.

Some plants have a juvenile stage in which leaves of quite different form to the mature ones are produced. The most notable groups in this respect are the eucalyptus and the cypress families. Eucalyptus often have roundish leaves when young and narrow sickle-shaped ones when mature. After a completely juvenile stage such plants may have clusters of juvenile leaves within a crown of mature ones. They can sometimes be kept indefinitely juvenile if they are regularly cut back to encourage young growth.

In some climbers much larger foliage is produced when the plant enters its reproductive phase. The ivy is the most familiar temperate case: it makes a bushy growth of rounded leaves when flowering, as opposed to the normal arrow-shaped climbing foliage on stems with aerial roots. The creeping fig normally has leaves about 2 cm long; when it is about to flower, stiff 10 cm leaves are produced on erect growth. In the aroid *Scindapsus aureus* the change is even more striking – the normal 6 to 8 cm undivided leaves are replaced by 60-cm leaves which may be lobed or deeply cut. Many aroids behave like this.

In all these climbers the leaves of reproductive maturity are produced when the climbing stem reaches a position of more light and air, or is unable to climb any further. In all of them, too, the mature flowering form can be propagated and grown on without reverting. One is reminded of the axolotl, that aquatic, gilled form of a salamander which can remain as it is indefinitely unless circumstances change and turn it into a terrestrial salamander.

Eventually leaves die. Those of annuals perish with the plant, of perennials wither with the above-ground stems. On deciduous trees and shrubs they fall annually in autumn; on evergreens and broad-leaved trees in tropical forests with no annual rhythms a few fall all the time, to be continuously replaced. The turnover of leaves in palms, bromeliads, century plants and the like, which are often very hard and stiff, is extremely slow.

If a leaf simply died and dried up it would remain on the tree as an unwanted relic. As it ages a layer of special cells is formed at the base of the leaf or its stalk, the separation or abscission layer. These are small cells which become externally gelatinous, a state in which any wind movement, or sometimes just gravity, will cause the leaf to become detached. At the same time, the cells on the stem side of the separation layer form a waterproof barrier so that the wound is effectively sealed. (Exactly the same occurs in the eventual dropping of fruit.) All these separations are caused by the action of chemical messengers, as described in Chapter 11.

The disposal of leaves annually is essential to the deciduous tree or winter-resting plant, which must not be active in winter; but on any tree leaves will age. Young deciduous leaves are bright green, which is due to the penetration of light right into the cells, where internal reflection makes every chloroplast operate fully. But the cell walls grow thicker as the months go by, and waste materials from the protoplasm build up in them, so that the green is dulled. Deciduous conifers such as larches and dawn redwood go so far as to shed entire branchlets. Sooner or later, in one way or another, the energy-producing units of every plant burn themselves out and must be replaced; the chemical mechanism involved is described in Chapter 10.

Once this begins to operate the protein content of the leaf declines, and waste products accumulate. These cause leaf yellowing in many plants, and in some deciduous trees those spectacular tints of gold, red and flame, purple and even pink, so memorable especially in the eastern United States, which give autumn its special bitter-sweet beauty before the bare branches are exposed to the long months of winter.

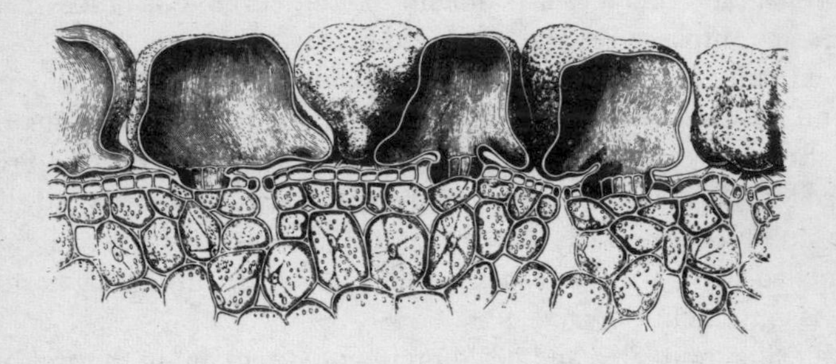

8 · *Eccentric and Bizarre*

'When my reason had ceased rocking on its seat, I rent the welkin with a cry of triumph . . . and in awe-stricken silence contemplated for the first time . . . the most wonderful plant in all the ranges of the Alps. It was an imperial moment, tremendous, breathless.' So Reginald Farrer, Edwardian plant hunter, in a typical lush passage; the event was his first finding of 'the Ancient King', *Saxifraga florulenta*, weird endemic of the Maritime Alps.

But it is quite uncharacteristic of plant hunters in general, most of whom are extremely laconic when recording their first stumbling on one of the many astounding manifestations of the plant world, which even today, seen in a garden, will cause one to catch one's breath. Even the usually magniloquent Schomburgk, when first confronted by acres of the huge leaves of the giant water-lily, could only write 'Lo! a vegetable wonder!', although he was less down to earth than a local botanist, Dr Santos:

> The aspect presented by the *Victoria* in its native waters is so novel and extraordinary that I am at a loss to what to liken it. The similitude is not a poetical one, but assuredly the impression the plant gave me . . . was that of a number of green tea-trays floating, with here and there a bouquet protruding between them.

He did admit that 'when more closely surveyed, the leaves excited the utmost admiration, for their immense and perfect symmetry'.

Be botanists as they may, there are many eccentric and bizarre variations on the basic structural theme of the plant, quite apart from the differences by which we classify plants into families, genera and species. Such variations are almost always highly specialized adaptations to various situations, although in cases like the giant water-lily they may just seem to be taking advantage of a perfect habitat.

We may start with roots, which combine the feeding function with that of providing anchorage and stability. Many tropical trees produce thin, plank-like buttress roots, sometimes starting fairly high up the trunk and descending obliquely to the ground, reminiscent of flying buttresses on a cathedral, sometimes emerging just above soil level and snaking along the soil surface for many yards. From them vertical roots penetrate the soil. Such trees include various figs, the kapok, cotton tree, and swamp cypress; many families have adopted this method of taking the strain off often rather shallow-rooted trees. In swamp cypresses the buttresses, which are narrow and sweep gracefully down the trunk, become hollow in old specimens. Though the word buttress is visually apt, these roots act more like guy-ropes, being subjected to pull rather than pressure.

Another bit of engineering for stability, again largely in tropical trees, is the growth of stilt-roots. Screwpines typically have a wigwam-like structure at the base which consists of a series of roots sticking obliquely downwards from the stem, more roots being produced higher up as the tree ages. This props up the rather weak stem and the large head of foliage. Sometimes the stilts may produce secondary props near their own base. If a screwpine is forced out of the vertical, it will make stilts all along the side of the trunk nearest the soil; a tree that has fallen nearly horizontal produces them under the trunk so that it looks like some grotesque centipede. Mangroves make stilt-roots to support themselves in muddy swamps and sea margins, and these form a tangled interlocking mass. In these trees the stilts carry most of the fine rootlets, and are able to make new ones as those below become silted over. While the roots are still in the air, the protective root-cap mentioned in Chapter 5 is clearly visible, resembling a brown thimble which is readily removed.

There is one very odd group of trees which makes stilts in tent-like shapes and has no central root at all; indeed, the base of the trunk often starts well above ground. The seedling roots in the normal way, but as the tree ages the stilts take over and the stem close to the ground dies back. One of these, the African umbrella tree, combines support with reproduction, for many of its stilts produce new stems where they touch the ground.

Sometimes stilts are combined with buttresses, but while the latter are formed by many forest trees, stilts are almost always restricted to plants growing in swampy or flooded conditions, and sometimes only appear if the plant is in a wet place.

Accessory roots of a different kind are produced by banyans, which are species of fig. As the branches extend they send out fine vertical roots which eventually touch the earth. If they are able to penetrate it they rapidly thicken up, becoming pillar-like. This allows the original tree to spread outwards more or less indefinitely and to occupy several acres. Banyans are sacred in India, where man often assists the air roots to penetrate the soil, and Buddha is said to have meditated under one; indeed, to be in the centre of an old banyan is like being in a sacred grove. There are a number of vast specimens about; a typical one in Ceylon provides shade for an entire village, with 350 large pillars and about 3,000 more slender aerial roots. Another Indian record, from the Andrha valley, is of a tree with 320 pillars with a crown 600 m in circumference. A famous specimen which grew in the Calcutta Botanic Gardens was only 280 m round but had 464 pillars.

In roots of these types – buttresses, stilts and pillars – the structure, at any rate where it is above ground, much more resembles that of a stem than a normal root. This is not surprising since the roots exist for support rather than conduction, although this is also carried out and must be important to vast trees like an old banyan.

A rather different form of rooting is carried out by 'stranglers', which are described in more detail later; these start life as epiphytes and send down aerial roots which, after reaching the ground, gradually grow around the original host tree, and eventually choke it to death. Many figs do this and also the rata vine. Another species related to the latter, which grows as a normal tree, sometimes produces masses of slender fibrous red roots from the lower branches, which do not reach the ground. This is an unexplained phenomenon, although it is possible that they aid in moisture absorption.

In this way they may resemble the aerial roots of epiphytic plants such as bromeliads, orchids, and climbers of the aroid family, which have to absorb moisture from the air and off the trunk of their host, and also secure the stems in position.

Some trees which grow in swamps or mud develop curious breathing roots, or pneumatophores. The mangrove tribe does this in particular, the roots sending up through the mud masses of erect tapering growths, the upper parts of which contain apertures and air passages and allow air to percolate down to the roots – an interesting proof that air is essential for the well-being of roots even though they are not involved in the process of respiration. Several other groups of tree have these breathing roots, including the palm *Raphia* and the wild nutmeg, whose root-lungs emerge in a contorted tangle from the floor of the swamp forest. In the swamp cypress, the roots of which are entirely submerged in the rainy season, the pneumatophores are in the form of knobbly 'knees', pyramidal structures sometimes up to 3 m tall, formed of soft spongy wood with spongy bark, and becoming hollow in old specimens.

Tree roots of separate trees often join up underground by self-grafting when they are pressed together, so that a wood consisting of one kind of tree may have a positive interlocking web of roots. This is one reason why cut trees often sprout so readily – their roots are powered, so to speak, by the leaves of their neighbours. Five years after some Malayan rubber trees had been cut down by the Japanese in 1945, their stumps were tapped: they produced as much rubber as neighbouring trees which had survived, because their roots were intimately grafted together. Certain diseases are spread through self-grafted roots, like viruses of citrus and wilt of oaks, and the only way of checking these is to destroy apparently uninfected trees around the diseased ones and to make sure any root connections are severed.

The sheer extent of roots is sometimes surprising. I have already mentioned the 622 km total root-length of a four-month-old rye plant. Even an alfalfa plant may penetrate 10 m down, and an apple tree will do the same. It has been calculated that on average around three-quarters of a typical 'higher' plant's growth is below ground. This is especially marked in seedlings, where nine-tenths of the growth may extend below ground as the young plant seeks moisture to start intensive growth; indeed, a linear measurement of all the root-hairs would probably result in a ratio more like 99:1. Roots extend further and further in proportion to top growth in arid conditions; thus baobab roots may spread for 100 m all round the tree, while tamarisks in desert conditions have been recorded as penetrating an astonishing 50 m down. The record for deepest roots is held by a wild fig in South Africa with estimated penetration of 120 m.

The distinction between root and stem is not always very clear, as the examples of the stilt-roots show. But basically the stem joins roots with leaves and flowers. It can vary in thickness from the half millimetre of some ephemeral annuals to the 17 m record of a sweet chestnut in Sicily, and in height from virtually nil to over 100 m. Even annuals can make relatively large stems, the 5·84 m of the common sunflower probably being the height record, while the policeman's helmet or Himalayan balsam thickens its hollow stem to at least 4 cm in a season. Some of the climbing annuals can exceed the sunflower's height, but they do not have to support themselves.

Although immensely thick tree trunks like those of redwoods are remarkable, very tall slender trunks are perhaps even more so, considering the strains they have to endure. Thus the Andean wax palm exceeds 60 m in height and is only a metre or two across at the base. The Guadeloupe palm reaches 17 m with a thickness of but 40 cm. The Douglas fir must be a strong contender for the relatively slimmest tree in the world, a 100 m specimen having a basal thickness of only 3 m. The slender 64 m flagpole in Kew Gardens was trimmed from an 82 m specimen (which was incidentally 371 years old): it is only 82 cm across at the base and 30 cm at the top.

Many trees are not just thick, they are fat, by which I mean of unexpected girth in relation to their height. The baobab can only be called grotesque with an average height of 12 m and diameter of 10 m;

examples exist where the height is less than the diameter. The baobab's trunk is filled with pulpy matter so soft that a bullet will pass right through it. Despite this, the trees are remarkably long-suffering; even if the interior is hollowed or burnt out (there is a famous South African example used as a bus shelter) the tree will survive.

Other trees have been given names appropriate to their shape, such as the flask tree, with a cylindrical trunk about six times as long as thick, and the similarly proportioned kapok; the bottle tree of Australia, swelling gently and then tapering to the crown; the belly palm of Cuba, which has a narrow trunk, then an oval swelling, finally more slender trunk. The Brazilian cotton tree looks like some monstrous turnip, its ovoid stem, about three times taller than wide, emerging from a small cluster of roots and giving rise to a scrawny horizontal crown. All these obesities are for water retention, and especially in deserts trees develop swollen stems as important water reservoirs; such include adeniums and elephant trees, which are odd also because of the smooth grey skin, looking very like an animal's, which is stretched over the muscular-seeming stems. The well-named tortoise plant or elephant's foot is a climber, thrusting up annual stems from a more or less rounded trunk, the corky bark of which is dissected into irregular polygonal areas by deep furrows. Idrias have stems which taper steadily from a relatively narrow but still thickened base to almost nothing; they can reach over 20 m, are leafless but thorny, and their fantastic appearance, especially when branched, has led to the name boojum tree. This is one of the most idiosyncratic of plants, never consistent in its stance. A rather enlarged version of this is the cucumber tree, to be seen in Socotra, which branches feebly at the apex of its conical trunk.

Other trees are bizarre to our eyes. Some of the Cuban wax palms, to quote the inimitable E. J. H. Corner,

> develop incredibly condensed, tussock-like crowns of stiff, barely stalked, crowded leaves below which the old and dead hang like the tattered petticoats under a new Andean poncho. *Copernicia macroglossa* and *C. rigida* are, indeed, guerrillas in the vegetable warfare of Caribbea; their leaves have astonishingly long ligules, and these tongues, one to three feet long, may remain on the trunk long after the blade has fallen, like the rude spouts of innumerable gargoyles.

In Australia the 'Blackboys' have a weird appearance. Their leaves, long and narrow, create a mop-head atop the dark, stubby trunk, and the plant's appearance is even odder when the narrow, vertical flower-spike thrusts upwards (which is frequently after a bush fire). Their botanical name *Xanthorrhoea* means 'yellow flow' and refers to the thick yellow resinous sap that is exuded around the leaf bases and hardens there. Every year, when new leaves are produced, more resin exudes, and so the trunk builds up with a central core of live conducting tissue surrounded by a resinous tube in which the old leaf bases are embedded. The resin is valuable in the making of varnishes, dyes and other things, and these living plastics factories are therefore cut down to extract it. It has been suggested that really old specimens, up to 6 m tall, may conceivably be 5,000 years old.

There is reason behind trunk thickening, which is usually water storage tissue, but there seems no very good cause for perforations in trunks. Some degree of perforation occurs in old olive trees as one can see in the Mediterranean, but there are some tropical trees which have large holes in their trunks. E. J. H. Corner described Adinas as 'archery-window' trees, the effect of an old specimen, which becomes hollow, being 'like a narrow wooden tower, or cylindrical shell, fitted with archers' windows'. One such tree, 24 m tall, could have imprisoned a man, but the perforations were only large enough for a hand to pass through. Similar trees in parts of South America are called mosquito trees because the perforations hold water and harbour these insects.

Certain trees and shrubs make more or less horizontal stems. There are prostrate junipers which make great circular mats sometimes 20 m across almost flat on the ground. Others make an impenetrable thicket of long horizontal stems about a metre deep. In Tasmania there is a notorious tree called the horizontal which grows both on its own or as an 'under-storey' in some forests. It has almost horizontal trunks, the branches from which bend over horizontally themselves, interlacing with their neighbours sometimes several metres above the ground. To quote Winifred Curtis in *The Endemic Flora of Tasmania*,

> In this way dense and springy platforms develop, often at a considerable height above the ground. They are dangerous to venture on, for a fall from a decayed, moss-covered branch traps the adventurer in a lower storey maze of

trunks and branches – a twilight world without landmarks, and from which it can be impossible to climb out.

There are tales of motorists who have driven on to the horizontal stretch alongside the road, with alarming results. In southernmost South America the southern beech makes a rather similar tangle of boughs which explorers have found very taxing to overcome.

Some prostrate shrubs root as they go and can cover vast areas, like the unique box huckleberry colony in the United States estimated at 40 hectares, and possibly 13,000 years old.

A quite different but equally remarkable style of growth is that of *Ficus benjamina comosa*, a form of fig which produces a single trunk about as high as a man, above which spreads an unsupported horizontal canopy over 50 m across and only a couple of metres deep.

The stems of climbing plants are remarkable mainly because they are so long and slender. The measured record for a species of rotang or rattan, which are climbing palms, is about 200 m long, and with a constant girth of not more than 4 cm, but there are almost certainly much longer ones. I cannot resist a quotation from Kerner and Oliver (*The Natural History of Plants*) describing, in one of their purpler passages, the forest home of the tropical climber or liana:

> Everything climbs, winds, and twines with everything else, and the eye in vain attempts to ascertain which stems, which foliage, which flowers and fruits, belong to which. Here the lianes weave and work green draperies and carpets in front of the stems of the forest border, there they appear as swaying garlands, or hanging down as ample curtains from the branches of the trees. In other places they stretch in luxuriant festoons from bough to bough and from tree to tree, forming suspension bridges, even actual arcades with pointed and rounded arches. Isolated tree-trunks are transformed into emerald pillars by the covering of woven lianes, or more frequently become the centres of green pyramids over the summit of which the crown spreads out in verdant plumes. Where the lianes have grown old with the trees on which they cling, and the older portions of their stems have been long stripped of foliage, they resemble ropes stretched between the ground and the tree-summits . . . Sometimes drawn out tightly, sometimes limp and swaying, they rise up from the undergrowth of the forest ground, and become entangled and lost far above the boughs. Many are twisted like the strands of a cable, others are wound like a corkscrew; and others again are flattened like ribbons, hollowed in pits or shaped into elegant steps – the celebrated monkey-ladders.

Lianas, or climbers, are in fact a very remarkable and presumably rather late piece of evolutionary adaptation, competing with immense trees despite their vastly lesser bulk, and making sure of their share of light and air by scrambling heedlessly over and through other plants. Nor are such climbers confined to the tropics as any temperate wood will show, with its old man's beard, brambles, roses and honeysuckles. Many very different families of plant have produced climbing members.

Climbers get up in various ways. The simplest is the thrusting, scrambling stem, very often carrying thorns or barbs like grappling irons, which grows upwards through or over whatever other plant offers unwilling support. Some have amazing powers of growth, like the Himalayan *Rosa gigantea* which flings out arching growths 10 or 12 m long. It may sometimes put out side-growths which help to make it irremovable once it is among other vegetation. Others twine spirally, making contact with supports by wide sweeping 'search' movements, often assisted in fixing themselves by bristles or barbs. Some climbers press themselves against a trunk and fix themselves to it with aerial roots. A further adaptation of this is to make many small branches which tend to fuse where they cross, so that a ladder-like lattice of growth is produced, as in the tropical *Clusia* family. This can sometimes entirely sheath its supporting tree.

Yet other climbers have special devices to help them cling. These may be tendrils which grow corkscrew-like around anything they touch. Such tendrils may grow out from the stem, as in passion-flowers and gourds, which can have tendrils 75 cm long. Other tendrils are the prolongations of leaves, as in the cup-and-saucer vine, a herbaceous climber which can grow 10 cm in a season. In some tropical climbing bignonias the leaf-prolongation resembles a three-pronged claw which searches for support and twists tightly around anything it reaches; the plant then puts out long aerial roots on to the surface concerned. In other climbers the base of the leaf-stalk replaces the tendrils, having the power of very rapidly twisting round and clamping on to anything it encounters, especially while the leaf is not fully developed. Clematis climb in this way.

Another development is that of tendrils with adhesive discs at the ends, which secrete a glue-like substance on touching a surface, and become fixed. This occurs, for instance, in Virginia creepers, allowing them to climb up absolutely smooth surfaces. Very often the discs enlarge on contacting a surface, spreading out flat. As with the corkscrew type, contact is usually followed by a spiral contraction in the tendril, drawing the main stem of the plant firmly down near the surface. Frequently such tendrils are light-avoiding, so that instead of aimlessly moving in the air they are on the shady side of the stem more likely to be close to a suitable support. They often seek out dark crevices, and in some species the tendril that finds a crack will swell up within it so that it is immovable.

In most of these climbers we can note a stimulus to surfaces which is followed by the production of specialized roots, by twining, by tendril-tightening and so on. A climber which does not find a surface is apt to collapse, although the more vigorous, like the rattans, will coil about on the forest floor and seek support away from their roots. Some, such as the hop, guard against lack of immediate support by making several growths at once which wind round each other and, rather like the rope in the Indian trick, can erect themselves to a considerable height without collapsing.

Judging by their wide distribution, and their equally extensive practical uses, the rattans are among the most successful climbers. They have very long internodes, often over a metre long, which enable them to grow quickly; a mature rattan can grow 3 m a year. In these palms it is the leaf which acts as the 'business end': it remains furled after developing a long, flexible barbed 'whip' at the end, which is pushed up by the extending stem below. Once support is reached the leaf opens, forming a further grip on the host like an opened umbrella. These rattan whips make passage through the jungles that they infest both hazardous and agonizing, and E. J. H. Corner goes so far as to suggest that the elephant and rhinoceros have developed their very thick skins partly at least for immunity to them.

Rattans and other climbers are remarkable also for the quantity of water in their stems and the speed at which they carry it. It is no traveller's tale that a refreshing drink can be obtained from a rattan: if a piece of the liana's stem, say 3 m long, is cut and tipped up it will yield two or three cupfuls of clear water. What is more, owing to the way the very long conducting vessels mesh within the stem (they may be from

3 to 6 m long), the stem can be cut again and will yield further water. Within the stem at least a third of a litre of water may flow per hour at a speed of 1 to 2½ m per minute, which is the fastest speed recorded from any plant. (Some comparisons include up to 2 cm per minute in conifers, 7 cm per minute in deciduous trees like beech and maple, and 70 cm in oak and ash.)

There are incidentally other trees that will produce water in this way. Thus the Australian pine or she oak, and the African tulip tree, have been called respectively bleeding tree and fountain tree because of this capacity. There is even a milk tree from South America which on being cut produces, to quote the famous explorer Alexander von Humboldt, 'a profuse flow of gluey and thickish milk, destitute of acridity and exhaling an agreeable balsamic odour'. It is, as von Humboldt points out, unusual for milky sap to be other than acrid and poisonous. A cheese-like curd is made from the 'milk' of what Edwin Menninger has called 'this snack bar of the forest'. Many palm trees are tapped for sugary sap, which is often fermented.

The 'monkey-ladders' of the quotation on page 86 are lianas such as *Caulotretus* and *Bauhinia* species in which the stems are ribbon-like. The central part undulates strongly, but the edges remain more or less straight, so that a series of steps with handholds appears to be formed. This is a device to ensure that if the stem has to undergo tension, by the host tree growing, for instance, the strong edges take the strain leaving the essential conducting tissues in the wavy centre unharmed. Some flattened lianas simply twist spirally, which also increases resistance to tension since a certain amount of unwinding may be possible.

Quite a different aspect of stems occurs when, in response to an arid habitat, the plant has vestigial leaves or none. The stem becomes green and photosynthetic, with breathing pores. Very often, as in brooms, the stems remain long, thin and whippy, when they are called switch-plants. The African passion-flower relation *Adenia globosa* combines a dome-shaped succulent base with a mass of green radiating stems clad in spines. In other cases, however, the stems become modified into leaf-like shapes. This occurs in the familiar butcher's brooms and in the fierce *Colletia cruciata*, in which the stems have become flattened and sharply pointed, the leaf-like parts (known as phylloclades) being arranged in an alternating cross-pattern. One can tell when a stem is acting as a leaf because not only does it not have a clearly differentiated stem but it will carry flowers, which a true leaf normally never does. Sometimes, as in the peculiar Madagascar tree called *Phylloxylon*, one leaf-like phylloclade sprouts out of the next in series, while the small purple flowers appear round their edges in season. Other plants develop photosynthetic spines instead of leaves; another colletia, *C. armata*, is an example.

Stems are not always above ground. One example is in the African genus *Parinarium*, whose tropical members grow as normal trees. In the South African *P. capense*, however, the tree has, as it were, sunk below the surface, having a woody vertical trunk and horizontal woody stems just below the soil, from which the bunched leaves and flower-heads emerge. Some alpine willows may also branch below the surface with a similar purpose of avoiding the cold.

The skin or bark of trees is often remarkable. It varies from the silky membranous material of *Melaleuca*, apparently endlessly peelable, to the stiffer parchment of birches, the peeling plates of eucalyptus and plane, to deeply fissured barks like oak and pine. The soft spongy bark of giant redwood (which is fireproof) can reach 60 cm thick, while the outer bark of the cork oak can be up to 20 cm thick. Commercial cork bark is the outer layer 3 to 5 cm thick, which can be peeled off in great pieces; I once witnessed an entire trunk about 5 m high and over 30 cm across having its outer bark removed in one denuding piece, exposing the glowing red inner bark.

The third important plant structure is the leaf. As I have suggested in the previous chapter, leaves show endless variety in shape and also in size. They range in size from scale-like organs almost invisible to the naked eye, via the 3 to 4 mm leaves of the lesser duckweed, tiniest of flowering plants, to the gigantic plume of the palm *Raphia taedigera*, with a petiole or stalk up to 5 m long and the blade carrying the leaflets up to 22 m long and 12 m across. This is the biggest leaf recorded and makes any other giant leaf seem puny. Mention should be made, though, of the genus *Gunnera*, which includes a species with leaves barely 3 cm across – *G. dentata* – and one with rhubarb-like leaves over 2 m across on 2 m stems – *G. manicata*, well known in Britain as an ornamental plant. Leaves of plants from topical forests tend to be large and of similar size; those of arid Mediterranean regions and of cold ones are almost always smaller.

Mind you, exotic plants do not have sole title to enormity of leaves. An ordinary edible cabbage over 3 m across and 120 cm tall was reported in the British press in 1971; while there is a once well-known strain known as the giant Channel Islands cabbage which grows up to 6 m tall, with a head 1½ to 2 m across. Twenty plants could produce enough food for one cow for over a year. The very hard stems were used for palisades, even rafters for sheds, and were also fashioned into walking sticks.

Nor can one omit the giant water-lilies, species of *Victoria*, whose circular floating leaves, of which a plant can produce forty or fifty in a season, may be well over 2 m across. They have vertical margins up to 20 cm high, while underneath they reveal a system of girder-like ribs which provide strength and also, being filled with air spaces, flotation. There are a series of large radiating ribs with flanges and a series of smaller cross-ribs. They are strong enough to support a weight of up to 90 kg. When mature the leaf area is around 2½ sq m, while in its early stages it can put on ⅓ to ½ sq m a day. It was indeed the structure of these leaves which gave Sir Joseph Paxton the idea for the curving, hollow metal framing of the greenhouse he designed to house the plant at Chatsworth (the first to flower in Britain), and a little later of the great Crystal Palace. He wrote, 'Nature was the engineer – nature has provided the leaf with horizontal and transverse girders and supports that I, borrowing from it, have adopted in this building.' The famous Palm House at Kew is built on the same principle. It is interesting to note that Dr Santos, already quoted, wrote that the underside of the leaf 'suggests some strange feature of cast-iron, just taken from the furnace'.

The sizes and shapes of leaves are correlated with their habitats, in particular with the amount of wind the plant may suffer. Thus the leaf of the Malayan *Trevezia* is 60 cm across, but though circular in outline it is cut into seven or eight deeply lobed segments radiating from the leaf stalk. A leaf of a single surface would suffer damage in wind which would probably, through decay, destroy the whole. The segments are less likely to be harmed and even if one is damaged the rest remain unscathed. The vast numbers of small leaves of most temperate trees are equally a means of spreading risk, just as are the deeply divided leaves with multiple leaflets of plants as disparate as ferns, the tree of heaven and palms. The banana has an originally undivided oblong leaf which is not particularly strong or rigid, but it is divided transversely on each side of the midrib by parallel strips of weaker tissue, so that in strong wind it tears into narrow segments which can continue to function. A leaf even more adapted to its function is that of the water hyacinth, a floating plant with a crown of radiating foliage which chokes tropical lakes and waterways. Its specific name, *crassipes*, meaning literally 'fat foot', refers to the swollen base of the leaf stem, filled with air spaces, upon which the plant floats.

Many leaves are deeply lobed, but in monsteras the foliage not only has cut-like indentations at the side but a fretwork of irregular holes between the side-ribs, sometimes almost more gaps than tissue, which gives them the popular name Swiss cheese plants. This is conceivably a device to protect against high winds since monsteras sometimes climb up exposed rocks. Another suggestion is that the holes, like the deep-cut lobes of other leaves, permit light to reach leaves below (they are carried more or less vertically on the climbing stems) which it would not do if they were solid. One can only say that this particular natural experiment does not extend beyond this genus, philodendrons and a few other relations.

The exterior coverings of leaves are often remarkable. This is especially the case with those which develop hairs as a protection against too much solar radiation and consequent transpiration. Many alpine plants are protected in this way, as are species from Mediterranean climates – in the Mediterranean, indeed, the impression of the flora is very much one of greyness because so many have hairy leaves. The hairs, which are hollow and air-containing, are usually very small, and one needs a microscope to appreciate their ingenuity and often beauty. They may be erect, or lie flat giving a silky effect; some are spiral, and a microscopic view may look like the gorgon Medusa; some resemble tiny thorns; many are star-shaped or carry 'shields' aloft on tiny stems. Others are T-shaped, some resemble miniature branching trees. There are, of course, plants whose hairs are clearly visible to the naked eye. Mulleins are very often thickly felted; some henbanes have very long silky hairs, and there are species of rhododendron with 'felt' so thick that the Tibetans used to employ it for lamp-wicks. In the small floating fern *Salvinia* the hairs are there to stop the leaf from being swamped with water.

A different method of surface protection is adopted by some succulents such as the South African rocheas where, instead of air-containing hairs, large roundish cells are packed over the leaf surface. These are again hollow and their walls are full of silica, making them flint-hard like a coat of mail.

One of the oddest protective coatings is that of the wax palms. Many live in Brazil in areas of scorching wind, and the leaves secrete wax so thickly that collecting it is very much a commercial proposition. This is done by drying the leaves and then flailing them to remove the wax. The leaves secrete wax in proportion to the extent of the dry-season drought, and around 1½ million kg (13,000 tonnes) a year are gathered: it is the hardest of all natural waxes and has the highest melting-point. Several other plants have thick waxy coatings.

As we shall see later in other contexts, leaves may become transformed or reduced according to circumstances, notably in succulents where the ultimate is a more or less spherical 'plant-body', sometimes composed of a pair of leaves which have almost or totally coalesced. Some succulents and many switch-plants have ephemeral leaves which may either be produced only on immature stems, or drop off in the dry season. The most remarkable leaf-transformations are those of insectivorous plants, in which leaves may be developed into trapping pitchers or have sticky, motile tentacles, as described in Chapter 24.

Some stems and trunks of trees, which become swollen to act as water reservoirs, have already been described. Roots may also swell up: the Mexican *Ceiba parvifolia* has the peculiarity of flowering and fruiting in dry seasons, and is enabled to do this because on its roots it carries miniature 'barrels', roundish growths up to 30 cm across, with soft, spongy tissue, in which water accumulates during the rains and is used up in drought. A species of custard apple from north-eastern Brazil has a subterranean root-stock over a metre across. When the dry period comes, all the top growth – branches and leaves – dies off and finally breaks away. Sometimes it may be two seasons before rain falls again, and only then does the root-stock send up fresh growth, making flowers and fruit.

The discarding of upper parts by this custard apple is reminiscent of the Malayan *Oroxylon indicum*, which is called the midnight horror because of its foul-smelling bat-pollinated flowers. Growing to a gaunt 20 m, it carries enormous compound leaves over 2 m long and wide, which crowd near the ends of the branches. Saplings, which seldom branch till they are 5 m high, look like monstrous umbrellas. When the time comes for a resting period, the tree literally dismembers itself. Each leaf falls apart in section, first each leaflet, then each side-stalk, finally the main stalk. In this way the trunk becomes surrounded with a heap of pieces which gives the tree another name, broken bones plant. The pole-like stems are then left carrying only a collection of huge pods, 60 to 120 cm long, from which eventually gauzy seeds are released to be carried away on the wind. A month or so after the stems have fallen new leaves begin to grow.

Most forms of storage are a combination of food and moisture, and tide plants over a resting period. Such stores may simply be swollen subterranean stems technically called rhizomes as in Solomon's seal and irises, or swollen roots of various shapes and sizes, generically called tubers, which can often be handled and transplanted quite dry. A special form of storage root is the corm, familiar from gladioli and crocuses, which is used up every year while a new corm is produced on top of the old one. Tubers can be enormous: those of edible yams may reach 45 kg and that of the aroid *Amorphophallus titanum* has been recorded at 54 kg and 60 cm across. Even our native black bryony will grow a tuber 60 cm long, which used to be prized as a 'false mandrake' reputed to cure almost any ailment.

Finally we have the bulb, composed of swollen leaves packed tightly around a central axis, which have abandoned their food-producing capacity for that of storage. In most bulbs next year's flower and operational leaves are formed within an embryo. Bulbs are usually subterranean but sometimes carried above ground, when they may have a green photosynthesizing layer. Some orchids have above-ground storage organs called pseudobulbs, thickened stems of various forms which hold moisture over long periods.

Miniature bulbs, called bulblets or bulbils, may be carried on stems: these are usually a distribution device, and the same applies to aerial tubers, which will form roots and act as a centre for a new plant if they come in contact with the soil.

Fleshy storage organs like corms are often associated with contractile roots. This is essential if the new corm is produced on top of the old one, or it would emerge above the soil. The contractile roots drag it down to the original corm's level. The pernicious weed *Oxalis* species makes a translucent conical tuber, at the top of which masses of tiny bulblets are produced. This contracts at the end of the season, dragging any bulblets which have not become scattered down into the soil. Such contractile roots are like those on runners, such as those of strawberries, creeping buttercup and tip-rooted brambles, which once anchored pull the young plant firmly down into contact with the soil.

The need for food storage is usually to overcome a dry season, as in the Mediterranean and Middle East where a great number of fleshy-rooted plants such as orchids, tulips, squills and so on spend the hot summer ensconced many inches down in the sun-baked earth. Sometimes such plants will miss a season and spend more than a year quiescent underground. There are also many fleshy-rooted plants which grow in woodland, where the soil remains reasonably damp all the time. The reason for their resting period is that they develop leaves and flowers fairly early in the season. Once the trees' own foliage makes a dense canopy there is not enough light on the forest floor for the bulbous plants to survive adequately, so they go to rest. Typical examples of such woodland plants are certain lilies and terrestrial orchids.

Many leaves, especially in warm climates, are not coloured green. They may be red or purple, carry regular markings, have veins of different colouring, or combinations of these. The average collection of house plants will produce plenty of examples. The arrowroot tribe, marantas and calatheas, excel in this respect; one of these, *C. ornata*, has a deep bronzy leaf on which the veins are finely picked out in an improbably light pink in a way which suggests that a not very steady paintbrush has been used. *Pilea cadierei* is peculiar because the silvery pattern is in fact the result of air spaces between the cuticle and centre of the leaf. Some of these plants have developed fantastically under cultivation: *Begonia rex* varieties show a wide palette ranging from deepest purples to pinks and silvers, often mingled with green, while crotons or codiaeums combine brilliant reds, oranges and yellows with an extraordinary variety of leaf shapes.

In some cases coloration is probably a masking pigment to reduce excessive sunlight damaging the chlorophyll; in others it may be to absorb the red light that filters through a forest canopy. But the regular patternings and vein-colourings are harder to explain. It has been suggested that they act as attractions to insects fertilizing insignificant flowers, but one can point to many plants with equally small flowers and plain green leaves, and to others, such as the familiar *Aphelandra squarrosa*, which has brilliant creamy-white veins as well as a bold yellow cockade of bracts to centre attention on the flowers.

Such colouring is not to be confused with variegation, which is usually due to chloroplast mutation and seldom persists in the wild, although it is prized and increased by us for ornament.

The final word on leaves must be a description of one of the most bizarre plants in the world, *Welwitschia bainesii* (*mirabilis*), which is virtually all leaf. It is one of the anomalous relics which stand somewhere between the gymnosperms and the flowering plants, although, perhaps fortunately for the future of floral evolution, it was basically an abortive experiment.

Welwitschia has been variously described as a giant turnip, the octopus of the desert, a plant coelacanth. It grows only in a narrow strip of desert close to the sea in south-west Africa, where sea mists and occasional night dews contribute most of its water supply, since rainfall is not only very sparse but may not occur at all for several years. A mature plant has a very thick, stubby tap-root up to 2 m long, somewhat swollen above ground, and a mass of lateral roots. Its water-conducting tissue (xylem) is like that of a flowering plant, its food-conducting tissue (phloem) like that of a gymnosperm. The shoot tip shrivels up soon after germination, so that only two adult leaves are ever produced at the crown and become longer and broader as the centuries pass (old specimens are thought to be 2,000 years old). The leaves fray away at the ends in the wind and literally scorch on the desert sand, gradually splitting lengthways into an irregular mass of ribbons. The biggest leaves recorded are nearly 2 m wide and 8·8 m long; in this case the living tissue was 7·3 m long, but this proportion is unusually high. Because of the need to absorb the sea mists rapidly when they occur there are millions of stomata on both sides – 22,200 per sq cm – which shut up exceptionally tightly in dry conditions. The leaf cuticle, though quite thin, has a central layer of calcium oxalate crystals which reflect sunlight. In their endless growth, at between 5 and 8 cm a year, these giant leaves remind one of a snail's tongue. Male and female flowers are carried separately in cone-like structures around a crater-like central depression which is, exceptionally, nearly 2 m across. It is not surprising that Sir Joseph Hooker called it 'the most wonderful plant ever brought to this country [England], and the very ugliest'.

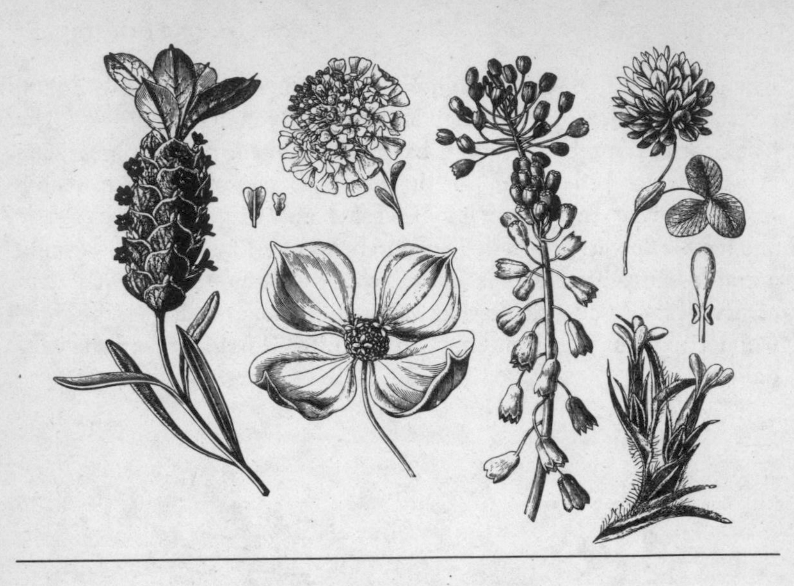

9 · *The Flower*

The flower is the most specialized piece of equipment the plant world has ever created. It is, as we have seen, the product or symbol of the latest group of plants to evolve. In the briefest terms it is a structure combining a protective exterior, created from modified leaves, with the organs for sexual intercourse which carry the sexual cells of the plant, the male pollen and the female ovules.

The typical angiospermous flower has an external ring or whorl, the calyx, whose basic function is to protect the bud and sometimes the open flower against nectar-robbing insects. Where there are distinct calyx segments they are called sepals. Within the calyx is the corolla, whose segments are petals. In the centre the sexual organs comprise the male parts, usually stamens carrying pollen in anthers, and the female parts, known collectively as the pistil. At the base of the pistil is the ovary, from which one or more stalks (styles) lead to stigmas; these are receptive surfaces on which pollen germinates in order to fertilize the ovules within the ovary.

All these parts can vary in many ways. The sepals, normally green and tough, may be ephemeral, or may take on the showiness in colour or shape of the petals in order to attract pollinating animals. They may replace the petals or, as in monocots, the sepals and petals may be indistinguishable. In some families these floral parts may be barely recognizable. Thus in eucalyptus there is a cup-shaped container surrounding the sex organs, the rim of which represents the sepals, while an expendable upper cap protecting the immature flower represents the petals.

Stamens are occasionally the most conspicuous part of the flower, as in *Calothamnus* and *Tristania*, relations of the Australian myrtle, where a small cup-shaped calyx conceals insignificant petals, while emitting four large, almost leaf-like growths carrying stalked anthers on their edges. The Mexican hand-flower is so called because from a blood-red 10 to 12 cm cup project five flattened stamens almost as large as a hand, and of the same form – a tree in full bloom is a sinister sight. In the Brazilian *Asteranthos* an outer ring of infertile stamens, joined by tissue like a parasol, replaces the corolla. Pollen is sometimes, as in orchids and asclepiads, carried in large pollen-masses, not on stamens. Conversely the stigmas may not project on styles: thus in poppies the stigmatic surface forms radiating lines across the top of what eventually becomes the seed-box.

Although flowers tend to be hermaphrodite, there are many in which only one sex is carried. Such unisexual flowers may look the same for the two sexes, but in some cases they are very different. The familiar hazel has a drooping catkin in which many male flowers are gathered, while the solitary female is a small red tuft. In the sandbox tree many insignificant male flowers are grouped in a spike, while the female flaunts a relatively huge lobed corolla. In the Italian cow-parsley relation *Pentagnia saniculifolia*, the large female flower literally carries two or three males dancing attendance on stalks arising from its exterior.

The size of flowers varies from a fraction of a millimetre in the duckweed *Wolffia punctata* to over a metre in the parasite *Rafflesia*. The shapes of flowers are almost infinitely variable, with numbers of parts ranging from a large indefinite quantity to one – thus the *Daphne* relation *Pimelea* has simplified its flowers to a single floral tube, one stamen, one style and stigma, and one ovary.

The ancient world knew about sex in the essential date palm, where male and female flowers are on separate trees, and the sexuality of plants had been experimentally proved in 1694. But it was the Swedish botanist Linnaeus (1707–78) who really brought it to the notice of the world. The basis of his plant classification was the number of the sexual parts; thus a family might be called *Diadelphia Decandria* or *Polyandria Pentagynia*. Linnaeus' view of sex was literal and positively modern. To explain his plant-classes he described *Monandria* as 'One husband in a marriage', and *Diandria* as 'Two husbands in the same marriage', ending finally with *Polyandria* where there are 'Twenty males or more in the same bed with the female'. Sometimes the whole thing got rather out of hand; he had one class called *Syngenesia Polygamia Necessaria* (literally 'Confederate Males with Necessary Polygamy'), as in the marigold with fertile ray and sterile disc florets. Here Linnaeus' exposition described the situation as 'the beds of the married occupy the disc and those of the concubines the circumference; the married females are barren and the concubines fertile'.

This emphasis on flowers' sex organs shocked many of his contemporaries. Thus the Rev. Samuel Goodenough, later Bishop of Carlisle, in a letter, wrote 'To tell you that nothing could equal the gross prurience of Linnaeus's mind is perfectly needless', and we find Goethe worrying over the potential embarrassment of the botanical textbooks to chaste young people. Linnaeus certainly showed the world that plants had a sex life which could, with little recourse to imagination, be equated with that of animals, with, as Wilfrid Blunt puts it, 'all its implications of polygamy, polyandry and incest'.

Erasmus Darwin, grandfather of Charles, brought the subject on to a rather more coy but no less explicit level in his epic poem *The Loves of the Plants*, in which he versified the Linnaean system. Thus the three stigmas and six stamens of 'fair Colchica' become:

> Three blushing Maids the intrepid Nymph attend,
> And six gay youths, enamour'd train! defend . . .

Of the Glory Lily, in which three of the six stamens mature after the first three, he wrote:

> Proud Gloriosa led three chosen swains,
> The blushing captives of her virgin chains.
> – when time's rude hand a bark of wrinkles spread
> Round her weak limbs, and silver'd o'er her head,
> Three other youths her riper years engage,
> The flatter'd victims of her wily age.

As for *Silene*, where 'The harlot-band ten lofty bravoes screen', perhaps the less said the better.

Flowers can be carried singly or in groups, widely or closely spaced. The greatest number in one group, or inflorescence, is probably that of the talipot palm, with 100,000 individual flowers. The inflorescence of palms is often remarkable, emerging from the tree's crown almost like a distinct plant. In some families the flowers are carried in composite heads with one external layer of overlapping scales – an involucre – protecting all, the result often looking like a single larger flower. The daisy family is called *Compositae* for this reason, one involucre often enclosing several different kinds of floret.

Although normally produced at the end of stems, usually on separate stalks of their own, flowers may, very occasionally, sprout directly from leaves, or more often from the more or less leaf-like stems of plants which have abandoned leaves, such as cacti. Occasionally, especially in some tropical trees, the flowers are actually carried on the main stem or trunk, bursting from the bark in an unexpected manner. The cacao tree is one such, and there are several figs which do the same, as well as more orthodox flowering plants.

In some cases flowers are actually produced below ground. Certain Malayan figs do this, making slender stems near the base of the trunk which pass downwards into the soil and there bear bunches of figs on rope-like runners. Some other trees are equally extraordinary, having an orthodox trunk with branching crown, but producing flowers on slender horizontal runners at soil level. These include the 30 m Malayan *Polyalthia hypogaea* and the 10 m West African *Caloncoba flagelliflora*, in which runners radiate for over 10 m round the trunk. In the Brazilian *Geanthemum rhizanthus* and the Malayan stemless palm *Salacca flabellata* the flower-bearing shoots are entirely underground.

The flowers of figs are among the oddest that exist. Numerous, minute and simple in structure, they cover the inner wall of a hollow receptacle which we later eat as the fruit, with access to the outer world via a tiny orifice at one end. D. H. Lawrence called it 'fruit of the female mystery, covert and inward'. In the related Dorstenias a half-way stage can be seen, in which minute flowers are embedded in the surface of a fleshy disc, sometimes lobed like an octopus, which in one way imitates a large flower to attract pollinators, in another seems ready to fold up into a fig-like bag.

Flowers vary greatly in their duration. Orchids can last eighty or ninety days if not fertilized; others are flowers of a day or less. The colourful Mediterranean rock-roses drop their petals in the afternoon; the morning glory closes and shrivels by evening. Even the flower of the giant water-lily only lasts twenty-four hours. When the Mexican tiger flower was featured early on in the *Botanical Magazine*, the author, after extolling its beauty and singularity, exclaimed 'we lament that this affords our fair countrywomen another lesson, how extremely fugacious is the loveliness of form; born to display its glory but for a few hours, it literally melts away'.

To the average human eye the flower, however ephemeral, is an object to be cultivated for decorative purposes, in garden or vase, or to cause admiration, amazement and occasionally revulsion when seen in the wild. To the taxonomist, it is the means of establishing the plant's classificatory status and its affinities in the vegetable kingdom. To the enquiring mind, every flower, however insignificant, has, if examined carefully, intricacy, ingenuity and unity of purpose which create a beauty and interest of their own.

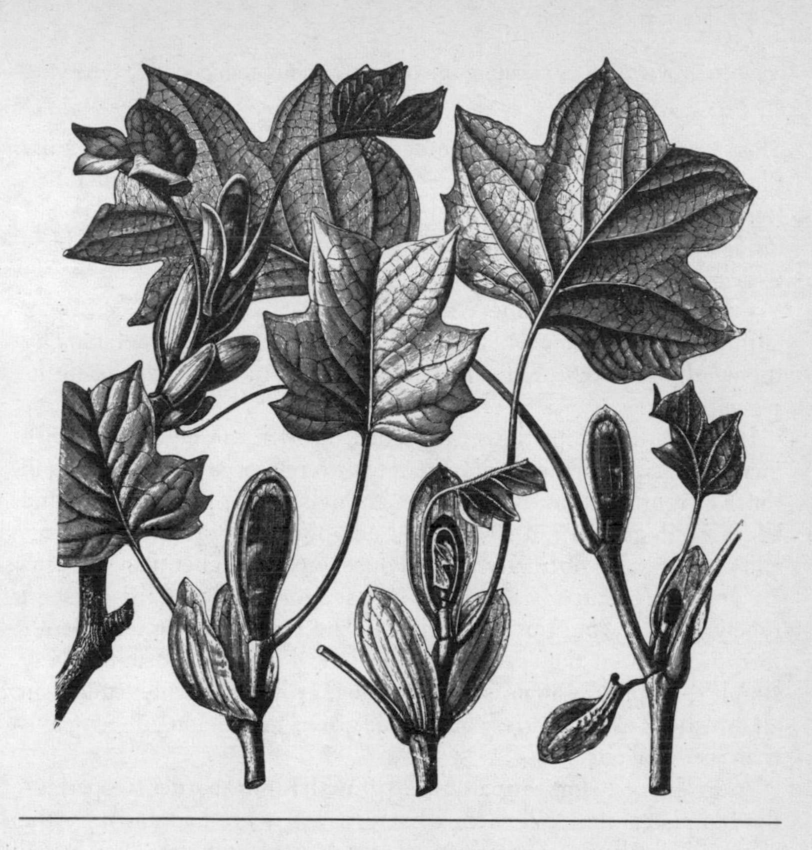

10 · *Growth and Rhythm*

A Chinaman travelling up the Yangtze Kiang river in the nineteenth century lodged at an inn which was in the centre of a grove of giant bamboos.

Before dawn he was awakened by an awful squeaking, whining, and faint screaming coming from the bamboo grove. Alarmed, he called his companion, who explained to him that the noises were produced by the growth of the shoots of young Giant Bamboos, as they pierced their way through the bracts and sheaths clothing the base of each shoot. On warm, moist mornings, bamboo grows at an astonishing speed, and friction produces these terrifying noises.

This fascinating tale was recorded and published by the botanist Blossfeld. One can sometimes detect the sounds of growth from much lesser plants; I recall one winter morning wondering what was squeaking as I worked, and tracing the noise to a bowl of hyacinths whose buds were pushing past the stiff foliage.

Even so, to quote one eminent botanist, Professor G. E. Fogg, 'although most people will feel fairly sure of what they understand by the word, "growth" has little precise meaning in the strictly scientific sense'.

In all plants except the single-celled, growth can be equated with increase of size and the development of different organs culminating in those of reproduction. But while in animals organs develop early and become integrated into circulating, nervous and hormonal systems, and each organ is normally formed once only, in higher plants organs are created repetitively and more or less indefinitely by the constant renewal of growing tips (meristems). The programming of a plant's form is based on the speed and direction of growth. To take a very simple example, the apple is round because growth continues equally in all directions as it matures; but a pear grows faster along its long axis than its radial one.

As in all organisms the patterns of growth for the particular species – the templates, jigs, call them what you will – are laid down by the genetical mechanism carried in the nucleus of each cell. The actual capacity to grow, develop and reproduce depends in many ways on the organism's powers of absorbing nourishment in the right proportions.

I have briefly discussed the major aspects of growth: the soil-searching by roots, the thrusting up of the stem, the deployment of leaves. The sensitive root tips thrust through varying kinds of soil, dealing as they go with problems such as stones or other impenetrable surfaces, or forcing their way into the narrowest crevices. The shoots push through the soil, perhaps protected from damage by crooking the neck so that the growing point has no pressures on it, or by strong, sharp, thick sheaths. That great and observant gardener E. A. Bowles described how snowdrops and crocuses emerge from the soil:

The two leaves are tightly bound round by a sheathing leaf so that their tips are pressed together to form a sharp point that cleaves the ground and makes way for the fragile flower, in much the same way that you put your two hands together and hold them in front of your head when diving into the water . . . It is marvellous what power lies in a growing shoot of a crocus. It makes light work of a hard, well-rolled gravel path. A single crocus leaf is a flaccid, weak instrument, but the whole series of leaves . . . tightly bound by the tough, sheathing leaves, and the sharp and toughened points of the leaves thus all brought together, form almost as sharp and strong a weapon as the underground shoot of one of the running bamboos.

Sometimes growth is less controlled than that; there are plants, like the African fireball lilies, whose new shoots tear through the layers of the old; there are some trees where the flower buds break through the bark on the trunk and branches, and in the shapeless succulent *Muiria* the flower bursts its way out of the soft living tissue in the most painful-looking manner.

But all the time one is impressed by the inexorable quality of plant growth. There is little a determined root cannot penetrate except solid rock, and if there is a tiny crevice roots soon force their way in, as many house owners who have planted, say, poplars too close to their foundations will know. Cracks in water and sewage systems are equally soon found out, the conduits rapidly becoming choked with a mass of fine absorptive root-hairs. Over the years the swelling of a root in a rock fissure can crack a huge block and lift it: Kerner and Oliver record an alpine larch which, over the decades, had split a block of schist, raising the upper part, whose weight was estimated at 1,400 kg, by 30 cm. The demolition of ancient stone-built stgructures in tropical forests, as in Central America and Cambodia, provides further examples of the power of tree roots.

Tree roots crack rock by imperceptible expansion. More spectacular results can be observed with fast-growing plants, the most remarkable I have seen being the lifting of a large concrete paving slab by a horse mushroom. Here the expansion is due to hydraulic pressure, for it is the result of preformed cells expanding rapidly as they take up water. But why the flimsy mushroom is not brought to a halt, or its base is not forced into the soil by the weight on top of the expanding cap, can only be a source of amazement. Shoots too can sometimes penetrate apparently impossible substances. I have records, for instance, of daffodils emerging through 8 cm of Tarmac and flowering successfully, and of suckering shoots of *Rosa hispida* coming through 20 cm of rubble foundations and top layers of gravel and Tarmac. At the other extreme of plant life the germinating spores of single-celled marine algae have been shown to burrow into and under the paint on ships' hulls, cracking it and exposing the metal to corrosion.

Leaves emerge from their buds in a miraculous unfolding and expansion of compressed tissues, lubricated by mucilage from special glands. Sometimes the leaf within the bud is literally crumpled into a ball, as in rhubarb; or it may be pleated or deeply folded, as in beech. Pinnate leaves with multiple opposite leaflets open like the pages of a book. Leaves with radiating lobes unroll their segments one out of the next like some complicated parlour trick. The large round disc of the Indian lotus unfolds like a flag.

The tulip tree keeps its embryo buds in little oval sheaths which drop off when the young leaf is ready to unfurl, revealing the next embryo in its sheath by the leaf stalk. Even more remarkable are the buds of the shrub *Amicia*, which can be likened to one of those intricate Chinese puzzles. The buds are oblong and purse-like; and if you open the outermost there is another purse within, and another within that, and so on for seven or eight purses and embryonic leaves.

Flower buds unfold in similar ways, and to see a very large flower like that of hibiscus starting from what resembles a rolled umbrella and open up into its solid beauty is comparable with watching a butterfly emerging from its chrysalis. The bud of a magnolia emerges from a furry sheath which has protected it all winter; the buds of Californian poppy have little dunces' caps which fall off as the flower expands.

Speed of growth varies enormously. I have given examples of sizes achieved by annual plants; one might add that the giant water-lily is usually grown as an annual in captivity. The fastest growing trees are eucalyptus; *E. deglupta* in New Guinea has been recorded as adding 10½ m in fifteen months, although in normal Australian conditions 2 m a year is to be expected. Some acacias and especially the lamtoro are other very fast growers, like eucalyptus being widely planted to re-green deforested areas in the tropics. However, these are eclipsed by bamboos, which can grow over a metre in a day, and achieve 30 m in under three months.

The slowest growth on record is of a sitka spruce growing at the tree limit in the Arctic, which reached 28 cm in height after ninety-eight years; its trunk was less than 3 cm across. Many trees, such as box and yew, grow very slowly at best. Cycads such as *Dioon* reach only 2 m after a reputed 1,000 years.

The size increase of which a plant is capable in its lifetime is also remarkable. One of the largest specimens of giant redwood has been estimated as having a wood volume of over 1,500 cu m and weighing well over 1,000 tonnes. Since the seed weighs less than 0·005 gm, this represents a weight increase of over 250,000 million times! 'Big trees' may live for nearly 4,000 years and this reminds us that, unlike an animal, a plant can combine in its continuously developing body parts of extreme antiquity at the same time as embryonic growth.

Individual organs may grow very fast indeed: stamens enlarge almost visibly in certain stages, those of wheat at 1·8 mm per minute for example, while a banana leaf-sheath grows at 1·1 mm a minute.

In every living organism the 'programming' which determines its shape and structure is contained in the genes lodged in the chromosomes. The form of organism produced is the result of evolutionary processes already discussed. In plants this programming determines whether the species is an annual or a tree, for instance, or if a climber whether it twines to the left or to the right. It also lays down the way in which shoots and leaves will be produced; this is always in a geometrical sequence.

This sequence is such that it can be expressed as a fraction, e.g. 2/7, which indicates that the seventh leaf of a series will be found vertically above the first, the intervening leaves having made two spirals round the stem to achieve this. If the different spiral-fractions found in the plant kingdom are collected together they can be arranged in three separate series. These are: 1/2, 1/3, 2/5, 3/8, 5/13, 8/21, 13/34 . . . ; 1/4, 1/5, 2/9, 3/14, 5/23 . . . ; and 1/4, 2/7, 3/11, 5/18 . . . These series share the characteristic that each fraction can be obtained by adding the numerators and denominators of the two previous fractions. In fact the most frequent fractions, or phyllotaxes as they are called, are 1/2 (in which the leaves are in opposite pairs), 2/5, 3/5, and 5/13.

These sequences of numbers are known as Fibonacci series after a thirteenth-century mathematician, and the primary and secondary spiral patterns that result in plants represent the best possible solutions to problems of structure, growth, and economy of space occupied. The sequence of florets in a daisy head say, are at an angle of 137.50776 . . .° which, to quote Robert Dixon, 'provides a uniquely flexible design for the harmonious and efficient spacing of plant parts repeated around a central stem even as they grow . . .' In geometrician's terms, the angular interval of 137.5 . . .° is the 'golden angle' of a circle, an expression of Euclid's 'golden section'.

Phyllotaxy creates striking, precise patterns more clearly seen in composite flower-heads, pine-cones, cacti and such compressed bodies than in a widespread leaf arrangement; nature's geometry providing the optimum solution for its purpose.

Flowers themselves are remarkable examples of geometry, as anyone taught botany with the aid of floral diagrams will know. Flowers often deploy their parts in multiples of a basic figure – three, four, five – but some have a remarkable disregard for such simplicity. Consider the passion-flower, where the flower bud is initially protected by three bracts, and opens to reveal ten almost indistinguishable sepals and petals, a 'crown' of an indefinite number of radiating filaments, and an erect central column carrying first five stamens and then, above the ovary, three stigmas. Flowers or flower-heads with an indefinite number of parts, like composites or cacti, usually carry them in a Fibonacci spiral. Otherwise flowers are either radially symmetrical, or zygomorphic, which means that they can be divided into two similar halves if cut down the centre line. It is in zygomorphic flowers that the greatest variation occurs in shapes of the different flower parts or of petal lobes, as for example in a snapdragon or an orchid. Some flowers are irregular, like those of milkworts, where the stamens twist to one side of the flower. Whatever their geometry, floral arrangements are concerned with pollination, as we shall see later.

Programming for a complex flower is remarkable enough, but at least a flower emerges from a bud and has an initiation point in the tissues. Predestination of quite a different order occurs in plants such as palms. I have previously described the complexity of the vascular system of the palm, where a cross-section of, for example, a coconut trunk 30 cm across may reveal some 18,000 vascular bundles. These bundles grow obliquely, being initiated near the outside of the stem and, after approaching the centre, develop sideways to form the conduit for a leaf. The fascinating point is that each such bundle has to start growing sometimes literally years before the tree is mature enough for the leaf concerned to be 'born' and its vascular connections to become functional.

The programming of a plant is often remarkably flexible. This is well illustrated by the powers of different organs to regenerate roots under special circumstances, powers which gardeners make use of when multiplying their plants so that this phenomenon has become commonplace. Roots are not normally produced by stems and even less by leaves; yet one can stick a willow branch in the ground and it will 'strike', or push an African violet leaf into a pot and it will make new roots and a new young plant. The tropical gumbo-limbo tree is widely used for fence-posts. Within a short time these make roots and leaves, and in a few years there is yet another line of the trees along the road or field. Many succulents like stone-crops will make roots at the internodes if they are hung up in the air, and leaves of these and other succulent plants will, if laid on the ground, put out roots at the basal end. A begonia leaf in contact with soil will make roots from its ribs, especially if they are cut.

A plant deprived of its leaves by some disaster very often makes new ones in unexpected places; trees can regenerate leaves on their trunks for instance. The removal of a trunk or shoot will usually stimulate the production of new growth, as anyone knows who has chopped down dandelions or docks, which can regenerate almost endlessly from a truncated taproot, or has removed the main trunk of a plum tree only to find new shoots appearing all over the garden from the buried roots.

Such regeneration reminds one of many animals which will make new parts or limbs if one is damaged, like starfish. Amphibian larvae and newts will regenerate lost limbs, lizards will regrow lost tails, but this power does not exist in otherwise more advanced creatures

Whatever the size of a plant, whether a tiny annual or a forest tree, the main features of its growth follow a general pattern. But on top of this pattern, and on top of the genetic programming which results in a species of characteristic aspects, the growth of a plant is conditioned by rhythms of various kinds, mainly those based on the twenty-four-hour day. As with animals there is no doubt that plants can 'tell the time' – indeed they are living clocks.

Such time-keeping has been known for a long time. Linnaeus worked out a floral clock based on the opening and closing of various flowers, which was reputedly accurate to within half an hour. Such knowledge of floral time-keeping is reflected in popular names for various plants, like four o'clock plant for *Mirabilis jalapa* (a tea-time opener) and John-go-to-bed-at-noon for wild salsify.

The most remarkable individual plant time-piece must be the evergreen Malayan shrub *Wormia suffruticosa*, locally called shrubby simpoh. First, when mature it flowers every day for its entire life of half a century or more. Its buds open at three in the morning and are fully expanded an hour before sunrise; the petals fall at four in the afternoon of the same day. The resulting fruits are ripe in exactly five weeks; they split open into several narrow rays at three in the morning of the thirty-sixth day.

The main kinds of rhythm to be observed in plants are daily and seasonal. As well as flower movements, there is scent and nectar production in flowers, turned on and off at distinct times; sap production in roots; and 'sleeping' of leaves at night (first recorded in 1729). Usually this involves the leaf drooping from an erect position and this occurs in a wide range of plants – one American species, *Madia elegans*, does it twice a day; but *Maranta leuconeura* is called the prayer plant because it folds its leaves together from a horizontal day position. The South American rain tree folds up its leaves not only at night but in cloudy weather, just as many sun-sensitive flowers do; the snow gentian closes whenever a cloud passes overhead and reopens in the sunshine, and will repeat this performance several times. Observers of a total solar eclipse, on Santorini, noted that during it 'small daisy-type flowers closed', but not larger ones.

Even very simple plants may show such rhythms: the flagellate alga *Euglena* and certain diatoms are phototactic, that is they move up and down in the mud in which they live to take advantage of daylight for photosynthesis, retreating downwards at night. Daily rhythm also controls spore-production in some simple algae, and in at least one case there is a rhythm in the rate of photosynthesis in a single-celled alga. There is incidentally no suggestion of such rhythms in bacteria or blue-green algae.

It is important to realize that daily rhythms are expressed from within; they are *not* imposed by external conditions, and they are self-sustaining in artificial conditions, say total darkness, for some time. These biological rhythms are called circadian, which means 'about a day' and the 'about' is useful in reminding us that they may not be exactly of twenty-four hours' duration, although growth rates are usually at their peak in a twenty-four-hour rhythm. The interval may become somewhat shorter or longer in artificial conditions where the cells are not 'reset' by daily light and dark, and plant cells maintained in continuous light finally lack all rhythm. Sometimes they do not follow day-length exactly. Among the examples of the latter are some seaweeds whose rhythms appear to be conditioned by tidal movement, so that their activities are governed both by solar and lunar days. Some phototactic diatoms also show movements correlated with tides, so that they only come up to the mud surface in daylight when the tide is low. This is a very useful adaptation considering the habitats of such plants since the microscopic ones would be swept away if the tide was in, but it is in any terms a remarkable piece of biological chronometry.

Although these last examples are clearly useful to the plant and correlated with its working day as it were, the daily circadian rhythm most often seen, that of the 'sleep' of leaves at night, seems to have little reason behind it. To anthropomorphic eyes it looks natural enough. Darwin believed that it protected the leaves from the effects of cold at night, which it has been shown to do in some measure in temperate climates; but this does not explain why so many tropical plants appear to slumber. It may possibly reduce evaporation from delicate leaves. Sometimes leaf movements appear to have no rational explanation, as in the telegraph plant – named after the old-fashioned semaphore with arms – which waves its leaves and leaflets in all directions.

Ritchie R. Ward has put forward the hypothesis that an organism with a circadian rhythm is like an electronic computer controlled by a master clock. Such computers will be handling a number of 'sub-routines', to use the jargon, some of which are slower than others. If the different speeds of the sub-routines get very much out of step, it may be advantageous to 'gate' them so that they can all start up together again when the 'gate' is opened. A night-long pause would hardly be considered efficient in a man-made computer, but to plants such pauses hardly matter, especially as only morning daylight can bring about a restarting of the photosynthetic operation. (In the night-long rest they are on a par with man.) Scientists have mastered the facts but, to quote G. E. Fogg again, 'So far, only the hands, as it were, of these biological clocks have been studied, and the nature of the clockwork itself is unknown.' However, there is a relationship between sleep movements and other rhythms, which is expanded later in this chapter.

There is some evidence that flowering plants are sometimes affected by lunar rather than solar periods, and one American investigator, Professor F. A. Brown Jr, is certain that a quarter-lunar rhythm is widespread among both plants and animals. One of his recent experiments shows that water uptake in beans is closely correlated with quarter-lunar periods. One must mention in passing the strong anthroposophical belief in lunar influence on plant growth. Some people consider that certain plants will produce leaves more quickly if sown when the moon is waxing, or strong roots if sown when the moon is waning; sowing with the waxing moon generally enhances subsequent growth and yield.

In any case it appears that seasonal periods are by far the most important to plants overall. Seasonal rhythms are those controlling such long-term activities as the production of buds, development of leaves, initiation and opening of flowers, falling of leaves in deciduous trees or resting in other kinds of plant. A gardener will know that these activities are not markedly affected by weather. A very cold spell may restrain leaf and early flower buds from opening for a time, but if it persists they will eventually open even if this means their immediate destruction. Equally a very warm spell early in the year will bring plants into leaf and flower earlier; but the variations from the average are usually to be measured only in days.

The year can be measured by the activities of plants, as W. W. Garner and H. A. Allard described in a historic 1920 paper:

One of the most characteristic features of plant growth outside the tropics is the marked tendency shown by various species to flower and fruit only at certain times of the year. This behaviour is so constant that certain plants come to be closely identified with each of the seasons, in the same way as the coming and going of migratory birds in spring and fall.

In some primitive communities the activity of a local tree in putting out leaves or flowers, or maturing fruit, is a signal for some seasonal human activity to start. F. R. Irvine collected several African examples, such as the explosion of the inflated pods of *Griffonia*, a signal to farmers to plant their crops.

In temperate climates there is a distinct pattern of growth based upon an annual pause. This applies mainly to annuals, which live within the year time-unit and spend the pause in the form of seeds; to herbaceous perennials, which make annual stems; and to deciduous shrubs and trees, which lose their leaves in the winter. In this remarkable and rapid transformation from a full-scale food-making factory to a complete shut-down situation, these plants very largely curtail their metabolism, although the roots are seldom entirely inactive and start work early in the new growing season in order to pump food into the thrusting forth of new leaves and fresh growth. Deciduous trees grown in tropical climates behave oddly. Thus pear trees planted in Java become evergreen, although their buds may show individual growth cycles not synchronized with each other.

Tropical trees are often laws unto themselves. If they rest at all, which means losing leaves for a short time, they are apt to do so individually, not synchronizing with their neighbours even of the same species. One kind may flower in ten-, another in fourteen-month cycles; some do so continuously, often one branch at a time; yet others may bloom only once every ten or fifteen years. This accords with their congenial, little-changing environment in which time-keeping is of little importance. As the explorer Henry Bates observed, 'In England, a woodland scene has its spring, its summer, its autumnal and its winter aspects. In the equatorial forests the aspect is the same or nearly so every day of the year: budding, fruiting and leaf shedding are always going on in one or another species.'

The fact that the young growth of temperate trees is sometimes damaged by late frosts might be thought an argument against the deciduous habit. Most conifers and other evergreens retain their leaves in winter; the conifers especially inhabit cold and harsh climates, and their leaves are in use whenever conditions allow. But here the leaves are adapted to stand desiccating conditions and they lose water vapour very slowly. This is especially useful when the roots are unable to take up moisture due to dry or frozen ground and this is the main reason why deciduous trees lose their much larger leaves: they would otherwise lose water faster than the roots could supply it. We may note that the coldest, northernmost forest of all, in Siberia, is of a deciduous conifer, the Dahurian larch.

However, there is very little evidence that plants are aware of the passing of a year – a demonstration of which would in any case be extremely difficult. But if this does not exist what is it that plants do measure? It is in fact the length of the night. The plant's awareness of this changing period controls the seasonal activities previously listed and in particular the initiation of flowers. It has been found that plants can be divided into three groups according to night-length, although the word 'day' is more conveniently used in describing them – short-day plants which flower when their lighted days are of twelve hours or less; long-day plants needing more than twelve hours of daylight; and day-neutral plants whose flowering is initiated by climatic factors other than day-and-night length. Thus in the equatorial Andes there is no change in day-length nor in mean temperature over the seasons, although in a daily temperature range of some 13°C the plants have 'summer' every day and 'winter' every night. In this climate the control of annual flowering is imposed by the rainy season. The sexual rhythms of most seaweeds are under the influence of water temperature, although like conifers these evergreens of the sea never rest completely.

Clearly this segregation is conditioned by geographic distribution of any particular species: in the tropics, where days remain around twelve hours long, it is only short-day or day-neutral plants that will be able to succeed, while in high latitudes we find long-day and again day-neutral plants. Quite closely related plants may be entirely different in their day/night requirements.

Examples of short-day plants are chrysanthemums, annuals of tropical origin, and most spring- and autumn-flowering species of temperate areas; long-day plants include wheat, potato, lettuce, beet and summer-flowering temperate species; while among day-neutral examples are maize and tomato and the many tropical plants which produce flowers and also leaves the year round, having in fact no specific annual rhythms.

We know that it is the leaf which controls the plant's flowering according to long or short days. Some fascinating experiments to show this have involved leaf-grafting. Plants can be conditioned by artificial lighting treatment to be ready either to flower or not to flower. If a single leaf from the former is grafted on to the latter, it will induce flowering in this plant. In one experiment a leaf from a positively conditioned plant was grafted in turn on to seven negatively conditioned ones and induced flowering in each! If a single leaf of the clot-bur is given the short-day treatment that the plant needs to bloom, the rest being kept in a long-day condition, the whole plant will initiate flowers. It is amusing to note that the flowering of parasitic plants depends on that of their hosts, which are seldom related.

Such a situation must mean that the leaf produced a chemical messenger under the right conditions; this can be called the 'flowering factor' and it is similar in non-related plants. But how does the leaf establish that this is to be done? The most satisfactory hypothesis involves the daily cycles already described. If we imagine the plant to possess, in effect, a twenty-four-hour clock mechanism as the circadian rhythms imply, we can suggest that during the twelve-hour period of the average night (which is exactly six a.m. to six p.m. on the equator) the plant is regarded as 'dark-loving', while in the other twelve-hour period it is 'light-loving'. Retaining our mechanical simile, we can further suggest that there is an on-off switch for flower initiation built into this clock. Light reaching the leaf-clock in the dark-loving phase, especially its early part, keeps this switch 'off' in short-day plants and 'on' in long-day ones. (This is a simplification: in fact long-day plants have a shorter critical dark-loving phase than short-day ones.)

At this point the sleep movements previously mentioned take on some meaning, because it is possible to establish from the position of the leaves whether the plant is in a light- or dark-loving phase. With the exceptional *Madia elegans*, which as noted earlier 'sleeps' twice in the twenty-four hours, we find that flowering can be started in either short or very long days, but not in intermediate-length days.

Many complicated experiments have gone to build up this theory, originally postulated by E. Bünning in 1936, which implies that there must be a plant 'clock' which measures the length of succeeding days and nights and triggers off an instruction to originate flowers when the day-length condition is suitable. The chemical messenger which carries this instruction has not been isolated, although there are various theories as to how it may operate.

Sometimes the plant's reaction is keyed to temperate conditions, especially those at night. A specially important temperature effect is that exerted on certain plants as seedlings. Winter-hardy cereals are among these: they will flower sooner if the seedlings have endured near-freezing temperatures. If they are sown in autumn this occurs naturally in winter, and they will flower in the coming summer; but if not sown till spring they will not flower in the summer. A similar effect occurs in beet and other biennial plants, where it is the rosette normally made in the first year that requires winter chilling for flowering the subsequent summer. This winter chilling is called vernalization, and is restricted to plants of cold and temperate regions. Rice, among other tropical plants, has a converse need for high temperature at a certain time in order to flower at a later one.

One should add that plants have to reach a certain maturity before being able to initiate flowers under any circumstances. One or two are exceptional, such as the precocious peanut which has no 'puberty' since the flower 'initials' are formed as soon as the seed germinates.

Having worked out that these rhythms and triggers exist, man has been able to manipulate them, especially those controlling flowering and hence fruiting or cropping, for his own ends, as I shall explain in Chapter 27.

Although daily rhythms, oscillations, periodicities, call them what you will, are quite clearly internal, or endogenous, they must have been imposed over the millennia by the actual alternation of day and night which plants had, for their own good, to make sense of as they ventured from the tropical forest into lands of colder winters and longer nights, and indeed of pronounced seasons. Only in those parts of the tropics where conditions remain consistent throughout the year can plants remain happy-go-lucky and flower more or less all the time.

There are some plant periods which are difficult to understand. Although many fungi have a twenty-four-hour rhythm, some show much longer ones. Fungi are very antique organisms, and a four-day rhythm may be an ancestral recollection, either of a very different day-length or of a fumbling towards a twenty-four-hour period. In fact recent geophysical research on the Pre-Cambrian fossils called stromatolites, which exhibit daily growth-rings, suggest that 2½ billion years ago the lunar month may have lasted forty to forty-five days, each day being only five hours long. The examination of fossil corals has further suggested a 425-day year in Cambrian times, 600 million years ago.

At the other end of advancement, many bamboos have rather extraordinary flowering cycles, measured often in decades; to say in some Asiatic countries that a man has seen the bamboo flower twice is to imply he is very old. Although older records suggest that bamboos flower after a constant number of years, this is not in fact the case, and they may do so at widely spaced but erratic intervals; thus *Arundinaria falconeri* is known to have flowered in 1876, 1890, 1929 and 1936 – intervals of fourteen, thirty-nine and seven years. Some bamboos die after flowering, so perhaps we should not talk about a rhythm here at all, but in terms of the building up of a plant until conditions are right for it to flower, as in other monocarpic plants like the giant agaves. Yet it is extraordinary that plants from the same stock, but in very different climatic conditions, flower simultaneously. Is there a rather haphazard long-term clock built into these stocks? It has been suggested that sun spots may have something to do with the problem; but such long-term cycles, resembling those of the cicada and locust, represent a puzzle we have not solved; one of the many unexplained aspects of plant growth.

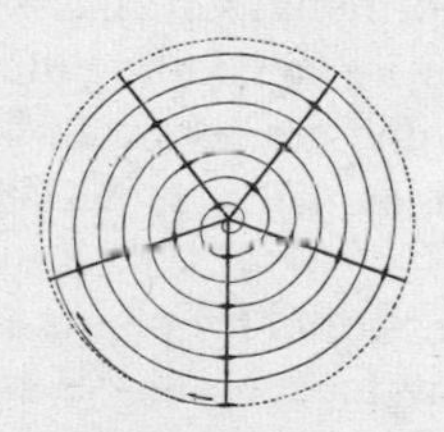

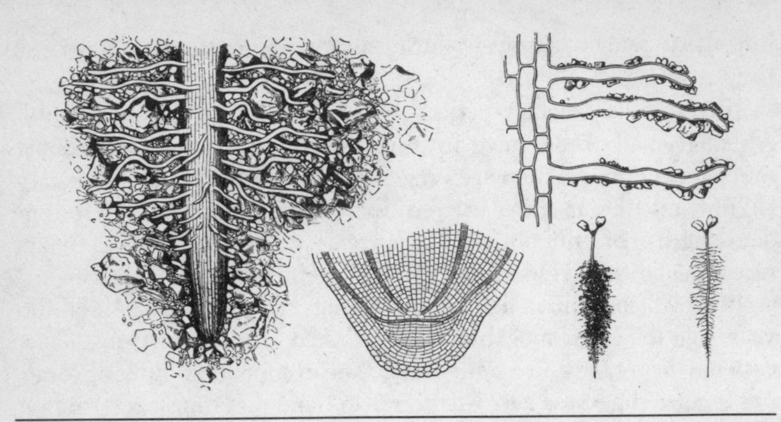

11 · Control and Communications

A factory has to have a manager to keep track of both day-to-day and long-term output of raw and finished products and, through foremen on the shop floor, of individual processes. Behind every aspect of plant growth there must likewise be control, from the simplest multiplication of cells to the building up of complicated organs; somewhere orders have to be issued for more roots, shoots or leaves, for flowers to appear and fruits and leaves to fall in season, and for controlling the rates at which these activities occur.

In the simplest terms certain chemical or occasionally physical operations in cells detect external situations, and cause other substances to be fabricated which, in infinitesimal amounts, act as messengers for these orders. These latter, the growth-control or growth-regulating substances, often loosely referred to as hormones, can be compared to the animal hormones which the endocrine glands secrete into the bloodstream, likewise controlling the growth and operation of organs and structures. Some are found in both plants and animals: auxins for instance in human urine. However, too close a comparison can be misleading.

Vitamins again are as essential for plants as for animals in enabling them to make use of raw food materials for growth, and must be considered as growth hormones of a different kind from auxins. There are also growth inhibitors such as abscicic acid, once more picturesquely called dormin. There are a great number of hormones, some of which have still to be isolated, although the plants' operations cannot satisfactorily be explained without their existence. Such are the hypothetical flowering-stimulator florigen, and vernalin, which encourage flowering after a period of cold.

In some long-day plants, oestrogens (sex hormones until recently linked exclusively with animals) build up as the plants flower, while in unsuitable flowering conditions, namely short days, no oestrogen is produced. In certain cases the application of this universal female sex hormone will cause a plant to flower much more rapidly than normal. It is known that other animal hormones connected with sex are widely found in plants, but at present their role is uncertain. Oestrogens in subterranean clover, incidentally, inhibit breeding in sheep!

All these substances interact with each other and the plants' metabolic processes, and their extremely complex operation is by no means completely understood. They exist in simple plants and in fungi, where vitamins usually operate instead of auxins.

The auxins in particular are created and are most concentrated at the growing points or meristems of a plant, whether these are of leaf, shoot or root. Meristems are areas of permanently embryonic cells, either dividing actively or capable of doing so under stimulus. From the meristems the growth regulators move to the other parts of the plant, where they are used up or destroyed; they are never recirculated. They move mainly in the food supply conduits of the phloem, and often only in one direction. Fruit growth hormones emanate from the developing seeds.

Each meristem controls a fairly restricted amount of territory, as opposed to the single animal gland that can spread its orders throughout the body in the blood. Even so, the meristems are in balance, and an excess or lack of 'hormone' at any point is made up by neighbouring meristems. One example familiar to gardeners demonstrates this. On a shoot of a fruit tree it is the end bud that normally develops, the side buds further down the shoot remaining dormant. If the tip bud is removed, the side buds closest to it immediately start developing. In slightly simplified terms, the tip bud produces auxin in high concentration, which promotes the growth of the shoot from the bud itself, and at the same time, diffusing down the stems, inhibits any development by the side buds. If this inhibiting influence disappears by removal of the tip bud, the lower buds start developing auxin on their own, their shoots are stimulated, and the surplus auxin diffused down the stem prevents yet lower buds from moving. A gardener can often stimulate a bud into growth by removing a small crescent of bark above it, which prevents auxin reaching it. Conversely he can stop it growing by removing bark, or merely pressing a knife into the bark, below a bud, which then accumulates auxin. This may be desirable for instance when training fruit trees, and is responsible for the old gardening rhyme,

> A dog, a wife, and a walnut tree,
> The more you beat 'em, the better they be

– the beating of the trunk having, presumably, the effect of stopping downward movement of auxin and hence increasing flower and fruit production. Auxins are also involved when branches are trained horizontally, which encourages a greater production of flowers and subsequent fruits.

We see then that these chemical messengers, carriers of orders, have various co-ordinated functions. Operating at a distance – and this is the vital criterion of a growth-controlling substance – they balance the external factors affecting the plant (nutrition, water supply, temperature, light), control the growth rate accordingly and the development of different parts of the plant (shoot growth as opposed to flower development), and, to borrow a phrase from that great expert on the subject, L. J. Audus, deal with the distribution of the plant's capital, namely the raw materials which manufacture protoplasm, cell walls and so on.

These substances also orientate the parts of a plant in the correct position, making roots grow down, shoots up; they adjust the direction of growth according to light, temperature, moisture or other external stimuli; cause tendrils or other fastening devices to operate and the tentacles of insectivores to clasp their victims; form wound-protecting tissue; cause flowers to develop, determine their sex (when they are unisexual), control flowering according to day-length, direct pollen tubes to the ovaries, stimulate the production of fruit, and make fruit and leaves fall in season; and inhibit seeds from germinating for various reasons. There is no part of plant development which they do not control.

The mechanisms involved are complex, if not baffling. Those which make leaves fall are a good example. Before this happens the auxin concentration in the leaf decreases, and ethylene is produced in the 'abscission layer' where the leaf stem becomes detached when adjacent cells part company. Abscicic acid is also involved; it appears to inhibit the synthesis of proteins, thus accelerating the run-down of the doomed leaf. Abscicic acid is also operative in the production of resting buds and storage organs.

Ethylene is a remarkable compound in this connection. It is a simple hydrocarbon (C_2H_4), which can affect plant growth in a concentration as low as 0·06 parts per million in air. I recall seeing a large carnation show in which a small exhibit of apples had also been staged. Before the afternoon was over every carnation had wilted because of the ethylene given off by the fruits. Ethylene affects almost every aspect of plant growth, including seed germination, elongation of cells and growing shoots, the onset of flowering and the ripening of fruit. It is normally in very delicate balance with auxin concentration. In the case of leaf dropping, the production of ethylene is believed to be triggered by another hypothetical material known as the senescence factor.

One of the more extraordinary plant reactions involving ethylene is its production when a plant is wounded. Even light stroking with a fine brush steps up ethylene production by up to thirty times the normal level within half an hour. Since ethylene controls the bending of stems, this is probably why tendrils curl around a support once they have touched it, but the reasons for extra ethylene production after wounding are not clear.

Fruit, incidentally, does not merely fall when ripe. As growers know, newly formed fruitlets may fall, and there is often a 'June drop' of immature fruits. These drops are the plant's method of regulating the amount of fruit to manageable proportions depending on the food and moisture situations at the time.

One of the most interesting growth-regulators is called kinetin, which activates cell division. It is found in coconut milk, and is the only natural plant product which will stimulate growth in fragments of plant tissue; it can also be concentrated from yeast and fish sperm, and there is reason to believe that minute quantities exist in all higher plants. Substances of this type are grouped as cytokinins.

I have said little about the mechanisms that trigger the creation of auxins in the first place. This is largely because the process is still shrouded in mystery. Taking as an example the movement of plant organs in relation to light, one has to postulate a light-receptive chemical whose reaction to light triggers the meristem to produce auxin. Possible contenders are several: they include carotenoids and riboflavin, both of which are involved in animal vision, the latter especially being universally present in plants. A reaction to gravity must certainly be linked with the physical movement of minute bodies, probably starch grains, within cells.

At some point cells have to stop extending; the whole plant may have an upper limit of size, and maturity is an important factor. To stop auxins over-acting, it seems likely that an enzyme is produced when auxin quantities exceed a certain limit. When auxins and this inactivating enzyme reach a certain balance, the enzyme destroys the auxin and the operation concerned ceases. Other mechanisms certainly exist for this essential turning-off process.

The complexity of the chemical messenger system in plants can be demonstrated by an example from a very simple organism, the water mould-fungus *Achyla*. In this there are male and female filaments and hormonal chemicals control their sexual congress. First the female makes a chemical activator and the male an inhibitor. When the concentrations of these are suitable the female activator causes the male filaments to produce short sexual branches. Once these have grown they in turn produce another hormone, which now diffuses back to the female plant and causes this to form the short, thickened branches which are the female organs. Now a third hormone is produced which, diffusing from the female to the male, induces the filaments of the latter to grow towards the former and, on establishing contact, to create a male sexual cell. Once more the mechanisms turn and a fourth hormone, manufactured by this sexual cell, passes a message which makes the female filaments put forth their sexual organs and the eggs. Finally fertilization occurs by the passage of male nuclei along a fine tube with which the male connects with the female cell. One part in 10 billion of initial female hormone or activator can make the male branches start the cycle.

Such complexity in the hormone system of an elementary organism – and equally complicated operations are recorded for some of the single-celled algae – strongly suggests that there must almost certainly be far more intricate ones for flowering plants than those so far worked out.

From the human point of view growth-regulators are extremely important, for we have found how to synthesize many of them and by this means to make plants behave as we want them to, producing roots on cuttings, or flower and fruit out of season. Such hormonal manipulation is described in more detail in Chapter 27.

The study of growth regulators is complicated by the increasing realization that electric potentials not only exist in plants but are responsible for many aspects of movement control, and probably other aspects of growth and metabolism. These 'action potentials' occur in animals also.

These potentials, which last about one-thousandth of a second, cause sudden depolarization of the cell's membrane potential, which is around one-tenth of a volt. Extrapolated from the tiny width of a cell, this represents a voltage gradient of some 10 million volts per metre – which in air would create a lightning flash. This depolarization is created by the pumping in of negatively charged ions and pumping out of positively charged ones, thus making the inside of the cell negative and the outside positive. After this metabolic action the ions leak back into the cells, often bearing a molecule of sugar or other nutrient and thus providing the cell with energy for its functioning.

These action potentials occur notably in young tissue still expanding. It is suggested that others are connected with cellular repair as well as with growth and movement, healing of the cell membrane being impossible with a high voltage potential across it. In the alga *Nitella*, action potentials make the streaming of cytoplasm stop, which may reduce the loss of essential cell contents after injury.

Such electrical charges transmit information and cause action in actively moving insectivorous plants such as sundews and Venus flytrap (described further in Chapter 24). Charles Darwin postulated this as long ago as 1875 and he persuaded Sir John Burdon-Sanderson, who was investigating the electrical properties of muscles, to carry out comparable tests on the Venus flytrap – the first experiments of their kind.

Action potentials also control the relatively fast movements of leaves and stems in the sensitive plant. In this the signals are actually carried through specialized cells in the phloem and, to quote Andrew Goldsmith, 'are perhaps the nearest thing to a nervous system found in the plant kingdom'. However, he emphasizes that the action potentials of plants differ from those of animals, which control nerve impulses; one can only suggest that in plants there may be the beginnings of, or at least parallels to, animal nerves, while emphasizing the vast difference between a transmission rate of 3 cm per second in plants and a one of 10,000 cm per second in human nerves.

Lately there has been some criticism of the scientific basis of the action of the chemicals which have been discussed earlier, and doubt has been expressed about the mode of their operation and whether they do, in fact, fit as neatly as has long been suggested into the physiology of plants. The origin of this doubt is the too-easy conception of 'plant hormones' as analogous to animal hormones, and the enormous difficulty of actually pinpointing hormone-like chemicals in plant tissues at the minute concentrations in which they exist.

Despite these doubts, a summary of the 'plant hormones' and their action, as presently believed, may be useful here:

Abscicic acid: controls dormancy of seeds and buds, growth inhibition, abscission, closing of stomata.

Auxins: stimulate growth, fruit development, root initiation and growth, cell differentiation and apical dominance.

Cytokinins (including kinetin): involved in cell division and differentiation, germination, and retarding of senescence.

Ethylene (ethene): involved in growth at many stages including germination, with wound and stress responses, flowering, and abscission.

Gibberellins: cell expansion and extension growth of stems; germination; flowering.

There are of course comparable materials in animals, not only the growth-controlling secretions of the endocrine glands, but also chemical transmitters actually within nerve tissues, and it seems possible, if contentious, that certain chemicals in animals convey general emotional messages, like the scotophobin which is claimed to instil fear of the dark into rats and to be transmissible by extracts of brain tissues – just as vernalin and florigen are apparently transmissible to other plants by grafting.

Finally, control and communication in plants is a highly complex operation, carried out by a combination of chemical messengers and electrical impulses, and still imperfectly understood. But it is not a nervous system: plants just do not need this.

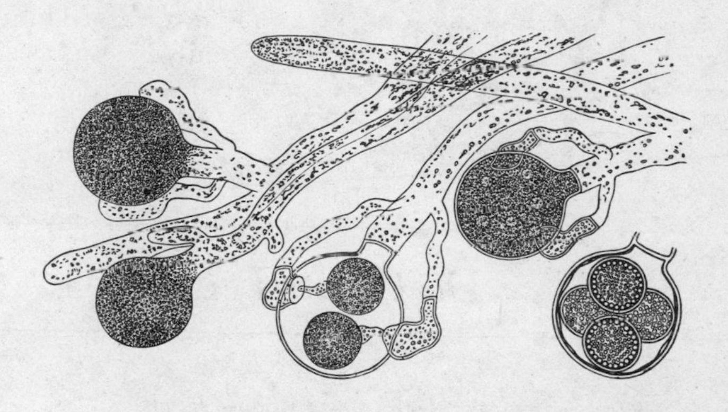

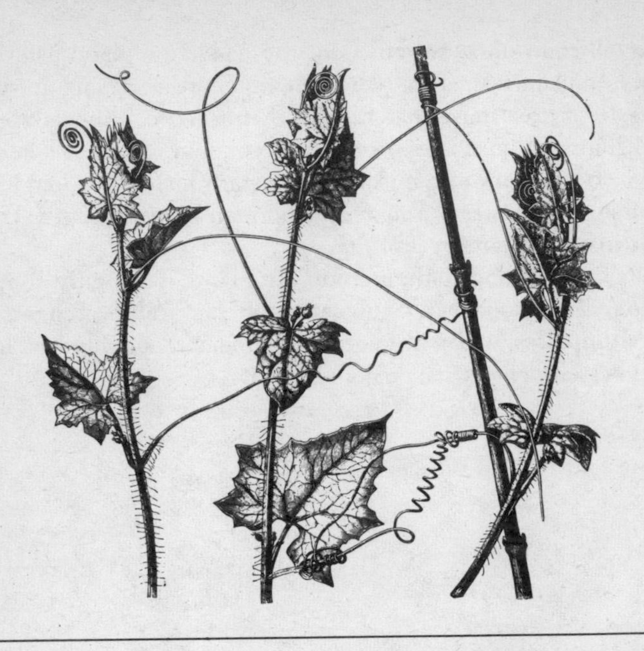

12 · *Do Plants Feel?*

Weak with nice sense the chaste Mimosa stands,
From each rude touch withdraws her timid hands . . .

So wrote Erasmus Darwin about the famous sensitive plant, which can be seen to 'feel'. Like many gardeners who are in constant intimate touch with plants he certainly thought of them in human terms. They do, after all, give birth, develop from infancy through growth to maturity, show signs of ageing, and eventually perish. In this development they are constantly moving, and these movements may show considerable determination. And for an organism to move in itself suggests some kind of basic feeling.

Larger animals feel heat and cold, wind and rain, fatigue and pain, and various degrees of trust, affection and sexual pleasure. Human beings add to these all kinds of emotional and intellectual perceptions and one might add confusions. In such terms plants do not feel; but if we apply one of the main definitions of the word – 'to perceive or be aware of through physical sensation' (Webster) – they most certainly do.

At the elementary end of the scale we find unicellular plants which move to and from light, and can also sometimes move according to chemical and temperature stimuli. The main stimuli which cause movements, or tropisms as they are technically called, in higher plants are gravity and light. Reaction to gravity, expressed either positively or negatively, makes roots grow down and seedling shoots grow up; it causes buried fleshy roots such as iris rhizomes and above-ground runners like those of strawberries to grow horizontally, and fleshy subterranean stems as in ground elder to grow at a remarkably constant depth beneath the soil surface. Gravity also influences the twining of climbers, and conditions the position of flower stalks. These last may change their position according to the stage of growth: thus the flower stalk of a peanut starts by being negatively geotropic, that is growing upwards; after the flower has been fertilized the reaction is reversed so that the stalk turns down and forces the fruit into the ground. In just the same way the ivy-leaved toadflax tries to push its fruits into cracks in the vertical rocks or walls it normally grows on.

Reaction to light normally makes plant growth develop towards the source or greatest intensity of the light, or controls the orientation of parts of the plant relative to the source. The one-celled swimming alga *Chlamydomonas* will move towards the light just as, in principle, leaves turn their maximum surface to it. The angle of the leaf stalk is conditioned by light, as in nasturtiums where the round leaf is carried at right angles to the light source. Adhesive tendrils of some climbers move away from the light in order to have the best chance of finding a surface to cling to, and aerial roots as in ivy are produced on the dark side of the stem. Sometimes light and gravity combine in affecting the posture of plant organs, especially in abnormal conditions; thus strawberry runners will grow upwards if kept in the dark. One would expect roots to shun the light but this is seldom the case

Roots, once below ground, are partly actuated in their direction-finding by available moisture. It is the root tip which alters direction both to seek water and to avoid obstacles, and Darwin was so impressed by its propensities that he wrote 'it is hardly an exaggeration to say that the tip of the radicle thus endowed, and having the power of directing the movements of the adjoining parts, acts like the brain of one of the lower animals'.

Seedling parasites searching for a host do so by reacting to chemical secretions from the host, though sometimes its moisture is sufficient.

A plant reaction to touch is perhaps less expected than one to light or gravity. It is, however, the vital one for many climbers. Those which make wide sweeping movements in order to find a support start twining round this at once when they touch it, producing a spiral of entirely different size in order to wrap round tightly. A tendril or clasping leaf-stalk shows similar very rapid reaction when it reaches a suitable surface; their sensitivity is often restricted to a short length near the tip. Twining stems act quickly enough when in contact with a support, but tendrils are even faster. The record appears to be held by *Cyclanthera pedata*, whose tendrils begin to curve around a support within twenty seconds of initial contact, and make their first complete coil round the support in four minutes. One interesting factor about all these touch-responses is that simple pressure is seldom responded to: there must be an element of rubbing or movement to stimulate the plant. This movement is also what activates the tentacle-like sticky, digestive hairs of sundews to arch over living prey.

The rapid reaction of tendrils after touch-stimulation is operated by the rate of growth on what becomes the outer side of the curve increasing by up to two hundred times its normal speed; the inner side of the curve may stop growing altogether. In many cases this differential growth-rate extends backwards along the tendril for several centimetres. Once a tendril has latched on to a support, it thickens for strength, while in certain climbers the end of the tendril produces sucker-like discs following contact. Once a climber has found support and has fastened itself by twining or tendril activity, its whole mechanical tissue develops; this both tends to strengthen the point of fixing by increased size of stem, and provides increased conduction for further growth higher up. In the parasite dodder the organs which penetrate the host (haustoria) are only produced if its stem is in contact with that of the host.

The classic plant with 'feeling', as already mentioned, is *Mimosa pudica*, the sensitive plant, also rather charmingly called humble plant. There are in fact several sensitive plants, mostly relations of the mimosa but including species of wood sorrel. Not only do these plants 'sleep' at night, as described earlier, but they are sensitive to touch. In nature this normally means raindrops, strong wind or contact by animals; at home one finds the warmth of a match flame will also cause collapse. The leaf of the sensitive plant is composed of several leaflets radiating from the top of a leaf-stalk, and each leaflet carries numerous small opposite segments. A tiny touch may cause just one of these feather-like segments to move; increasing disturbance will make a whole leaflet fold its feathers, and this normally makes the adjoining leaflets follow suit. If the shock is large enough the drooping extends to the leaf stalk and to other leaf-groups on the stem. Thus there is a definite process of transmission, and this occurs in all the other sensitive plants.

The mechanics of the operation are well known, and are in fact merely an extension of the movement carried out by the pulvinus – the swollen area at the base of the leaf-stalk – in other plants which controls 'sleep' movements, or the much less dramatic ones necessary to orientate leaves satisfactorily in relation to light. In sensitive plants the pulvinus contains cells with very thin walls, which surround the flexible lower strands of the otherwise stiff leaf stalk. These cells are able to fill rapidly with water or to lose it. Only the operating cells in one half of the balloon-like pulvinus respond to the electrical action potentials invoked by contact; the resulting shape alteration partly collapses the pulvinus and hence moves the leaf. This mechanism is similar to but much faster than that which controls the opening and closing of the stomata or breathing pores. After a period the pulvinus cells take up water again, stiffness is restored and the leaflets and stalk return to their upright positions.

It can be shown that the speed at which collapse occurs over a whole plant is up to 3 cm per second. Transmission of the impulse can occur through sections of stem which have been killed, and also across small gaps under water. Once the plant has recovered the mechanism of collapse will operate again and again – though there are tales of 'nervous breakdown' exhibited by shedding leaves if plants are made to respond too often.

It is not only the mode of operation but the point of it that is in question. Although it can be shown that in freak circumstances like a hailstorm leaves of sensitive plants avoid damage which is suffered by their neighbours, there are equally delicate species growing in similar conditions which survive perfectly well without this device. As a common scrub weed the sensitive plant gets grazed a great deal. The approach of cattle causes the leaves and stems to droop so that where the startled cow thought it saw a vivid patch of succulent green there seem now to be only thin greyish twigs – the plant has made itself invisible. Other notable sensitivities are displayed by sundews, Venus flytrap and some other carnivorous plants, which are given more space in Chapter 24. But in general such quick sensitivity must be regarded as one more experiment carried out in the vegetable empire and not taken very far.

Much more useful movements are those performed when the plant 'feels' excessive light or heat. Plants as different as the silver lime and wood sorrel allow their leaves to droop in bright sunlight in order to avoid scorching, and in others like beans and leguminous trees the leaves take up a vertical position in the heat of the day. The opening and closing of grass leaves to avoid excessive transpiration in arid conditions is especially remarkable. The leaf is elongated and basically flat, although the inner surface is typically ribbed, with the stomata concealed within the resulting grooves. In moist air the leaf will remain flat so that photosynthesis carries on at a good rate. In dry air the leaf rolls up to resemble a tube, with the tender photosynthetic tissue tucked away in the centre, very little air movement occurring through the stomata, and the exterior well protected by armour consisting of a thick skin and exterior cells.

This movement is not controlled by pulvini, but by the taking up or losing of water by groups of cells around the bottoms of the grooves on the leaf underside. Each rib typically has a layer of strong, non-photosynthetic tissue across it, and this acts like the centre rod of a hinge as the cells on either side open the leaf by expansion or close it by losing moisture. The structure of a wide grass leaf with a large number of ribs much resembles that of those flexible garage doors composed of metal strips which roll up like a blind.

A large proportion of plant movement is carried out, as explained for tendrils, by differential growth rates. Perhaps the most remarkable example is that of the Venus flytrap where the extremely fast reaction which clasps the trap over an insect is not carried out by a pulvinus as in sensitive plant, but as a result of very rapid cell expansion – in effect, high-speed growth. Hydrogen ions are released and the cell walls acidify; this activates enzymes that allow expansion and increased flexibility in the cell walls.

On a slower scale, differential growth rates also apply as, for instance, to those flowers which open and close each day. Opening is typically the result of the petal 'feeling' a temperature rise. A temperature increase of 10°C normally makes growth increase between two and five times, but the inside of the petal steps up its growth twenty or thirty times beyond normal under temperature stimulus; it turns back and thus the flower opens as the sun warms it. At night the growth of the outside of the petal relative to the inside causes closure. Some plants open and shut their flowers several times a day if cloud and sun alternate, like the showy gazanias, Livingstone daisies and similar South African plants, and small alpines.

Flowers have other movements too. The house lime has a central boss of radiating stamens. If an insect alights on this, or it is touched, all the stamens immediately move outwards from the centre, presumably a device to distribute pollen on the insect visitor. Again, an insect visiting barberry flowers and touching the stamens makes them spring inwards. And there are many other comparable trigger mechanisms in flowers by which they 'feel' a visitor's presence and act accordingly to dust it with pollen, notably those in orchids where the movements are sometimes sudden and relatively violent. One of the fastest movements of all occurs in the Australian trigger plants, so named because of the long column, bent down between and under the petals, which bears the stamens. When touched, this 'trigger' whips through over 180° in 1/100th of a second to cover a visiting insect with pollen and often to force it into contact with the stigma.

Such responses to touch and less direct stimuli show that plants have percipience and react to what they feel. Some also respond in a way which looks like a positive act to totally unfamiliar situations, such as producing leaves on tree-trunks if existing leaves are destroyed, and roots from stems and leaves if these are severed from their normal roots or stems respectively. A herbaceous plant which has been bent or broken down so that its stem lies on the ground will, as long as its conduction system is still functioning, bend the end of the stem back into the vertical position so that the flowers resume their correct position relative to the ground. This is carried out by a portion of stem which has, as it were, no previous experience of the need to bend. Likewise a tree brought horizontal by accident will often send up vertical shoots and occasionally down-growing roots from the prostrate trunk in positions where they do not normally occur.

The arrangements for movement in sensitive plant, Venus flytrap, grasses, flowers and leaves, all operating on some kind of hinge, are perhaps the nearest a plant gets to having joints which can be activated.

One of the odder plant reactions, discovered by florists, concerns cut tulips. These usually become very limp when put in water after cutting, because the water in the stems is at high tension when growing, and cutting creates an air-lock. They normally need a long period in water after re-cutting the stem before they will stand erect. But if a pin is inserted just below the flower the stem stands up rapidly. This result of plant acupuncture has not been scientifically explained. The subject of chemical and electrical communications in plants, examined in the previous chapter, has resulted in much work designed to show that, effectively, plants do feel.

The Indian J. Chunder Bose carried out hundreds of experiments on this thesis earlier this century, proving to his own satisfaction that plants had a nervous system, which he located in the phloem conduction strands, and that 'the transmission of excitatory impulse in plants is essentially similar to that of the nervous impulse in animals'. In his experiments he pricked and cut plants, which inhibited growth for various periods; cutting a plant might produce a 'convulsive contraction'. Bose also injected plants with stimulants and depressants. According to him alochol made plants weave like any drunk, caffeine made them perk up, chloroform calmed them so much that he was able to transplant a large tree without the usual precautions.

Similar work has been carried out more recently by diverse researchers. In 1970 reports were published of work carried out in the Agricultural Academy of Moscow, which suggested that plants not only have electrical impulses similar to those of nerves in man, as described in the last chapter, but went so far as to suggest that plants had a recollection centre at the 'root neck', which was reported to pulsate like an animal's heart. Newspaper accounts, even in *Pravda*, not surprisingly developed the electronic recordings of tiny electrical charges into plants speaking and shrieking. The latest reports purporting to come from Russia suggest that this research is designed to aid the space programme: if electronic communications fail it is hoped that astronauts might send messages to each other via plants carried in the space-craft.

The Russians also claim to be able to work out electronically which plants have desirable breeding characters, without any genetical information.

In the United States during 1972 reports appeared of a lie-detector or polygraph expert who carried out similar experiments. His instruments led him to believe that plants he had tended remained in communication with him when he left them, and showed excitement when he decided to return home. An experiment involving a friend on an air journey showed strong reaction from her plants to the stress she felt every time the aircraft took off and landed on a trip which ended nearly 5,000 km away.

This expert's further experiments, to quote an account by a staff reporter of the *Wall Street Journal*, 'seem to indicate that besides some sort of telepathic communication system plants also possess something closely akin to feelings or emotions . . . They appreciate being watered. They worry when a dog comes near. They faint when violence threatens their own well-being. And they sympathize when harm comes to animals and insects close to them.' In one experiment polygraph electrodes were connected to three vegetables: one was picked up and dropped into boiling water. *Before being touched* the selected vegetable registered a violent upward movement on the polygraph chart, followed by an abrupt straight line – interpreted as 'fainting'. A comparable experiment of scalding a cabbage leaf had been carried out much earlier at the Royal Society, and convinced Bernard Shaw that plants might feel. Another piece of recent research involved a student destroying one of two plants in a room by shredding it up. A polygraph attached to the surviving plant showed no reaction when other students entered the room afterwards, but a 'fainting' reaction when the 'killer' did so.

An IBM research chemist who duplicated these experiments reported that when his students were discussing their work the plants did not react, but if their conversation turned to sex or ghosts the plants had the polygraph needle jumping about. An ITT engineer actually made his empathic plants record wild reactions when he achieved orgasm with his girl-friend several miles away.

In another experiment a philodendron is reported as giving a 'sulking' reaction when telepathically insulted, while a Japanese scientist connected a polygraph to a cactus with acupuncture needles and made it successfully add up a list of numbers, its answers appearing as a series of surges on the pen recorder. Yet another experimenter, both an electronics engineer and an ESP investigator, has reputedly reversed the direction of a toy electric train by relaying his orders via a plant.

The Japanese doctor mentioned above invented a device to transform the electrical surges from his plants into vowel-like sounds; and in 1980 a London factory was turning out similar 'Bio Activity Translators'. These convert a plant's electrical impulses via a synthesizer into musical sounds which indicate whether a plant is happy or depressed – to use the inventor's type of approach – which may be due to availability of warmth or food, weather conditions, and the presence and intentions of humans or animals within range. So far the device cannot actually spell out exactly what the plant is reacting to!

It is easy to dismiss such experiments out of hand since some of them appear ludicrous, they involve plants in telepathic communication, still an unexplained phenomenon in humans, and appear also to involve them in value judgements. In any case, recent British experiments suggest that most of this work can be duplicated if a damp cloth replaces the plant! Yet electricity certainly plays a part in plant cells, at a molecular level, as it does in animal cells, while the basic photosynthetic process involves movement of electrons as light energy is transformed to potential chemical energy.

As described earlier, electric potentials are detectable in the leaves of flytraps and the sticky leaf-tendrils (*not* the leaves proper) of sundews when they are activated, and the visually recorded pulses of such potentials resemble those of vertebrate peripheral nerves.

This activity – based on existing cells, which are not structurally anything like nerve cells – is reminiscent of similar activity in elementary animals such as hydrozoans and possibly sponges; the former have elementary nerve fibres, the sponges none. Such potentials very probably exist in many plant cells and systems; something similar has been detected in stigmas following pollination, and elementary mobile cells with cilia or flagella, not only planktonic individuals but mobile sex-cells in less advanced plants, are probably activated in this way. Plants have, in fact, what researchers have described as 'a primitive neuroid system'. What they do not have is a nerve *centre* of any sort, any more than a hydroid has, and we must equally recall the quantitative difference between the processes – the time-course of a potential in a plant being measured in seconds, not the milliseconds of an animal nerve. Nor has anyone suggested that sponges and hydroids utter silent shrieks when deracinated or cut up.

There seems no reason why such elementary impulse transmission systems should not exist both in plants and elementary animals, which after all began evolutionary existence in parallel. One can also point out that a nervous system in advanced, mobile animals confers sound survival value, while no such value would seem to be involved with plants. Their need for some kinds of feeling has arisen in entirely different circumstances, and to develop pain-feelings seems to be against their whole life-style, where they can so easily be damaged by external activities ranging from storm and lightning to the biting of caterpillars and the grazing of herbivorous animals, against which they can take no evasive action.

One has finally to say that the imputation of fainting and telepathy in plants sounds like wishful anthropomorphism – that wish on man's part to make other organisms resemble him in as many ways as possible because, presumably, it makes them less alien and perhaps less frightening. It sounds also like science fiction; one recalls a story by Roald Dahl about a man who could hear grass screaming as he mowed it; and there are John Wyndham's brilliantly imagined triffids, capable of walking, elementary communication, and a certain amount of prescience, apart from their capacity of striking at people almost unerringly and always fatally with their flailing stings.

And if plants do feel, scream and faint, I doubt whether it would put most people off eating vegetables or cutting flowers. As one correspondent wrote in *Garden News*, following the account of the 1970 Russian experiments, 'People are going to say, "If you feel so strongly about plants' feelings why do you eat them?" My answer is: "Because I'm hungry and my brain is stronger than the plants, so I'm the dominator."' A race which eats animals having killed them, and can tolerate mass-production of chickens, calves and pigs in rather unpleasant ways, is not going to worry about the soundless shrieks of a carrot or a cabbage.

There are, of course, vast numbers of gardeners who are absolutely positive that plants respond to their talking to them – encouragement, blandishment, threats, even singing. Some further extracts from *Garden News* correspondents' letters illustrate this: 'Of course plants like being talked to. How many times have I known an ailing plant respond to an encouraging chat? Most women will instinctively say "Sorry", if they accidentally damage a plant.' 'See how tenderly any real gardener handles his "babies". See the gardener fret if he has to go away from home and trust someone else. See how the plants fret for their owners.' 'All dedicated gardeners believe that plants have feelings and talk to their plants, not always aloud but in their thoughts.' In a recent 'talking' experiment, identical sets of seedlings in identical conditions were talked to 'nicely' and 'abusively'; the former grew faster and better.

A recent book suggests that, one day, it should be possible for a plant to open the garage door when the owner reaches home, and that once we learn to communicate with plants properly there will be a new Garden of Eden. This book describes a Scottish garden at Findhorn whose owners believe that plants are full of 'archetypal formative forces – nature spirits, fairies, gnomes and elves', which is on a rather different plane from the lie-detector and Russian astronauts. This seems more akin to some earlier Californian research by a professor of natural science, after two years of which he concluded that he could accurately assess the emotional disposition of plants. Potatoes, tomatoes and cabbages reacted best to kind words and even flattery, while orchids and gladioli were described as temperamental and nervy. However, amazing results have been reported from Findhorn including gigantic cabbages and wheat over 7 m tall: a reliable and knowledgeable expert wrote that 'the vigour, health and bloom of the plants in midwinter upon land which is almost barren sand cannot be explained by the application of any known cultural methods of organic husbandry'. There is indeed a mystery here; the growers claim that the results follow their entering into communication 'with the spirits that animate their plants'.

Prayer has also been invoked as a communication medium with plants, and there are exponents who believe that they can cause plants to thrive or to die through prayer, rather as some primitive tribes claim to be able to destroy their enemies' fruit trees through ritual. A *Sunday Times* survey in 1972 showed that 14 per cent of people believed that the power of prayer can materially affect the growth of plants. In a recent American experiment, carried out under controlled conditions by a scientist studying plant growth rates, a group of people concentrated prayer upon a plant 1,300 km away for five minutes. A recorder showed an immediate growth response which had within twelve hours increased to over eight times above normal!

One might comment that such feelings and beliefs tell us more about the human beings concerned than about any nervous system in plants; and wonder what is the aural organ in a plant, and how it learns our language. Some gardeners certainly have a very strong sympathy with plants which is reflected in the way the plants grow for them – they have 'green fingers' in short. But this is surely no more than a knowledge, born out of varying proportions of experience and intuition, comparable to a person who is good at electronics and can play musical instruments or learn foreign languages without apparent effort. It is a matter of what one may call rapport, and it probably does the talker more good than the plant.

But although one may smile condescendingly at ladies who talk to their African violets there may be a germ of truth in the process. It has nothing to do with communication, but it does seem very likely that plants are sometimes stimulated to grow through external vibrations such as are created by speech. Once again, the way the experiments have been carried out has sometimes resulted in derision. The easiest way to produce sound waves which create aerial vibrations is, after all, to play musical instruments or gramophone records. And the image of the Australian gardener playing violin concertos to her plants for half an hour every morning, or of the Indian lady research worker executing a traditional dance among potted marigolds for fifteen minutes every day and claiming a 60 per cent increase in height over control plants, does tend to make most of us smile, and wonder whether plants prefer Beethoven or Stravinsky (perhaps best for cacti) or if pop is better than the classics. One experiment suggested that flowers prefer the violin, vegetables music of lower pitch, another that plants did in fact respond best to Bach and classical Indian sitar but, subjected to 'acid rock (*sic*), they cringe, lean sharply away from the sound, and die in a few weeks'. Will there eventually, one wonders, be special agricultural workers to programme concerts for specific crops?

This all reminds one of the ancient Indian practice of *dohada*, which involved singing, dancing and stamping around trees, and sometimes actually striking them, which was supposed to encourage growth – like the old English walnut tree of page 120.

Joking apart, there have certainly been some remarkable results. Indian research in 1958 involved vibrations generated by an electric motor, and claimed a resultant 30 to 50 per cent increase in height, acceleration of flowering by up to two weeks, and four-fold increases in crop yield, over control plants. This was followed by apparently productive experiments in which music was broadcast over rice crops. In 1972 it was reported from the United States that exposing field crops to harsh noises significantly increased their yields. It has even been found that very loud sound can speed up seed germination.

Some of the most convincing evidence that vibration does enhance plant growth comes quite accidentally from workers in an entirely different field observing an otherwise unexplained discrepancy in results. Propagation experts often use closed frames in which the soil is warmed from below by heating wires. These can either be low or high voltage. Different workers recorded rather varying results in speed of rooting and subsequent growth under apparently similar conditions. It was found that the better results were consistently achieved above low-voltage warming wires which vibrate, whereas high-voltage ones do not.

Certain contrary claims, however, have been made about the beneficial effects of vibrations upon plants. Solomon Islanders claim to be able to kill trees – their enemies', one supposes – by creeping up on them before dawn and suddenly uttering piercing yells close to the trunk. After a month the tree is said to die as a result of shock from such frequent and violent awakenings. A more scientific experiment, reported in 1972, arose from observations that trees on the edge of forests were typically much shorter and stockier than those sheltered within. Identical seedling trees were grown in the same conditions, half of them being shaken violently for thirty seconds each morning. At the end of a month those shaken had grown an average 4·3 cm, those left alone 20 cm. A rather similar result followed another experiment in which a scientist rubbed the stems of various plants for ten seconds twice a day, making sure that this was light enough to avoid mechanical damage. Some of the plants, though not all, stopped active growth within three minutes, and did not start again for three or four days after the treatment. Growth reduction and consequent stunting was accompanied in sensitive plant by prevention of flowering.

Such shaking and rubbing are of course quite different in degree from the vibrations of musical instruments or broadcast noise, which apparently has more beneficial effects on the protoplasm or on cell division.

A final fascinating point which may help to explain some 'green fingers' is that three people in a hundred are known to exude substances like plant growth hormones in their sweat.

J. Chunder Bose's experiments on anaesthetizing plants, mentioned earlier, followed earlier ones made on all kinds of organism ever since the first modern anaesthetic was demonstrated by an American dentist in 1846. In recent years more scientific experiments show that sensitive plant, Venus flytrap and other plants can be stupefied by chloroform, ether or morphia. The anaesthetics appear to be blocking the action of the ion-driven electrical impulses described previously. Although somewhat contradictory results have been obtained on plant metabolism, it does seem that low levels of anaesthesia increase respiration and slow down photosynthesis, and higher levels inhibit both. And, as described in Chapter 15, anaesthetics have remarkable powers in overcoming seed dormancy. Barbiturates slow down cytoplasmic streaming and can, for instance, hold back the sprouting and growth of pollen tubes; and they have been shown to make seedlings weak, probably by affecting their photosynthesis.

These experiments with narcotic substances, apparently working on a plant like its natural 'hormones', suggest future possibilities in controlling plant growth.

One concludes that plants do feel in certain ways, and that there is much more to be discovered about the ways in which they do so, which may indeed conceivably help the world produce more crops in the future. Their reactions are comparable to those of many animal organisms, especially the simpler ones, from which human beings – who are seldom in such close contact with, say, amoebae or even worms as they are with plants, which they keep around them for pleasure as well as for use – have not found it necessary to invoke emotional responses or telepathic capacities. Nor do they have souls as has been imputed to them (their soul was supposed to reside in the pith). Plants thrive when conditions are suitable, but there is no other implication of pleasure or enjoyment. Where a man can say 'I think, therefore I am', a plant will suffice on 'I grow, and thus I am'.

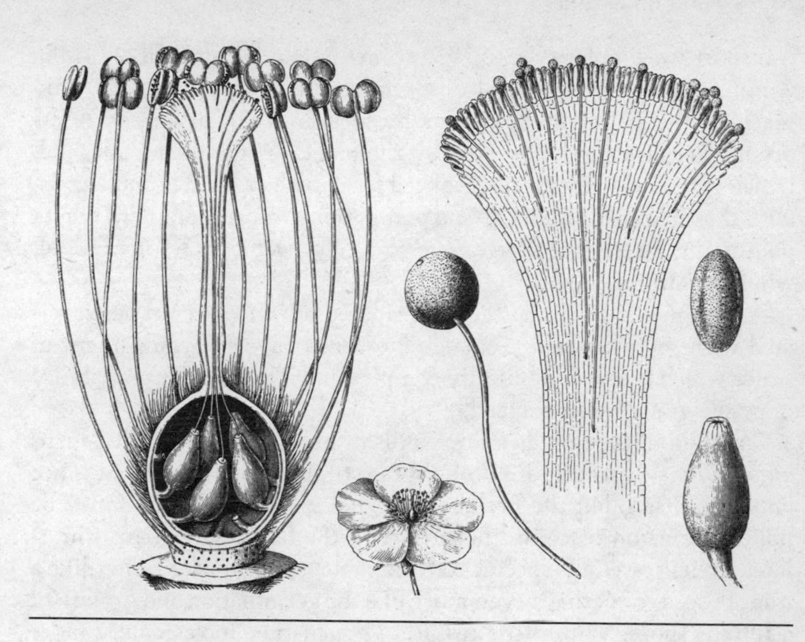

13 · *Sex*

Tennyson wrote of 'The white flower of a blameless life', and Blake found 'All Heaven in a Wild Flower'. Flowers have indeed often stood, in literature, poetry and symbolism, for innocence, freshness and ephemeral beauty. But as I have described in Chapter 9, flowers were shown by Linnaeus to be nothing more than sex organs, usually blatantly displayed, which act, to quote Croizat, 'like husbands and wives in unconcerned freedom'.

Mating is important to most plants, for it evens out weaknesses: the more vigorous offspring are likely to survive better than the frail, and sexual interchange is far more likely than non-sexual methods to produce variations with new capacities or the ability to cope with changing conditions.

In the more advanced and generally more familiar plants male gametes are produced within pollen grains which are designed for transport. In simpler types such as pines pollen is launched on to the wind, sometimes assisted by special air-sacs for buoyancy, and reaching the static female organs by luck born out of an enormous production. This is the fee paid by the plant for this chancy method, which is used also by more advanced trees such as birches and hazels, with their familiar catkins. Wind pollination is most prominent where pollinating insects are scarce, due to a cold, wet climate or exposed, windy conditions.

Although wind pollination is considered primitive, it can be associated with sophistication. Female pine-cones have an arrangement of scales which is aerodynamically complex and increases the probability of pollen grains being trapped.

Pollination itself needs a few words of explanation. It is the means whereby male gametes are brought into contact with females: it is not an end in itself, but the prelude to sexual union or fertilization. The pollen-grains on reaching the surface of the female organ are stimulated by its secretion (specific to each plant) to germinate rather like a minute seed; or perhaps even more like the germination and growth of a fungus spore. Within the resulting pollen-tube (occasionally more than one is produced) the male gametes develop. In conifers these gametes are released in a cavity above the egg-cells; in flowering plants the pollen-tube has first to force its way through the tissues of the female organ (pistil), which nourishes it as it does so. Not till the tube closely approaches the female gamete within its ovule are the male gametes liberated, and sexual union follows.

Because self-fertilization nullifies the exchange aspects of sexual reproduction mentioned earlier it has been suggested that in evolutionary terms self-pollinated flowers are doomed, however well they may seem to survive at present. Darwin himself wrote that 'Nature . . . abhors perpetual self-fertilization.' Yet he was not altogether right. It may occur where pollinating insects are scarce and, as explained further in Chapter 21, there can be real advantages for self-pollination in certain classes of plant, which is one of the reasons for the success of some weeds. It is interesting that quite often closely related species are self- and insect-pollinated. Continuously self-pollinated plants of one species in limited habitats may produce such pure-breeding 'lines' as to create virtually new species, and the specialized regional botanist can often recognize which local population a plant specimen comes from.

Having said this, one must note how many hermaphrodite flowers have elaborate arrangements for preventing self-pollination. The simplest involve the stamens ripening and the stigmas becoming receptive at different times. In some such flowers, however, there is a fail safe device, such as a final movement of the stamens, to ensure self-pollination as a last resort – better one's own seed than none. The normally bat-pollinated cup-and-saucer vine is an example, fertilizing itself just before the corolla falls.

Another safety device is cleistogamy, meaning literally 'shut-marriage', by which flowers which never open can fertilize themselves without external help. Our native violets and wood sorrel do this when in summer the leaf canopy of the trees under which they grow makes too much shade for their pollinators to visit them, while in spring they produce normal flowers. In some acacia forests in west central Africa it is much too hot for insects to operate at ground level (70 to 80°C), but in the tree-tops the temperature is below 40°C and a large wasp and many birds act as fertility agents for every species of tree. Underneath, the flowers are almost all cleistogamous, despite the fact that in less difficult environments most are open-flowered. One of these plants, a *Commelina*, goes so far as to produce closed flowers filled with slime, which actually develop below ground.

A simple method of avoiding self-pollination is an alternative arrangement of stamens and stigmas, as in the primrose made famous by Darwin, in which there are 'pin-eyed' flowers with the stigma at the top of the flower-tube and the stamens half-way down, and 'thrum-eyed' ones with reversed positions. As Darwin showed, it is much easier for an insect to cross-pollinate two different flowers than two similar ones. Other flowers may even have three distinct variations in length and position of stamens and stigmas, as in the purple loosestrife. In such cases there are often different sizes of pollen which 'fit' best one type of stigma, the one which provides maximum out-breeding. In certain plants, such as flax, pollen can never germinate on the stigma of the same bloom.

In avocados the hermaphrodite flowers each open twice. At the first opening, only the female organs are receptive and mature; at the second, the pollen is ripe. Although fresh flowers are produced throughout the season, the timing of openings is such that one tree's pollen can never reach its own flowers. There are two types of tree – one in which female readiness occurs in mornings and male readiness in afternoons, and the other in which the situation is reversed, so that cross-pollination is ensured, and commercial growers interplant the two types.

Some plants produce separate male and female flowers, and these may show varied climaxes. Thus in the hardy orange the early flowers only carry stamens; later there are male, female and hermaphrodite blooms; the last have only stigmas. This seems to be going a little too far for efficiency, and so does the carrying of male and female flowers on entirely different plants. Not only does this complicate matters, because male and female plants cannot get together like animals, but it seems a waste of half the representatives of a species, made worse when, as in many trees, they may take ten years or more to reach sexual maturity.

In any case one must stress that even decidedly out-breeding plants are very likely to be pollinated by close congeners, for pollen does not on average travel very far. With trees sex relations are probably with the nearest neighbour. But, even so, solving the problem of pollen transfer is one of the great technical achievements of the plant kingdom. As we have seen earlier, evolution can be mutually beneficent to very different groups of organism. Thus the development of insects led to the elaboration of the simple pollen-producing flowers, so that the insects would be persuaded to visit the flowers and, in exchange for a suitable reward, carry the pollen from one flower to the female part of another. Birds, bats and a few other animals became linked with pollination in similar ways. The bewildering variety of specialized flowers is echoed in their pollinators. Where the relationship is basically one flower to one pollinator two American workers have aptly called it a 'lock-and-key' situation.

Such animal pollinators make it far easier to ensure that flowers on different plants are visited; the problem was always to persuade the pollinator to visit the same kind of flower and not just to wander unselectively, when its activities are little better than the wind's. But animals can have instincts built into them so that they will return regularly to the same flower species, and the flower imprints these instincts by developing an appearance and other attributes as distinct as possible from those of its neighbours. The animals may combine both sight and scent in finding their reward, whatever it may be.

If pollination is to be carried out by animals the pollen itself has to alter. Wind-carried pollen is as small and light as possible, each grain separating from the others to ensure maximum dispersion. Pollen can, exceptionally, travel 5,000 km, often above 2,000 m altitude. When animals are concerned the pollen must not only be strategically placed where it can contact some part of the visitor, but must be adhesive, stringy or even in a compact mass. There is a change too in the whole basis of the flower. Wind-pollinated plants have many males producing much pollen, but few females; the latter will often have an extensive stigma, like the feathery tassels of maize, and only a few ovules, or even only one, since only a single grain of pollen may reach them. Pollen production is astronomical in wind-pollinated flowers; a single male plant of the annual dog's mercury, only a few centimetres high, has been estimated to produce over 1,300 million grains, and trees like birch and hazel produce respectively about 5½ million and 4 million grains per *catkin*.

Conversely in animal-pollinated blooms there will usually be relatively less pollen but the female can have numerous ovules, because as long as the postman knocks much fertilization is likely. Their stigmas will be relatively strong and stocky so that they neither impede the animal nor are likely to be damaged by it. Insects are the most numerous animal pollinators, and we can recognize flowers designed for beetles, bees, bumble-bees, moths, butterflies, wasps and flies. All types and sizes of insect are known to pollinate flowers, from enormous moths to minute owl-midges and even wingless, crawling species, while springtails may – if they land luckily in their leaping – find themselves in another bloom of such plants as golden saxifrage whose pollen they eat.

When the insect or animals needs to hang on to or push into the flower, as a bee does for instance, the flower will have a 'landing platform' or something solid to grip; where the creature can hover there is none. One of the oddest 'grips' is that of the elephant-head lousewort, whose flower really does resemble an elephant's head. The bee sits astride the projecting 'trunk' to forage in the main bloom. The curved end of the trunk, from which the stigma projects, pokes up under the bee's wing between the thorax and abdomen, looking as though it was tickling it, where it comes into contact with pollen previously accumulated by the insect.

The earliest flowers we know of, such as magnolias, are pollinated by relatively primitive insects, usually beetles. They tend to have large, strong female organs. The custard apple family, *Annonaceae*, relies on small beetles which eat the stamens, while in the giant water-lily cockchafers chew up the stamens. Enough pollen remains on the feasting beetles to be rubbed on the stigmas. Some flowers, for example the giant water-lily, close up on the gourmets to ensure that they pay for their meal by doing this as they seek a way out. In such flowers pollen is an end in itself. But many other plants, such as species of *Cassia*, offer bogus food-rich pollen as well as the operative kind; relations of *Tibouchina* have two distinct kinds of stamen, one with real pollen which catches on the insect while it is enjoying the food-pollen. Flattish, open flowers like roses and anemones are likely to be beetle-pollinated. Even the bizarre relic *Welwitschia*, in its very difficult habitat, has its pollen-eating bugs in profusion, although there is some doubt as to whether they are the true pollinators. But beetles so often eat pollen wholesale, and lie about in the flower as if at a Roman orgy, without moving to other flowers, that they are the least efficient animal pollinators.

Sometimes pollinators, notably beetles, are provided with a repast of a different kind. Parts of the flower itself may be attractive: the giant aroid *Amorphophallus* is full of small beetles which, quoting Proctor and Yeo, 'remain stationary for days on the lower part of the spathe chewing away at special cells filled with starch and oil'. In the related *Typhonium trilobatum* the beetles are provided with yellow hook-shaped 'baits' on the spadix.

The bee is the great pollen-collector, of course, visiting flowers very frequently to bring pollen to its protein-hungry larvae; and, although it is more extravagant in terms of energy for a flower to produce excess pollen than the nectar described later, the evolution of bees clearly forced flowers to do so. Bees are far quicker and more efficient as pollen collectors than the beetles, so natural selection has favoured flowers attractive to bees.

How do flowers allure their visitors? The most obvious means to our eyes is direct glamorization – they become larger, showier, brighter, welcomingly colourful. They may produce outgrowths which wave like banners in the air, like the slender spiralling 10 cm trails of the tropical cucumber relation *Hodgsonia heteroclita*, or the even longer but less numerous growths from certain orchid flowers. Such 'tails' appear to be particularly attractive to flies, sometimes by producing strong odours as in arum relations so that one might call them 'smell-banners'. Sometimes the banners are external to the flowers themselves, like the scarlet floral leaves (bracts) of the poinsettia, and the often coloured bracts around other *Euphorbia* flowers; the pocket-handkerchief tree is so called because its small boss of true flowers is signalled to visitors by one very large and one much smaller white bract. Showy bracts appear also in dogwoods and bougainvillaeas. These bracts, while not technically part of the flower, are inseparable from it. In other cases small fertile florets are surrounded by larger sterile ones which act as signals to insect visitors: the lace-cap hydrangeas are an example. In some cow-parsley relations, for the same reason, the florets on the outside of the flower head are much larger than the inner ones.

In certain bromeliads, such as nidulariums and neoregelias, the centre of the basically green rosettes flushes red when the small flowers appear in the centre, and many other bromeliads, such as vriesias and tillandsias, carry small flowers in a spike-like or club-like head in which most of the brilliant colour comes from external bracts.

Flowers also carry a pattern of spots or lines which provide the insects with guide-lines to the nectar, and hence to the sex organs, almost like the flare-paths or homing beacons of an airfield. Examples are horse chestnut, monkey musk and eyebrights. Such guide-lines are not always visible to us, because many insects, notably bees, are able to perceive ultra-violet light; indeed, illumination of flowers with ultra-violet has revealed to our eyes many previously unsuspected patterns of this kind. Flowers 'know' what their visitors can see. The central part of the flower, sometimes the stamens, is 'lit up' in this way for insects. Flower colour may appear very differently to man or insect –

thus a dandelion apparently is purple to a bee. Bees in particular appreciate vivid colour contrasts while insects in general can discern small variations in colour shades.

Insects can also be affected by the shape of flowers. It has been shown that honey-bees prefer radially symmetrical blooms, while bumble-bees like those symmetrical in the vertical plane. Honey-bees also prefer flowers with a cut or even ragged outline to circular or lightly lobed ones.

Besides visual guide-lines there may also be tactile ones. These often begin on the landing platform with various knobs, hairs, fringes or grooves, designed to place the insect in the right position; nearer the centre of the flower there may be tongue-guides, especially if nectar is placed within a narrow orifice or spur.

Although flowers of different families usually have a very distinct appearance, there do appear to be some mimic situations. A small eyebright which lives scattered thinly among ling has a similar colouring and flower arrangement, which presumably 'reminds' insects of its existence. There are many flowers resembling the buttercup, with five or more petals forming a shallow bowl; such flowers have perhaps developed a similar 'advertising style' because their not very specialized pollination factors resemble each other's. The insect mimicry of the bee orchid tribe, the most remarkable known, is described in the next chapter.

Further animal-attraction is created by producing odours from various parts of the flower, sometimes attractive to human beings and sometimes the opposite. There are scents we find pleasant, like the delicate lily-of-the-valley, the heavier perfumes of lilac, jasmine or frangipani. The silvery leaved *Elaeagnus angustifolius*, called zinzeyd or zungeed in Persia, has a fragrance once thought so intoxicating that Persian men were wont to lock up their women when the tree came into bloom. Odours distasteful to us include those of lords-and-ladies and their relations. These include the stinking great dragon of the Mediterranean (*Dracunculus*) with a dark red spadix or spike and outer spathe, the *Helicodiceros* of the Middle Eastern deserts with fissured red-brown spathes looking like congealed blood, and finally the 2 m vertical enormity of their tropical relation *Amorphophallus titanum* in which an enormous conical spadix up to 2½ m tall thrusts out of a vase-shaped, ruff-edged spathe, emitting an odour which has been likened to a mixture of rotting fish and burnt sugar. The beetles which pollinate this monster are prevented from leaving before pollination by a sharp overhang; the flower begins to wither before they can escape. Apart from its flower, over 3 m tall, this giant stinker has leaves 2 m long and tubers over 1½ m across.

The African stapeliads are succulents with usually starfish-shaped flowers up to 40 cm across, which combine a smell and an appearance of decaying flesh to attract blowflies and their allies. As the famous aquatint 'The Maggot-bearing Stapelia' in Thornton's *Temple of Flora* shows, flies are sometimes so taken in that they actually lay eggs on the flowers. It would take a Salvador Dali to imagine a stapeliad had nature not done it for us, with its combinations of colours and textures, often covered or fringed in hairs that resemble mould growing on rotting matter, coupled with the smell. In some cases minute hairs move in the slightest breath of air, making the flowers look as if they were seething with tiny flies (a device shared with some of their relations and a few orchids). This in a family which also gives us stephanotis and wax vines of delicious scent!

It is interesting to note in passing that carrion flies prefer yellow and orange if there is no accompanying odour. As soon as a stench of carrion or faeces is present, their preference changes to dark purples and browns. The biggest fly-flower in the world is that of the tropical parasite *Rafflesia*, over a metre across. It is dark red to purple-brown, with irregular, raised whitish areas scattered all over it.

I have mentioned the illusion which sometimes causes flies actually to lay eggs on *Stapelia* flowers. This does the insects no good, for the resulting maggots are not provided with any food. But in some cases there *is* a food bonus for the young insects, as in cocoa, where the decaying seed pods provide breeding grounds for the pollinating midges; in *Alocasia pubera*, an aroid in whose spathes flies breed and pupate; and in the breadfruit relation *Artocarpus heterophylla*, where the pollinating flies breed in the fruit-smelling fallen male flowers when these have served their purpose. Such provisions ensure a continuous population of pollinators near the plants.

There are two classic cases of pollinator/plant relationships which are virtually examples of symbiosis or 'living together' (further expanded in Chapter 22), or at least what the biologist calls 'mutualism'.

Fig trees bear tiny one-sexed flowers on the inside surface of hollow, fleshy receptacles with one tiny aperture to the outside world. The fig fruit we eat – 'Involved, Inturned, The flowering all inward and womb-fibrilled; And but one orifice . . . a fruit like a ripe womb', to quote D. H. Lawrence – is in fact an inflorescence. There are three distinct types of flower, male, female and neuter. The pollinator is a minute fig-wasp only a millimetre long; almost every species of fig appears to have its own species of fig-wasp.

The female fig-wasp lays, with a very long ovipositor, within the neuter flowers (really bogus females) inside 'gall-figs'. While so doing she injects a drop of special liquid which causes the flower to turn into a gall, the enlarged fleshy tissue that results feeding the developing larvae.

These hatch when the fig is ripe and produce both male and female fig-wasps. The males are wingless and die shortly, without leaving the fig, having inseminated the females in a small dark orgy within it. The females push out through the fig's orifice, where they brush past male flowers and become dusted with pollen.

Some of the females fly to gall-figs and repeat the cycle. Others find true female figs; but the flowers here are a different shape, and frustrate the flies' efforts to lay eggs. In their bafflement the flies wander all over the flowers and transfer pollen to them, resulting in the ripening of seed-bearing figs. Whichever process occurs, the female fig-wasps die within the figs, having not eaten anything since birth.

Before hybrid edible figs were bred, which set fruit without the need for pollination, it was the custom, as recorded by Theophrastus, to hang a branch of wild fig in the fruiting tree so that the gall-wasps would cross-fertilize the two; otherwise the edible figs fell off before ripening. The wild edible fig is known as a caprifig, presumably because of a goat-like odour. In some parts of the world this 'caprification' is still practised.

The second classic pollination symbiosis is that of the American yuccas with certain moths, most of which are each restricted to one yucca species. The female yucca-moth, attracted to the flower by its basically nocturnal scent, climbs up several of the stamens in turn and scrapes their pollen together into a ball. It then transports this pollen-lump to another flower, which it carefully inspects to make sure that the stigma is receptive, and that another moth has not forestalled it. If all is in order, the moth lays from one to four eggs in each of the three cells of the ovary, and then applies the pollen to the stigma.

The ovules into which the eggs have been pushed grow abnormally large and form the food of the emerging moth larvae; but there are plenty left to develop into seeds. As the latter ripen the moth larvae, now mature, climb down the plants to pupate in the soil.

Not only do the adult moths emerge when the yuccas in their area are in flower, but the pupae spread their emergence over three seasons, just in case the yuccas have an off-year without bloom. The plant ensures that the moth-larvae have food and shelter; the moth guarantees the pollination of the yucca. They are completely mutually dependent. There is incidentally a bogus yucca-moth which copies the true species only in laying its eggs in the ovules.

Such associations seem so complex and improbable that one is forced to wonder at their existence and their development; but exist they do.

But I must return to other methods. Another major flowery attraction is nectar, a variously balanced sugar secretion (exceptionally up to 74 per cent sugar) which is of course food of a delicious kind to many insects (including ants), as well as bats and birds. Many plants secrete sugar at various points in nectar-producing glands, not always in the flower; in ancient times this may have attracted primitive insects which simply clambered about, knocking into the flowers from time to time. More subtly, nectar is made only in the flower, placed in such a way that the insect's tongue is used to extract it. Sometimes, as in cow-parsleys, palms and other plants with many small flowers packed together, there is just a little nectar in each flower, so that insects go from one to another for a square meal, and pollinate more actively. It may be remarked, however, that the 'large floral tables' of such flowers, to quote Proctor and Yeo, are ideal for insect idlers; they have not achieved a specialized relationship and some of the very numerous insect visitors may be of little use to them. Indeed, so many insects feed and rest on umbels that they are often visited by others which prey upon them. Alternatively, nectar may be more profusely produced, so that fewer flowers are need. The South African *Melianthus*, literally honey-flower, produces so much nectar that it positively rains sweetness if shaken.

It is all a matter of specialization and the degree of natural 'trust' built up by evolutionary adaptation. The more hidden and inaccessible the nectar the longer the insect's mouth-parts are likely to be and the more highly adapted the association. Nectar in a tubular spur, as in the familiar columbine, usually indicates a single insect pollinator, and indeed when Darwin saw an orchid called *Angraecum sesquipedale* found in Madagascar, with a spur up to 30 cm long of which only the last 4 cm were filled with nectar, he postulated that there must be a moth with a correspondingly long proboscis. Although entomologists of the time ridiculed this, the moth was in due time discovered, proboscis and all. What is more, the reverse situation occurred in South America where an appropriate moth was discovered, with a proboscis over 30 cm long, and after many years the appropriate flower with a 30 cm spur was found, a terrestrial *Habenaria* orchid.

Moth flowers are likely to have narrow lobes radiating from the central entry, and hence indicate the target for the hovering insect's proboscis; bees do not mind flowers with more rounded parts nor even those of for instance Himalayan balsam or foxglove into which they must penetrate entirely, because they can clamber about on the blooms. A word might be added here on the extraordinary rapidity with which some insects operate: a humming-bird hawk-moth has been observed to visit 106 flowers of the alpine *Viola calcarata* in four minutes.

Many flies are attracted by glistening droplets of fluid such as exude from moulds on decaying matter. Certain plants have acquired this signal, for instance the grass of Parnassus. Within the outer petal are five hand-like 'nectar petals', fringed with little stems (staminodes) topped with a round button which secretes fluid and glistens with this moisture droplet. They also produce scent. The flies do in fact receive a reward, for the nectar-petals also have nectar, placed near the centre of the flower so that the supping fly will brush against the sex organs. Some exotic species have a positive fringe of these droplet-hairs.

Nectar-petals on their own are not uncommon, forming for instance a ring of small pockets within the flowers of hellebores. Among the most remarkable nectaries are those of the crown imperial, at the base of whose pendant bell-shaped flowers are six glistening drops in white 'bowls', some 5 mm across, which hang in defiance of gravity. Long ago the herbalist Gerard likened them to 'faire orient pearls', commenting that 'the which drops if you take away, there do immediately reappear the like' and that 'they will never fall away, no not if you strike the plant until it is broken'.

Even more visually remarkable are the tropical marcgravias, epiphytic shrubs whose dangling flowers look like something out of a circus. The numerous flowers are carried in a wheel-like horizontal whorl below which five elongated nectar-pouches hang vertically. Humming-birds visiting these sugar-tubes can leave only by rising and touching the flowers with their heads, thus ensuring pollination.

Many insects (and also birds) have ways of robbing nectar without doing their 'duty', notably by biting or piercing holes through the flower itself, or by pushing their tongues into slits not designed for them. Some bees have even developed pollen robbery of flowers not actually designed for fertilization by pollen-gatherers. In some flowers there are protective devices, such as hard or swollen calyces, against such robbery. The simple inversion of a flower, so that it hangs, will encourage bees but put off flies. Sometimes false nectaries are provided to stop unwanted insects short of the flowers; sometimes sticky hairs trap crawlers hoping to attain them by way of the stem.

Heavy production of such attractants is obviously important in the competitive flower world. Thus a study of alpine flowers in Canada showed that pollinating insects are abundant when the earliest flowers open after snow-melt during May, and compete actively for the relatively scarce pollen and nectar available. But in early June what are known as 'cornucopia' species, such as willows and dandelions, with unlimited nectar/pollen supplies, take over, and the insects abandon the low-yielding early flowers. Once the cornucopians are over, further flowers compete for insect attention.

Besides attractive scent, pollen and nectar, some flowers also create warmth. Although the warmth is typically a by-product of stink-making (remember that stink can be insect-attracting), it is not without its purpose. Thus the faecal smell produced by our native lords-and-ladies to attract pollinating owl-midges (which breed in decaying excrement) is produced actively within the spadix (the erect, inner spike of the flower), raising its temperature by up to 15°C. The enclosed flower-chamber below is only a little warmer than the exterior, but its shelter is enticing to many insects. Shelter is equally provided in a curious Malayan relation, *Cryptocoryne griffithii*, which is an aquatic plant. Its slender flower tube, 30 to 45 cm long, just projects out of the water; a faint odour and a little warmth encourage a kind of beetle 'to overnight in an underwater cabaret', as E. J. H. Corner nicely puts it, and so carry out pollination. A trap-door prevents any escape, however, till pollination has been carried out.

Something comparable occurs in the palm-like Costa Rican shrub *Cyclanthus bipartitus* which is closely related to aroids and has flowers with a somewhat similar structure, though without detectable odour. As the female flowers mature the inflorescence attracts night-flying scarab beetles which collect in an enclosed chamber beside the whorled flowers. Over a period of many hours the male flowers mature and the beetles – by now engaged in long-drawn-out copulation – become coated with pollen. Eventually the inflorescence opens, the beetles disperse, and some find other inflorescences. To quote Dr Peter Moore, the plant provides neither inducement nor reward, 'and relies solely upon the beetle's requirement for a site protected from its predators in which to carry out its clandestine orgies'.

To return to lords-and-ladies, the main device here is in fact a trap, the flower-chamber formed at the base of the large boat-shaped, vertical spathe enclosing the spadix. The flowers are carried in this around the base of the spadix, a group of females in one band at the bottom, a group of males above. On top of all, at the neck of the flower-chamber, is a band of hairs, which bend downwards so that insects can pass down but not up. The midges go to the strongest-smelling area first, which is just above the flower-chamber; the surface of this and the lower part of the spathe produces oily droplets, so that the insects cannot help sliding in past the lobster-pot hairs, which are also greasy. Once the insects are in, the stigmas are receptive to any pollen they may already be carrying. Next day the stigmas dry up, and as they do so each offers a minute drop of nectar, a token payment for the insect's temporary incarceration. Then the stamens above become ripe: each opens two slits and showers pollen. Once this has happened, usually on the third day, the hairs shrivel and the pollen-covered midges can extricate themselves, to be attracted to another receptive flower.

Normally only a dozen or so insects are trapped in one flower, but sometimes there are many more. One record is of no fewer than 4,000 midges in the 6 cm chamber. Needless to say in such cases most of the visitors perish, as well as insect tramps, mainly beetles that have sought shelter. The highest spadix temperature known in an arum, and the most remarkable case of heating in the vegetable kingdom, recorded from *A. orientale*, was no less than 43°C in an external temperature of about 15°C.

A quite different flower displaying a marked rise in temperature is the giant water-lily. Within half an hour of its evening opening its temperature rises dramatically; measurements have shown a difference of 9·5°C, the flower centre registering 32°C in an air temperature of 21·5°C. The temperature falls about four hours later as the flower closes, to rise again the following day, when the flower opens in the afternoon. On the third day the flower sinks below the surface, after which the seeds ripen and eventually float out. Beetles which pollinate the giant water-lily are attracted by scent, become trapped by the nocturnal closure of the bloom, and are released on the second day well doused with pollen. Nearly fifty beetles have been found in a single flower.

Like the aroids, the giant water-lily increases its flower temperature by internal chemistry. Certain Arctic plants do so without such expenditure of energy. Like modern solar-heat collectors, they are bowl-shaped and these parabolic reflectors follow the sun to focus its radiant heat in their centre, stepping up the flower temperature by as much as 8°C. Mountain avens, visited by nectar-eating mosquitoes, and Arctic poppy with nothing to offer but warmth, thus create cosy havens for chilly pollinators. Focusing heat in this way is also used by the plants to help the seeds develop to good size.

Many arum relations create heat and have insect-attractive odours, but the trap mechanism is not so common. *Arisarum proboscideum* is called the mouse-tail plant because its purplish, hooded spathes have very long thin extensions which look just like mice tails among the leaves. Many aroids have these long extensions, especially the Himalayan arisaemas; they are usually on the hood, but occasionally on the spadix itself, and act as scent-beacons to the insect visitors.

I have already mentioned the dangers of suffocation to pollinators. The cobra-like arisaemas take this rather further. They have separate male and female flowers, pollinated by fungus gnats. If things work correctly the insects enter the pitcher-like male flower and simply slide down inside on a slippery surface, brushing themselves with pollen as they fall, to be let out through an exit at the bottom. Exactly the same mechanism operates with the lady flowers except that there is no exit; the ruthless plant lets its pollinators perish once they have done the necessary for it. *Arisaema* flowers are half-way to the leaves of insect-eating pitcher plants in their construction. Many of these 'cobra-headed' aroids have special tissue which, by reflection and refraction, concentrates light on the often darkly outlined entrance to the tubular spathe, thus creating a beacon.

Very similar trap mechanisms exist in the unrelated birthworts or Dutchman's pipes. Here a landing platform like that of a space-station faces a long tube, often curved as in a pipe, entirely fringed inside with inward-facing hairs. In arrangement it is exactly like an eel-trap. An insect-attractive smell – in *Aristolochia grandiflora* it is like rotting fish – lures small flies down to the sexual organs where, as in arums, the females are ripe first, the males shed pollen later, and the hairs then wither to allow the insects to escape. In many cases this is accompanied by the flower tube bending down, so that the insects are positively pitched out. Birthworts are however even more likely to suffocate their visitors than arums – they have become almost too clever.

To us the smell of birthworts, where we can detect it, is unpleasant. Some species which grow on the goat-ridden steppes of Western Asia not only have flowers that merge visually with goat droppings but smell exactly like them. Yet another ingenuity in some birthworts is a transparent area in the wall of the tube, towards which trapped insects blunder hoping for an exit; in fact the light merely leads them to the stigmas.

Many plants have such 'windows'. They exist, for instance, in the *Cryptocoryne* mentioned earlier, and in ceropegias of the asclepiad family, some of whose flowers look like elaborate Edwardian lamp-shades, fringes and all, with apertures below the closed outer end of the bloom and scent-attractive areas around or within these areas. In these flowers there is a slippery one-way chute; the prison below is often translucent around the sexual organs. The bewildered insects batter around this area, where they find some nectar and become pollen-covered. They have to wait, sometimes for several days, until the flower tilts on withering so that the chute now aids their escape.

Some trumpet gentians have translucent elongated windows, but these are kinder plants: the light exists for the pollinating bees to find the nectar in what is known as a 'revolver flower', at the base of which a ring of narrow tubes, like the chambers of an old six-shooter, face the insect so that it feeds in several different positions and collects the maximum amount of pollen. Columbines and bindweeds are similar.

A trap of quite a different kind to those already mentioned is that of various asclepiads such as the milkweeds and swallow worts. Here the paired pollen masses, combined with a sticky gland, are placed just above slits which lead to the stigma. Insects are attracted by nectar in large cups, and as they move about while drinking from these one of their legs inevitably goes into a slot, or sometimes their proboscis. In theory, the butterfly pulls out its leg with the pollen clamped around it, so that on reaching another flower the pollen is thrust on to a stigma. But the slit is so deep and tight that weaker insects cannot extricate themselves, and flutter in this too-efficient gin-trap until they perish. *Araujia* is known as cruel plant because it so often traps moths by their probosces in this way.

Many flowers take positive action rather than set a passive trap. There are several in which the stamens are so arranged that the touch of an insect makes them move rapidly so that pollen is flicked over the visitor; the Australian stylidiums are known as trigger plants for this reaction. Several barberries have similar trigger-stamens. The Brazilian *Hyptis* have small hinges on the lower lip of the flower, which hold down the anthers like pistol triggers; lip and stamens are under tension. Once the lip is moved the stamens are released and the pollinators – humming-birds or bees – are showered with catapulted pollen. Some bees, objecting to this explosion, have learnt to hold down the anthers while obtaining nectar.

Many pea relations have ingenious mechanisms: some, including broom and alfalfa, have a spirally coiled 'stamen tube' incorporating the style (the stalk carrying the stigma) which an insect 'trips' on forcing its way into the flower. This strikes the insects on the back, causing it and the stamens to jerk and pollen to be dusted on to the visitor's underside. Honey-bees are known to rob lucerne flowers of nectar by piercing them, which may simply be a way of gaining food without suffering the discomfort of constant sharp blows on the abdomen.

Mountain laurels have explosive stamens; so do various nettle relations – *Pilea muscosa* is indeed called artillery plant because of the puffs of pollen emitted at a touch. Other stamens are less violent: those of the house lime, which form a semi-circular boss, will move gently outwards if touched, so brushing their pollen on any visitor. Stamens of cacti carry out a reverse movement, concentrating on a central spot if disturbed. Cornflowers and knapweeds have an ingenious, repeatable method of ejecting pollen when sensitive filaments are touched.

Turning now from insects to other creatures, many birds are highly adapted pollinators, just like insects. They usually love nectar, but are not attracted by scent. Bird-pollinated flowers tend to have vivid, pure, even harsh colours, often in red shades, sometimes in garish associations of, say, green, yellow and red, and often with guide marks. They tend to be strongly made, and to have plenty of rather watery nectar, which sometimes forms the main source of liquid for quite large birds. Thus coral trees, which make so much nectar that they are called cry-baby flowers, are widely visited by birds in the dry season when they bloom. Each flower of the spear lily produces sufficient nectar to fill a liqueur glass. Proteas carry many small flowers in a 'cup' of tightly fitting bracts which hold nectar by the spoonful. Bird-pollinated blooms tend to be tubular or conical, suitable for holding plenty of liquid. Some bird-flowers such as fuchsias hang, or like trumpet vines stand erect, so that humming-birds can suck while poised in the air. The enormous spikes of the giant *Puya raimondii* offer the Andean humming-birds 20,000 metallic-lustred sucking points. Humming-birds, incidentally, are as sensitive as bees to colours humans cannot detect.

Other bird-flowers are strongly made so that the visitors can perch on them or, more often, on a bare stem, stalk, bract or bud in a suitable position. Some have developed literal perches, like the South African

Antholyza ringens, where the flower stem elongates above the flowers allowing sunbirds to cling to it and thrust their beaks into the upturned blooms. Sunbirds are particularly acrobatic as they probe for nectar. With the gaudy bird-of-paradise flowers the bird stands on a stout trough-shaped platform and extracts nectar from the centre of the flower, little knowing that the stamens are pressing pollen on to its feet ready to be transferred to the stigma of another flower in the same position. The several flowers produced arise out of the trough in turn, a fully developed bloom looking like a partly opened fan.

Tropical mistletoes thrust their brilliant flowers out of their host's branches so that they cluster or hang below the branches and are perfect for bird visitors. One group in Indonesia has buds with small apertures which birds of the *Dicaeum* group have learnt conceal a source of nectar. They thrust in their beaks and the flowers literally explode in their faces, showering them with pollen. This is another very close animal–flower relationship, for as we shall see the same birds eat and distribute the seeds.

Very often we find tropical members of a family bird-pollinated and their temperate relations dependent on insects; and since evolution of the flower almost certainly began in the tropical forest it has been suggested that bird pollination began before that by insects. Current thinking, however, is that it did all begin with insects. Birds incidentally often have a second food supply in the insects directly associated with flowers; they are also great nectar-thieves, their sharp beaks allowing easy piercing of the sides of flowers. Sometimes it seems that they do this to their habitual nectar-flowers simply to avoid having pollen brushed or ejected on to their feathers.

Many insects are nocturnal and hence the flowers they visit open at night and usually have a very strong scent which may be 'turned on' or at least increased only at night; they also have nectar to ensure a stay. Sometimes the habits of the insects are peculiar. Thus the night-blooming moonflower is in certain places visited by death's-head moths. A friend in Malta told me that he could almost set his watch by them: they arrived at 6.15 in the evening, when dusk was falling, and after 6.30 were no more to be seen although the white flowers remain open most of the night. Like this moonflower most crepuscular-pollinated blooms are white or pale-coloured.

At night the bird equivalents of insects are largely replaced by bats. The flowers they visit open at night; often the initial attraction is a powerful odour, sometimes quite revolting to humans; it may be foxy, musty or fishy, but odours of cucumber and sour milk have been noted. Bats are colour-blind so the flowers are often black or white, or in a special range of colours, purples, browns and dark reds, when they are associated with a smell of decay. They may have large openings for the beakless, clumsy bats, while they must be very strong to withstand the creatures' claws, and typically hang on long stems, since bats can clutch but are less good at neat landings. The extreme example is the tropical tree *Mucuna gigantea* which reaches 20 m and whose flower stalks may dangle to within a metre of the ground well below the branches. When flowers on such long stems are followed by large fruits, as in the aptly named sausage tree, the effect is doubly bizarre.

Many bat-flowers are spherical or, like the baobab, have round stamen-masses which the creatures clutch. Bat-pollinated flowers must make masses of nectar for their greedy visitors, and equally much pollen, for this may be eaten too. The baobab, cotton tree, durian, and many trees of the bignonia group are bat-pollinated. The cannonball tree has a special off-centre 'brush' of stamens to take advantage of its bat visitors. The famed night-blooming cereus cactus or queen-of-the-night (another flower chosen by Thornton for his *Temple of Flora*) is visited by the small vampire bat, and the giant saguaro cactus by a long-nosed bat. These are examples of bats which tend to hover rather than clutch. Some bat-pollinated flowers are very large: in the balsa tree they are 12 cm long and 8 cm across; in the sausage tree 7 cm long and 12 cm across.

Other small animals literally have a hand in pollination. Thus the agile bush-babies also visit the pendant night-open flowers of the baobab. One plant is pollinated by rats: this is the Hawaiian screwpine *Freycinetia*, whose cone-shaped flowerheads are surrounded by a number of orange-red leaves, fleshy and sweet at the base. Besides the birds and bats which normally pollinate these flowers, rats eat their floral leaves and in so doing transfer pollen on whiskers and paws from male to female cones. Other screwpines have similar succulent floral leaves which attract flying-foxes, which are bat relations. The Australian banksias are pollinated by 'dibblers' – marsupial mice – and honey possums, both small creatures adept at climbing, with long tongues adapted for nectar feeding and soft fur which picks up pollen.

The oddest pollinator is the snail, which is reputed to carry pollen from one lurid, cup-shaped aspidistra flower to the next; these are produced at ground level so it is certainly possible. The lily relation *Rohdea japonica* is also pollinated by these gastropods, which eat the fleshy outsides of the massed flowers (which smell of bad breath) and take pollen from one stigma to another.

Flowers usually close in rain because the pollen may be harmed, or pollinators may be deterred. A few plants take advantage of this, as for instance in the Faroë Islands, where pollinating insects hardly exist. In the creeping buttercup rain drops knock pollen from the anthers, the liquid running into the cup-like centre of the flower and swirling around the stigmas on which pollen is deposited. In bog asphodel the rain is trapped on long hairs on the stamen filaments; the water forms a bridge between anther and stigma over which pollen can float.

Truly aquatic plants have special problems, but there are in fact few which carry their flowers below the surface; these are mainly those remarkable species such as the grass wracks which have re-invaded the sea, to compete with the ancestral seaweeds. In such circumstances the flowers no longer need colours, scent or nectar, and often become very small. The main adaptation here is to produce water-resistant pollen in long strands, designed to wrap around the long thin stigmas after release.

Most aquatic plants produce flowers in the air and are wind or insect pollinated; pollen falling on the water usually decays, although occasionally true pollination on or even below the water surface occurs, as in some pondweeds. Here the pollen has a spiny surface which traps air and keeps the powder dry and afloat.

The aquatic plant most remarkable in its flowering is perhaps the European tape grass. This makes strap-shaped leaves up to 80 cm long, and lives in water of about this depth. The female flowers are borne on thin stems emanating from the base of the leaves, the buds being carried safely upwards in an airtight bladder of enclosing bracts. Once at the surface, the flowers open into three boat-shaped lobes, within each of which is a brush-like stigma. The male flowers are also produced in bladders, but much lower down. When they are mature, the bladder opens, and the buds detach themselves, floating up to the surface like so many divers, each with a bubble of air firmly enclosed within the flower segments. Once in the air these segments reflex so that they act as floats for the two stamens, which project obliquely upwards. If the little floats touch a female flower they are likely to jam between two of its lobes, when the pollen is transferred to the projecting female brush. This amazing plant has one more trick: once pollination has occurred, the female flower is pulled down, back to the base of the plant, by the spiral coiling of the stem.

Tassel pondweeds release their pollen in little bubbles of gas when it is ripe, and on reaching the air it spreads over the surface film where the stigmas are exposed.

In every plant the flower plays no further part after pollination and eventual fertilization, and finally withers or drops its corolla. Some are able to signify to late visitors that they have been pollinated, so that these will turn to other flowers and not waste their time. The peyote cactus actually closes its flowers within five seconds of fertilization. Blooms of *Magnolia virginiana* go brown within half an hour. The germander speedwell changes from blue to purple when its nectar is expended, while the shrubby milkwort flushes red when fertilized. Scented flowers usually shut off their insect-attracting odour very rapidly after pollination. In plants with several flowers in a group the changed colour of pollinated blooms may act as an insect-lure to the yet unpollinated ones; thus in the alsike clover the new flowers are white, the pollinated ones a strong red colour. Pollinated flowers in a head can also be seen to grow away from the unpollinated ones so that insect access to the latter is always clear.

What happens if the flower loses its pollinator? Our native bee orchid seems to have no pollinator in Britain and has resorted to self-pollination. Lobelia relations such as *Cyanea* are known to have lost their pollinators, birds called Hawaiian honeycreepers, when these became extinct. They, too, have been able to revert to self-pollination, which is a fairly remarkable feat in evolutionary terms, since it had to be done quickly.

Once again, certain flowers hedge their bets in this matter. Dandelions have very sweet nectar coupled with nutritious fatty pollen much enjoyed by honey-bees, and they have an ingenious method of ensuring cross-pollination, in which the stamens appear first and then, like a piston, the female organ pushes through the male part to spread its stigmas well above. Despite this, dandelions and the related hawkweeds very frequently, and in some groups always, resort to apomixy, in which viable seed is produced without fertilization, although pollination may need to occur as a stimulus. Apomixy obviously destroys the parent-sampling or recombination of characters which gives a species adaptive flexibility; but dandelions and hawkweeds, and other unrelated groups, such as lady's mantles and brambles, have intrinsic variability which is passed on to their progeny, as well as a high rate of 'sporting' or mutation, in which characters become permanently altered, so much so that species of these genera are now numbered in hundreds or even thousands.

Apomixy occurs most typically in open, unstable habitats and probably began in the difficult conditions characteristic of many areas during the Pleistocene era.

Occasionally plants produce fruits called parthenocarpic in which no seeds are formed. The pear variety Conference quite often does this, making sausage-shaped fruits; bananas too are parthenocarpic. These are special cases where an accident of nature helps us but not the plant, although bananas grow very readily from suckers. As I shall describe later, it is sometimes possible to stimulate plants artificially to produce fruits without pollination.

Some plants evade expending the large amounts of energy which others put into pollination – in wind-pollinated plants the creation of the inducements to visit and associated rewards. The trapping devices outlined earlier curtail energy expenditure, and the mimic orchids described in the next chapter reduce energy output to the minimum by perpetrating gross deception on their insect visitors. As in so many aspects of their lives, plants show all kinds of experimentation, in this case to balance the energy needed with ensuring that pollination is adequately carried out.

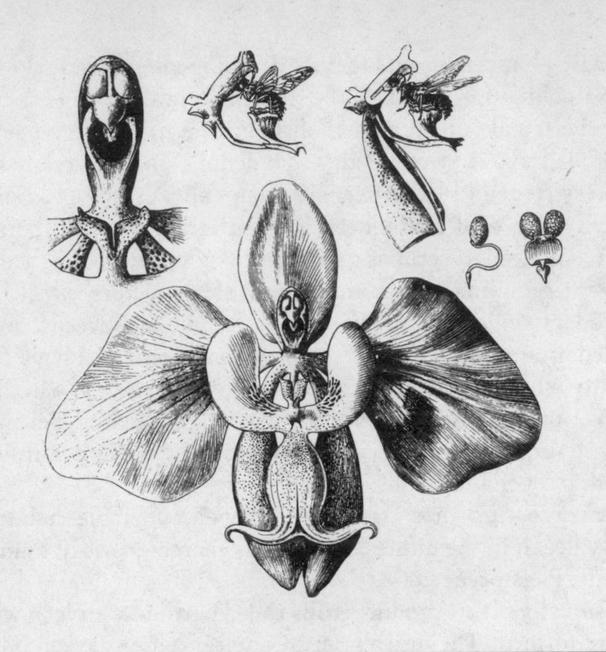

14 · A Floral 'Kama Sutra'

If nature ever showed her playfulness in the formation of plants, this is visible in the most striking way among the orchids . . . They take on the form of little birds, of lizards, of insects. They look like a man, like a woman, sometimes like an austere sinister fighter, sometimes like a clown who excites our laughter. They represent the image of a lazy tortoise, a melancholy toad, an agile, ever-chattering monkey. Nature has formed orchid flowers in such a way that, unless they make us laugh, they surely excite our greatest admiration.

So wrote the seventeenth-century German botanist Breynius. But he was not quite as observant as one might imagine, for this 'playful formation' of orchids conceals a sex life so elaborate that it reminds one of the *Kama Sutra*. While many other plant families are linked with pollinating insects, the orchids have consistently gone to more extra-ordinary lengths than the rest, and the orchid flower can indeed be considered a device for taking advantage of insects. In anthropomorphic terms, indeed, we can suggest that although the insect usually has some kind of reward it obtains this only by being made uncomfortable or positively ridiculous.

Although it has nothing to do with the plants' elaborate sex life, it is somehow appropriate that the word orchid derives from the Greek *orchis*, meaning testicles. This is because the terrestrial orchids found in the Mediterranean typically have a pair of rounded tubers which indeed resemble testicles. Another classical name is satyrion. In medieval England orchids were called dogs' stones or bulls' bags, and Shakespeare wrote of Ophelia's 'long purples' 'to which the liberal shepherds give a grosser name'. The doctrine of signatures linked man and nature: if a plant resembled a part of the human body, it must have some curative or tonic effect upon that part. Not surprisingly this doctrine, which flourished well before Christ, made orchid tubers greatly in demand for aphrodisiac potions, and the orchid was a symbol of sex in works such as the Unicorn Tapestry.

But that is by the way. The orchids form one of the largest of plant families with perhaps 30,000 species, one of the more recent in evolutionary terms and one which is still probably actively evolving. Many people think only of them in terms of the showy man-made hybrids seen in florists' shops – 'bosom orchids' as one gardener called them – but in fact they vary enormously in size from relative monsters to flowers a few millimetres across. The plants usually carry numerous flowers in spikes or on pendulous stems. One grower records a specimen of *Orcidium carthaginense* which produced a flower stem nearly 4 m long carrying several hundred flowers. The biggest orchid known is the Far Eastern *Grammatophyllum speciosum*, with pseudobulbs up to 8 m long and flower stems to 3 m, carrying hundreds of sizeable blooms. Many orchids live epiphytically perched on trees, others on the ground; some are saprophytic, basically leafless, living on decaying vegetable matter; and there are two which spend their entire lives underground, pollinated by tiny flies. Most, as shown in a later chapter, have a curious love–hate relationship with fungi. Here, however, we are concerned with what is in fact sexual engineering, and a few examples will have to suffice: they could be multiplied to cover almost the whole number of orchid species.

Orchid flowers differ from all others in form, especially in the mode of carrying the stamens and stigmas. These are fused into a 'column' which carries two functional stigmas and one or occasionally two functional anthers, while one of the six flower segments, the lip, is usually relatively large.

Orchid pollinators are almost entirely insects, and many individual orchid species rely on a single species of insect, so highly adapted are they. A few other animals, including small snails, may also occasionally pollinate them. Primary attraction for the insects comes frequently in the form of scent, which usually happens to be very attractive to humans. Some orchids change their perfume according to the clients they may hope to attract at different times: one moth orchid offers lily-of-the-valley scent by day and rose at night; a dendrobium offers heliotrope in the morning and lilac after dark. Some orchid odours, as in other flowers, are offensive to us; one kind smells of moth-balls, and some are carrion-like. Most orchid flowers last a long time if unpollinated – three months is not uncommon, and I have heard of some lasting for six in a greenhouse; but some are short-lived, as in the weird, pendulous, insect-like stanhopeas, whose flowers last but three days. This orchid achieves fertilization by emitting so strong an odour that insects from a wide area are quickly attracted. Once pollinated, orchids shut off their attracting odour; some change colour and most wither very rapidly.

Orchid scents are usually extremely specific for one kind or group of insect. There are at least fifty aromatic compounds found in the orchid family, each species making permutations of these in limited blendings. This must help to prevent the cross-breeding to which orchids are so prone, even among genera, as has been shown in cultivation where up to seven distinct genera have been brought together to create totally new flowers. It certainly helps to prevent crossing even between closely related species pollinated by related insects.

Some orchids have fly-attracting smells of rotting meat, and resemble its colour, like the carrion flowers of the last chapter. *Bulbophyllum macranthum* exudes a substance highly attractive to flies, which they lick. The flies land at the joined tips of the outer petals; but as they crawl along they find that the slippery segments spread further apart. They therefore transfer to the strong-looking lip, not realizing that this is delicately pivoted and will throw them downwards and backwards to make contact with the sexual organs. One 'bad-meat' orchid, *Masdevallia fractiflexa*, is a complete fraud, for it simply inveigles the flesh-flies into the suitable pollination position by smell, without actually providing any reward for them.

Scent, however, is almost always accompanied by a food reward. This may be straightforward nectar in a straightforward spur or cup or, as in the twayblade, the nectar may run down a groove in the centre of the lip. Nectar or food exudation can also occur on or in the tissues, where the insect must probe for it, or some part of the flower may be attractively edible. Some orchids have special protein-rich hairs, others crests or projections which insects gnaw, or insects may simply chew the basic structure – an offering of the flower's own flesh to its sexual agent.

Orchids also attract insects by their appearance: colour, pattern, hairy fringes, masses of quivering hairs, or long thin erect or dangling parts, the most sensational being the 75 cm ribbon-like petals of the Chinaman's whisker orchid, which often trail on the ground. Frequently, too, they have guide paths for insects which can be visible, physical or nectarious ridges or channels; and indeed the whole apparatus has been described as a 'cake-walk' or one-way obstacle course in which the insect is infallibly brought into a position where it cannot fail to collect pollen; or, if it has already done so, to deposit it on the stigma. Besides forcing the chosen insect along a definite path the cake-walk is designed to prevent unsuitable visitors from entering the flower, sampling the reward, or at least from contacting the pollen.

An added reason why orchids are so interesting in their sex habits is that they go to great lengths to avoid self-pollination, although in most cases both sexes are carried in the same flower, and a few will self-pollinate if all else fails. Where the sexes are separate the respective flowers are so different that botanists originally placed them in different species. This error has only been revealed, as with *Cycnoches*, the swan-neck orchid, and *Catasetum*, because of rare cases in which both male and female flowers appear on the same plant

The first bastion against self-impregnation is a structure called the rostellum which forms a physical barrier. In other cases there is a little flap or lid which closes over the stigma as the pollen is being removed by an insect. The pollen, incidentally, is usually carried in solid, unshatterable masses called pollinia. These are very often attached to a sticky disc by a stalk-like growth, or sticky material may be applied to them by the rupture of a small sac.

In many cases, as can be observed with the common British early purple orchid and its relations, the viscid disc clamps on to the proboscis of the insect: in this case a humble-bee probing the flower's spur. The disc has a glue-like material under it which sets very hard in a few minutes. But the position of the horn-like pollinia is now nearly vertical above the proboscis, and if the insect visits another flower it will merely push the first pollinia on to the next ones, not on to the stigma below them. To counter this the orchid has a trick up its sleeve: within half a minute of attaching itself to the insect the stalk of the pollinia bends forward almost at right angles. So poised the pollinia will be rammed on to the stigma of the next flower visited. As Charles Darwin noted, you can test this intriguing device with a pencil.

In the pyramidal orchid the long thin spur has a very narrow opening, but there are two adjoining ridges which, like a Victorian needle-threader, guide the proboscis of the searching moth or butterfly. Accuracy in guidance is essential because the pollinia must be clamped symmetrically upon the proboscis to operate successfully. As in the early purple the pollinia change attitude shortly after becoming fixed. Sometimes insects can be seen with several pairs of pollinia stuck to their probosces (Darwin once counted eleven pairs) which looks uncomfortable for them and can indeed eventually cause them to perish if these pollinia prevent them reaching the nectar.

In some orchids, such as the butterfly orchids, whose white colour and powerful nocturnal scent attract night-flying moths, the pollinia become glued to the sides of the moth's head, occasionally to its eyes!

In most orchids the insect alights on the flower segment called the lip. Although this is in effect one of three petals, it is usually much larger than the other parts, and often of very peculiar form. Sometimes the lip plays an active part, as in *Pterostylis*, where a hinge moves the lip up to prevent the insect leaving, forcing it along the inner cake-wall, indeed hurling it into the recess at the back of the flower. The lip goes back to its original position, ready for a new visitor, in about half an hour. In the Australian flying duck orchid visiting insects are again boxed in for a time, after being tossed into the centre of the flower by a jerk of the 'duck's-head' lip. Bulbophyllums again have lips so mobile that an insect which lands on these is literally tipped into the centre of the flower. In the marsh helleborine and related species the hinge is in the centre of the lip. The insect alights on it, depressing its outer end markedly; it crawls forward and sups nectar. But it has sprung the device: as soon as it tries to take off the end part of the lip springs sharply against its abdomen, and it can only go upwards. The back of its head breaks the glue-bag and its continued upward movement forces off the pollinia, which now stick on its head. In the feeding position on the next flower the insect is in the right place to smear the stigma with pollen.

Another mechanism is that of the pendulous *Gongora*, where the bee, attracted by a sweet scent promising nectar, finds it cannot hold on to the flower because of its highly polished surface. It endures a short sharp descent on its back down the curved chute of the column, at the base of which it encounters the pollinia. With these projecting horn-like from its backside, it visits another flower and the same helter-skelter descent rams the pollinia on to the sticky stigma also near the column base.

The pouch-like lip of lady's slipper orchids has nectar under its outer edge. Bees entering the pouch find that they cannot crawl out where they came in; it is too slippery and overhung. But translucent windows guide the insects towards the column, beside which are some stiff hairs to provide a foothold, and the insects push past the stigma and then the pollen-mass which project into the only feasible exits on either side.

The most extraordinary lip mechanism of all, smacking strongly of Heath Robinson engineering, is the 'invention' of the Central American bucket orchid. The lip, which can be many centimetres across, has literally turned into a kind of bucket, suspended by a rigid right-angled bracket and with a narrow trough-like spout. Just above this spout another rod-like part of the lip makes a right-angled bend, bearing on its under-surface the stigma and pollinia in niches. Higher up this rod are a pair of glands that secrete drops of liquid.

The stage is set. Humble-bees attracted by the colourful, strongly fragrant flowers (which hang below the plant) are drawn to an area of edible tissue from which they also collect a scented fluid. This literally makes the bees inebriated, so that they fall into the liquid in the bucket. But the sides are steep and there is only one way out, via the spout. The insects have to wriggle their undignified way between the trough and the right-angled rod, and in doing so can either collect pollen on their backs or place previously acquired pollen on the stigma.

The first bee to enter a virgin flower has a hard time with its escape because the rostellum tends to pin it in position. Sometimes the bee takes thirty minutes before it manages to wriggle out. After the pollinia have been removed, however, the rostellum is no longer so awkward and subsequent bees, some in theory with pollinia from other flowers, can pass through quite quickly. There is usually a positive procession of groggy, wet-winged bees extracting themselves from the bath.

The flower appears to have a scent shut-off device if a bee is too long escaping. One observer noted a bee taking forty-five minutes on its Houdini act; by that time there was no scent and nothing would persuade the insect to go back into the flower. Next morning the flower scent had returned: this may well be a device against self-fertilization.

Other orchids also make drunkards of their pollinators. Such are *Gongora*, *Stanhopea*, *Catasetum* and the swan-neck orchids, in each of which the pollinia are shot on to some part of the insect. There is, as usual, a reason for all this. Bees are normally brusque and clumsy. By intoxicating or drugging them the flowers can make them 'do their will' more easily, so that the pollinia are placed or removed with maximum precision.

The intoxicant is obviously highly pleasing to the bees concerned, for they collect it and store it in spongy tissue-pouches on their legs. One theory for this is that the male bees use the oil to mark out their courtship territories, as some animals do with their own scent glands or urine. Other evidence suggests that it is released at some suitable spot where it attracts bees of both sexes.

In catasetums the rostellum develops two 'antennae', which function as triggers; curiously enough in some species only one of the pair develops sensitivity. If a trigger is touched, however lightly, a catapult mechanism is set off which can fling the whole pollen-mass, with its sticky disc foremost, at least three feet out of the flower. No other shock sets off the catapult. Catasetums are visited by relatively large insects which gnaw the flowers and this rather violent ejaculation makes sure that the pollinia are securely fixed upon them.

Some orchids resemble remarkably the insect traps of the arum family, both in shape and colouring. Examples are *Pterostylis falcata* and *Masdevallia muscosa*, where a pitcher-shaped receptacle with a hood is formed by the interlocking and overlapping of the floral parts. The much-reduced lip combines trip-wire and trap, and can remain shut for about thirty minutes. As in *Arisaema*, the entrance is often relatively bright. Slipper orchids are also trap-flowers of a slightly different kind.

Perhaps the most unexpected pollination method is that of the *Ophrys* tribe, the European and Mediterranean bee orchid fraternity. Their habits are shared with the Australian *Cryptostylis*, the South American *Paragymnomma* and possibly one or two others. These orchids have no nectar, no oil, no edible area; yet they are readily visited by bees, flies and related insects. Indeed, the similarity between the flowers and insects was noted long ago. Because the bee orchid can self-pollinate in Britain (where it has lost its pollinator) one authority suggested (in 1831) that the flowers were designed like insects in order to scare away visitors. The truth might not have pleased the prudish nineteenth-century biologists: self-pollination as we have seen is only a last resort in the orchid world, and observation gradually showed that male insects carried out pollination by attempting to mate with the flowers; it was noted that with the insects concerned males either habitually emerged from pupation before the females, or there was typically a preponderance of males.

In fact we have here plants which can only be called prostitutes. The flowers are designed in their general visual appearance to attract male insects; they often have a reflective patch to assist the insect to 'home' as on to an airfield beacon; they have shiny eye-like spots, much reduced petals which mimic antennae, and often side-lobes on the lip which resemble the folded wings of an insect. But the main attraction for the male insects, which brings them close enough to the flowers to 'home' upon them, is an odour like that of the female of their species. So strong is this odour that the males will try to find flowers wrapped up in paper, and in at least one species the males prefer the orchids to their real mates. Once upon the lip of the orchid, the insect finds the right curves, projections and hairiness to persuade it that it has really found a female, and to attempt mating, an activity known as pseudo-copulation, carried out at length and with vigour.

In fact the orchid is never quite snug enough for the amorous insects, for although these may have their sex organ sufficiently stimulated to be extruded and rubbed upon the flower, and carry out mating movements, they do not ejaculate. Even so, they do fly hopefully from flower to flower, carry out a mating attempt, one hopes enjoy themselves, and very effectively transfer pollen. Once in a while the bee does appear to realize that all is not as it should be. Its pseudo-copulation is longer and more violent than normal and it may even bite the lip viciously.

These orchids vary greatly in lip shape and pattern according to their specific sex-partners; some have, to be anthropomorphic, 'shoulders' or even 'breasts' protruding from the lip, and in the horned orchid there is a pair of thin projecting horns over a centimetre long in the 'shoulder' position. Hairiness varies also, the most remarkable examples being perhaps the Venus's mirror orchid whose large shiny reflective patch is entirely fringed with thick brown or reddish hairs which extend over its pronounced 'arms', and its sub-species with narrower, hair-tufted lips which look exactly like blue-bottles.

O. lutea is a curious exception. It has a dark, raised area with two shining bluish marks in the middle of the bright yellow lip. Its visitors always place themselves head-down on the flower, their posteriors up by its sexual organs. It seems that the mimicry here is that of a female bee resting on a yellow flower, head facing away from the column.

Prostitutes these *Ophrys* may be, but they might do better if they arranged for a more varied clientele; some are entirely dependent on a single species of insect, and if this insect is not available they cannot normally exist in the area concerned, however suitable the climate and terrain may be. Still, the mimicry of animal shape and hairs in these *Ophrys*, as well as the production of female-insect odour, is one of the most remarkable pieces of evolutionary adaptation on record, so remarkable that one can hardly imagine the stages between the present-day species and an original 'basic' orchid with, perhaps, scent and nectar attracting any bee. After lecturing on this topic once I was indeed approached by a man who said he had never heard and seen a better demonstration of his belief in 'instant creation'.

It is a striking fact too that this special pollination method evolved independently in several places, involving different genera of both orchids and insects.

A rather different example of mimicry is that of certain bog orchids, including *Pogonia ophioglossoides* and *Calopogon pulchellus* in North America and *Disa pulchra* in South Africa. Their cup-shaped pink flowers have neither nectar nor pollen usable by insects. But they flower at the same time as unrelated pink-flowered plants in similar habitats – in South Africa, for instance, the *Disa* always grows among clumps of a pink watsonia which it closely resembles. So insects attracted to the pink flowers with nectar or pollen on offer are deluded into visits to the orchids' blooms alongside simply by mimicry.

One hardly needs to stress the really remarkable adaptations displayed by orchids. Let me quote again from *The Pollination of Flowers* by Proctor and Yeo:

> A simple explanation of any part of the biology of a species or family is always hazardous, because any one feature, such as seed-production or pollination-mechanism, influences and is influenced by so many other interacting factors. None the less, it seems fair to say that the Orchids have taken one particular solution to the problems of living in a competitive world very near to its logical conclusion.

Finally, it gives cause for thought that the orchid flower consists only of adaptations from the standard set of floral parts already remarked upon. To quote Stephen Jay Gould, 'orchids were not made by an ideal engineer, they were jury-rigged from a limited number of available components'. Perhaps this makes them all the more remarkable.

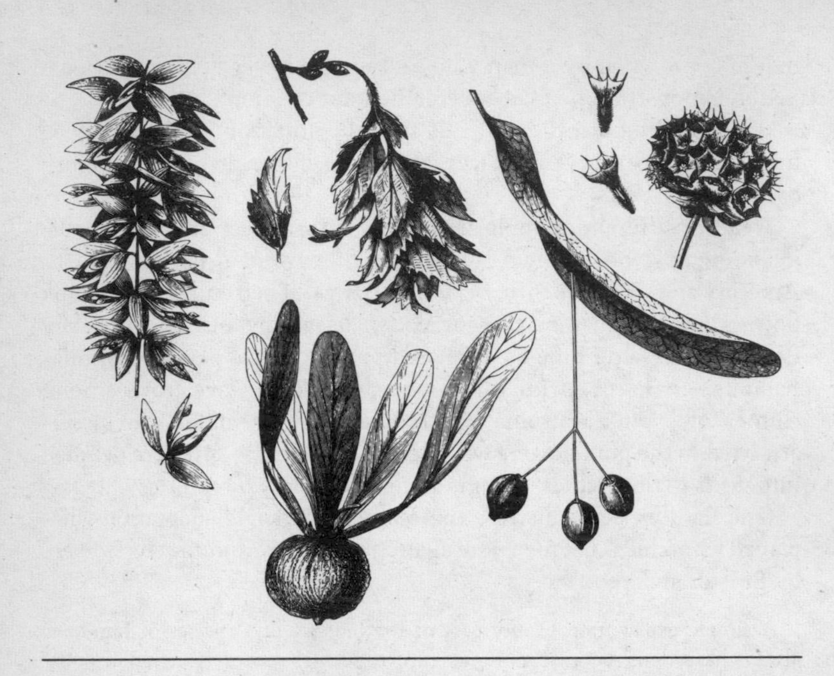

15 · The End Purpose

Nature is vastly prodigal with reproductive material, especially among some lower plants. An ordinary edible mushroom has been calculated to produce 16,000 million spores which are released at the rate of 100 million per hour, while an average-sized giant puffball, say 30 cm across, may produce 7 billion spores. The spore count for the biggest specimen ever recorded is almost incalculable: the puffball, 24 cm high, was 160 cm across at its widest part and 135 cm at its narrowest. It has been estimated that puffball spores could conceivably be carried in air currents for nearly half a million kilometres, while the rather larger spores of ferns may be restricted to perhaps 20,000 km. This compares with an average 10 km for the floating seeds of a dandelion, although in storms such seeds might travel 200 km.

Long-distance dispersal is not, in fact, necessarily desirable; it may take spores or seeds out to sea, to unsuitable habitats, or away from the plant's pollinating animals where these exist.

Spores (typically the offspring of the non-sexual phase of lower plants) are released in many different ways. Fungi such as puffballs are simply bags which, when ripe, wither and burst to release their dust-like cargo. Geasters or earth-stars have two layers, the inner being the spore-bag, which in some species is forced upwards when ripe by the splitting and turning inside-out of the outer. Gilled fungi such as the mushroom carry their spores on the sides of the gills, from which they are successively shot off with a very carefully calculated explosion for about one-tenth of a millimetre, a distance conditioned by the tiny gap between gills, between which the spores then fall. The dung fungi have gills which deliquesce, the cap gradually being 'self-digested' and turned into a fluid which contains the spores.

The tiny birds-nest fungi are a fascinating small group in which spores are carried in bodies called peridioles which do not release the spores till conditions are right. *Crucibulum* looks like a small almost vertical-sided basket; when ripe, its lid dries up to reveal the egg-like periodioles within. *Cyathus* is similar but more open and with slightly fluted sides. The peridioles are initially fastened by a thread which, in *Cyathus*, can be up to 18 cm long (the whole cup is at most 2 cm across). These fungi have discovered how to make use of the power of a raindrop. Their shape is such that a waterdrop of average size and velocity is forced back up the side of the 'nest', sweeping up the peridioles with it; in *Cyathus* they are carried at least 60 cm and sometimes up to 200 cm. The attaching thread unwinds as the peridiole is ejected, rather as a strand can be pulled out of a piece of knitted material; it finally breaks at its base and, being tacky, will stick to or tangle with any leaf or stalk in its course, where the peridioles hang until conditions are right to release the spores.

A striking catapult mechanism occurs in the fungus *Sphaerobolus*, where the spores are carried in a ball, held as it develops within a cup of many layers. Tension builds up between these layers so that the cup is finally and suddenly turned completely inside out, ejecting the ball to distances of 5 or 6 m.

Many fungi, including the dread ergot of rye, create drops of liquid below each spore-ball, from which nectarious spores emerge to attract dispersing insects.

One notorious mode of spore dispersal is in a group of fungi aptly known as *Phalloidaceae*. They produce spores externally in a film of mucus which has a faecal or carrion stench, as in the familiar stinkhorn. Botanists like calling a spade a spade, and this plant's Latin name, *Phallus impudicus*, is very apt, for from a soft, buried, egg-like mass it produces a fruiting organ shaped remarkably like a human penis, formed of a brittle material rather resembling expanded polystyrene. Not the least remarkable is the speed at which this occurs: the emergence of the organ may take only one and a half to two hours. Of course this is expansion of a preformed structure, as in a mushroom; it is not growth.

It is on the conical cap that the mucus is produced. Flies gather round this appetizing feast just as they do round the stinking flowers of aroids or stapeliads, devouring the mucus, spores and all. The spores pass through them unharmed and are excreted 'in another part of the wood'. One excretion by a fly can contain over 20 million spores. Slugs also enjoy the slime, and they can locate a stinkhorn by smell from 6 or 7 m. Once the slilme has gone, the fungus has a relatively sweetish odour.

Another British member of this tribe is *Mutinus caninus*. Mutinus was another name for Priapus, Roman god of sex, and *caninus* refers to the startling resemblance to a dog's penis, especially as the narrow cap is red. The smell of this fungus, though much less strong than that of its relation, is indescribably nauseating.

There are many tropical relations, some of which have a perforated skirt-like veil beneath the cap. Not surprisingly primitive people were struck by the resemblance to their private parts and used them for fertility rites, one such involving the rubbing of ash from incinerated fungi on to the vaginas of infertile women.

Other relations have no cap but look like thick-tentacled sea-anemones, and one extraordinary kind, once to be seen fairly often in old stove-houses where it lived on the fermenting tanbark which produced the heat, and still recorded from Britain and Europe, is *Clathrus ruber*, in which a striking red or pink hollow lattice-work sphere bursts from an outer coat.

More precision in spore dispersal is found among the mosses, where spores are carried in neat capsules, sometimes with a beautifully made little lid. In sphagnum moss the lid, and the spores within, are blown off and out by air pressure mounting within the capsule as its lower half dries out. *Andreaea* has a capsule with slits which only open in dry conditions, ensuring that spores are only released in dry weather, when they float on the air more readily. Variations in atmospheric moisture are again used by those mosses in which the capsule has under its lid a series of teeth. Changes in humidity affect internal tensions in the teeth, resulting in their twisting about abruptly, so that the spores are stirred up and thrown out. *Polytrichum* has a ring of little holes in the rim of the capsule on which the lid rests till the spores are ripe; the capsule then acts like a pepper-pot, and strikingly resembles the seed-dispersal mechanism of poppies.

Spores of horsetails are attached to the centre of a long X-shaped device. On release the arms of this contraption are wrapped around the spore, but once in the air they open out, changes in humidity causing sudden alterations in shape which provide considerable movement.

Tensions resulting from dryness are also utilized by various ferns, often merely to open slits in the spore-bearing body, but in *Dryopteris* to operate a fascinating catapult mechanism like a Roman ballista. A strip of special cells along one side and the top of the capsule exert such a force that the top of the capsule containing the spores bends gradually away from its base. A sudden release of tension flings the top forwards again and the spores are hurled out of it. This sudden movement is caused by the instant conversion of water into vapour in the special cells.

When we come to the higher plants spores give way to seeds. The seed has many more cells than the spore, and in principle contains an embryo and has a protective outer coat. The degree of development of the embryo varies, and so does that of the coat; some seeds have a food reserve while others, like those of orchids, do not – some orchid embryos have as few as 120 cells to start with. In some exceptional cases, as in *Ginkgo*, the seeds may not even be fertilized before they fall off the parent tree

Seeds vary enormously in size. The smallest, again in the orchid family, are less than ¼ mm long and are so light that 3 million make a gram. Such minute seeds are often produced in equally unbelievable quantities; a single capsule of the cradle orchid or the swan-neck orchid contains nearly 4 million seeds, and the ordinary *Cymbidium* of the florists may have 1½ million. Such powder-like seeds have almost the dispersal potential of spores. Seedsmen naturally find tiny seeds like these, or those of begonias and calceolarias, very difficult to handle, and often put a small envelope inside the packet with a barely visible pinch of seed from which in fact you can raise a greenhouse-full of plants.

At the other end of the scale there are plants producing a few very large seeds, the biggest of all being the Seychelles or double coconut. Also called coco-de-mer, because it occasionally floated across the world by sea (originally commanding enormous price as a supposedly aphrodisiac rarity), this is in fact a palm, and the giant seed can be up to 45 cm long and over 30 kg in weight; it takes over ten years to mature. It is remarkable for its amazing resemblance to a woman's pelvis, both back and front; and since its male flowers are carried on a narrow spike-shaped spadix over a metre long its sexual symbolism is naturally strong.

Technically this giant seed is a fruit, but this is a matter of terminology, since the entire outer casing of the 'nut' is filled with a single embryo. In this chapter I shall use the term fruit more loosely in its household meaning of seeds embedded in a much larger mass of edible tissue. The biggest orthodox seed is probably that of the South American *Mora excelsa*, which is flat and about 12 by 7 cm.

Size of seed is no guide to the length of time it takes to mature. Some orchids take up to eighteen months before the seed-containing capsules are ripe. The quickest production of seeds is, naturally enough, in annual plants with a short season from germination to flower and seed production; or perennial plants where the exterior conditions impose a very short growing time. The record is, I think, held by a Norwegian specimen of the alpine buttercup *Ranunculus nivalis*, which flowered within five days of emerging from its winter snow cover, and produced ripe seed seventeen days later. The giant Mexican cactus *Pseudomitrocereus* matures its seed within twenty-seven days of blooming.

Annuals are designed to produce seeds and die, but if cut back, even after ripening one crop of seeds, they will often send up more flowers and repeat the process; as all good gardeners know, dead-heading is essential to keep any plant flowering as long as possible. Other kinds of plants are termed monocarpic: they grow for a term of years, flower and die. The great agaves of America were termed century plants because of a belief that they only flowered – and then perished – once in a century; some certainly take many decades to do so, like the similar but unrelated giant bromeliads such as *Puya raimondii* which can wait for 150 years before erupting into flower.

Flower and seed production are often the final resort of a plant which 'senses' that it is going to perish, and here again gardeners take advantage of this by keeping certain plants – fleshy-rooted clivias and bulbs like hippeastrums and nerines are good examples – in very small pots. Given plenty of root room these plants would simply produce lots of leaves and be unlikely to flower; cramped, the instinct to bloom is encouraged. Dryness at the roots at a certain time is often a good stimulant to flower initiation; this is again applicable to many bulbs (which frequently have a dry resting season in their natural habitats), and is the cause of lettuces running to seed in a dry spell.

Once a plant has produced seeds, whether few or many, it must take measures to distribute them as far as it can. Just as fertilization may be passive and haphazard, or the result of evolutionary adaptation with animals, so is seed dispersal. It is interesting to see how in many cases plants with very highly developed pollination mechanisms – orchids are again a prime example – may well have no special adaptation for seed dispersal. The tiny seeds of orchids are blown about in the wind, but in this case there is a good reason for their landing close to the parent, because it is there that they are likely to encounter the fungus without which, as we shall see later, they cannot usually germinate.

Wind is in fact a prime mover of seeds, some of which have reverted to the dust-like proportions and weight of spores and can thus be carried long distances. A great many plants have developed winged seeds of various kinds, although their flying range is often very limited. These aerodynamic inventions may be a simple membrane in one plane around the seed, or develop into aerofoils like the familiar ash key or the spinning sycamore seed, so reminiscent of the old autogyro.

Some flying seeds have a single membrane or wing, occasionally a biplane; others, like *Dryobalanops*, a cluster of several wings to allow a relatively large, heavy seed to drift at least a short distance from its parent. The biggest wing belongs to the Brazilian *Centrolobium robustum*, 17 cm long, while in *Cavanillesia* from the same area several wings up to 9 cm across develop on the sides of the fruits, causing a rotating movement. Bignonias usually have winged seeds neatly packed into a long capsule or pod.

Some conifer seeds are winged; others, oddly enough, remain embedded in great woody cones until these rot or are opened up by fire. In *Pinus attenuata* the cones are grown over by the bark, and cannot be released until the tree has died, fallen and decayed! The biggest pine cone, incidentally, is that of *P. coulteri*, up to 35 cm long and nearly 2 kg in weight.

Scabiouses and thrifts develop little frilly membranes derived from the dry calyx, looking like shuttlecocks, which carry the seeds; many trefoils have inflated calyces. Such devices are effectively balloons, although their progress is likely to involve much tumbling over the ground. Bladder senna has large inflated pods, and *Cardiospermum* is known as balloon vine because of its spherical inflated seed containers; in both cases these become detached from the plant and blow about.

These shuttlecocks and balloons are half-way to the parachute, a device familiar to us from the dandelions and goatsbeards, but adopted by all manner of other families, including the South African silver tree, where the large seed is suspended from a feathery head on a delicate stem. Alternative to the parachute are plumes of the fine hairs, as in the alpine feather grass where the feather (technically an awn) may reach 30 cm long. The Antarctic parasite *Myzodendron* has a triple plume which is in fact a much modified stamen and can be several centimetres long, giving these plants the name feathery mistletoes. Very many plants such as willowherbs, clematis, valerian, cotton grass, willow and bulrush or reedmace have seeds borne aloft by a mass of very fine hairs which can carry the necessarily small seeds for miles on air currents. Plumes and parachutes are much more efficient dispersers than wings.

Some plants when in seed dry up, are uprooted and blow about shedding seeds as they go. In the Book of Isaiah there is a reference to the 'rolling thing before the whirlwind'. The prophet was a good observer of nature, for there are many kinds of 'rolling thing' which botanists now call 'tumbleweeds', especially in steppe or desert areas.

Most tumbleweeds have stems which form a rough sphere, and when dry are sufficiently hard to allow the sphere to blow about. The spiny saltwort, found both in America and on the Asian steppes (it is also a British native), is the typical tumbleweed; enmeshed masses of it as big as haystacks are sometimes seen and no wonder they are called wind-witches. Many thistles behave in the same way. In the Near Eastern *Gundelia tournefortii* – a kind of poor man's artichoke – it is the prickly heads which break off, mesh together and form large rolling masses. This is probably the 'rolling thing' of Isaiah. One account describes how 'at the proper season, thousands of . . . these vegetable globes . . . come scudding over the plain, rolling, leaping, bounding with vast racket, to the dismay both of the horse and his rider'. In Indonesia and Australia the radiating seed-heads of spinifex grass mesh together into 'wind balls'.

A literal 'wind ball' is the bitter gourd or colocynth of North African and Near Eastern deserts. When ripe the 10 cm melon-like fruits separate from the dry stems and become dry and light. Their smooth spherical surfaces allow them to be bowled across the sand in a wind. On coming to rest in hollows, or against shrubs or rocks, they eventually burst open and scatter the black seeds in the heat of the sun.

Water is of course another method by which seeds are transported. The spread of the introduced Himalayan balsam along British water-courses in the last few decades is a prime example, although this method presupposes a buoyant seed that will not germinate simply because it is wet. Such seeds usually have unwettable or water-repellent surfaces, as in the yellow flag iris; they contain air, sometimes in small pits on the surface, or in air spaces inside, or have buoyant corky tissues. The Indian lotus has a conical float-like seed-head with openings in the upper surface in which the separate seeds can be seen rattling. It breaks from its stem and floats till it disintegrates, when any seeds still trapped in it can float off separately, prevented from sinking by microscopic waxy outgrowths.

Hodgsonia heteroclita is a scrambling cucumber relation growing on river banks. Its large woody gourd fruits are covered with a greyish felt which prevents them from becoming wet when they fall into the water; what is more, the oily pulp within coats the seeds so that if they get into the water individually, they can float unharmed.

Seeds which are carried in the sea rather than in rivers must tolerate salt and be much stronger to withstand rough treatment and long periods of floating. The most familiar sea-transported fruit is perhaps the ordinary coconut, which not only has a thick fibrous husk for flotation but a very special food reserve, the milk, which allows it to germinate and establish itself on arid shores. But there are many much smaller seeds which are distributed in the sea; some Australian members of the *Veronica* tribe, with quite small seeds, have colonized Chile because of their floating persistence. Two species of the pea relation *Caesalpinia* have very hard pebble-like seeds which can float for literally years without losing their powers of germination; thus these plants have spread almost throughout the tropics.

The oddest method which a floating seed uses to reach a place for germination is perhaps that of persistent floaters around the Cocos Islands. Here frigate birds normally feed by threatening boobies and forcing them to disgorge the fish they have swallowed. When the boobies depart after their young are mature the frigate birds have no easy source of food, so apparently they swallow large quantities of pea-like floating seeds, which give them the impression of having a full crop; but on reaching land they simply vomit them up!

Mangroves are plants of tropical watercourses and sea-coasts, some of which bear a large pear-shaped fruit. Long before it separates from the tree, however, this pear produces a long thin shoot. This is not a true root; it is technically called the hypocotyl and is at the end of a tubular 'seed-leaf' or cotyledon. This shoot grows for seven to nine months until it is about 50 cm long and 3 cm across, with added thickness at the lower end. Just below the point of emergence of the fruit there is a ring of cells which, when the hypocotyl's weight is sufficient, rupture and allow the fruit to fall. If all goes well the thickened hypocotyl plunges dagger-like into the mud below, thus neatly planting itself. At the top end leaves unfold, roots form at the base, and a new mangrove starts to grow. In fact, many of the seeds do not land straight, or they fall into water, for these are plants of tidal areas, and then the buoyant 'pear' allows the seed to be carried by currents till it lodges in a suitable place for development.

There are other plants which more or less sow their own seeds, and some of these can be observed nearer home, for instance the wild herons-bills and related stork-bills – popular names derived from a long projection of awn on each seed, which at maturity are grouped in a beak-shaped mass. When ripe and dry they separate, first at the base where the seed is, and finally, twisting as they go, at the apex of the beak. At this stage the spiral 'tails' are light enough to assist the seeds in being blown about; eventually they lodge in debris or among other plants. The tail is very sensitive to changes in moisture, and the end carrying the seed is waved up and down and round as the weather changes. If the seed touches the ground, this motion results in the seed being thrust into the soil, where it germinates. It is barbed with stiff hairs like an arrow, and therefore cannot be pulled out readily once it is in the soil. The longest 'stork-bills' or 'granny's needles' are those of the Mediterranean *Erodium gruinum* which can reach 5 cm. Wild cereals have similar barbed awns which plant the seeds very rapidly after they are shed, and the long awn of the feather-grass also acts like a drill handle on the spirally twisted seed-end.

A few plants, some thirty species in all, literally force their seeds into the ground by downward elongation of the stems after flowering. The pea family has developed this trait most strongly (the peanut is a classic example) but examples occur in several quite unrelated families. At first glance it may seem unsatisfactory to have a mass of seedlings all growing and competing in one place, but with desert plants in particular it helps to form colonies in the best growing conditions of such usually inhospitable areas. The Negev *Gymnarrhena* has the best of two worlds, making one set of fruits underground, which are extremely drought-tolerant, and another above ground, capable of being blown about by wind, but not by any means so resistant to dryness.

With plants in really arid conditions it is obviously most important that seeds are liberated only when there is rain. This occurs in *Anastatica hierochuntica*, a member of the mustard family often called rose of Jericho, referring to a legend that all the plants opened up and bloomed when Christ was born. It is a plant of spreading stems which when dry roll up into a basket-work ball, when damp open out and release their seeds. As such it is sometimes sold as a curiosity under the name resurrection plant, like one or two others, one being a species of *Selaginella*. They are of course quite dead despite their capacity to open and close. Sometimes the rose of Jericho is pulled out of the soil by wind and behaves as a tumbleweed.

Moisture plays its part too in the Cretan plantain which can literally pull itself out of the ground. When the seeds begin to ripen the stems curve outwards and downwards, and push against the soil sufficiently to loosen the roots from the dry earth. The plant, by now ball-shaped, is soon gone with the wind. The highly succulent members of the South African *Mesembryanthemum* tribe develop very beautiful seed capsules whose intricacies almost resemble those Chinese globes within globes. In dry weather they stay very firmly shut, but moisture encourages them to open out their radial valves in a starry pattern – another hygroscopic device. But this is only the beginning. The capsule when open is so designed that, as in the little fungus *Cyathus* described earlier, a raindrop striking one of the valves exerts a thrust on the water already in it which forces the water out, carrying the seeds with it, for distances up to many feet. Different species have elaborated this application of hydrodynamics in various ingenious ways.

Not every seed is washed out; some are always so placed that it is difficult for them to be ejected, and in one group (*Apatesia*) some seeds are produced in special cavities away from the primary seed-producing valves so that they cannot possibly be forced out. This appears to be an insurance policy. It is very easy for the majority of seeds to be washed out in a quick deluge; but this may be short and followed by a hot spell, so that the resulting seedlings may perish. The few seeds left in the capsules remain there until these eventually break up, to germinate in a rainy period like most seeds. There is one group (*Hymenogyne*) which has gone beyond the intricate engineering needed for these symmetrical capsules, and carries all the seeds in a circular roll in which they are enclosed in narrow box-like structures. These separate quite rapidly, but the seeds remain imprisoned; the cell walls have holes in them into which water can penetrate and from which roots eventually grow out, usually a year after the seed has been shed.

In the cress family there are species with horizontal pods which, when struck by a raindrop, act like levers to eject the seeds: a throw of 80 cm is recorded for the pennycress *Thlaspi perfoliatum*. In the dead-nettle family one lip of the calyx is often extended to act as the raindrop-lever: in the sage *Salvia lyrata*, for example, seed throws of up to 2 m have been recorded.

Many seeds are moved by rainwater running over the soil surface: this happens extensively in deserts where the rare, violent rainstorms may create large temporary flows and floods. The seeds of certain 'living stone' succulents (*Lithops*) contain an air pocket which allows them to float for several days; apart from rain-wash transport it seems possible that some species are river-transported. Certainly the most widely distributed species occur in the areas of highest rainfall.

Some seeds exude mucilage when wetted which makes them stick to soil when they come into contact with it so that further movement cannot take place when germination starts.

A great many plants have capsules, pods, carpels, and so on, which open when the seeds are mature, spilling the seeds out in a mass or, more typically, by degrees as the seed-head is blown about in the wind. The 'pepper-pots' or 'censers' of poppies, with a ring of escape holes round the upper edge of the seed-head from which seeds can be widely broadcast, form a classic example. The larkspur has a cluster of pod-like follicles which open at the upper end and act in similar 'shaker' style.

Animal transport also plays its part in spreading seeds. At its most elementary this involves seeds which have fallen on the ground being picked up on a muddy foot or hoof. This is especially effective when the foot is a bird's, because of the distances it can fly. Birds are often important in 'planting' islands: thus there are thirty-five higher plants on Macquarie Island, 950 km south-west of New Zealand, which can only have arrived by bird, either on their feet, or by hooking or sticking on to their feathers.

A great number of seeds are hooked, barbed or covered with fine bristles so that they attach themselves readily to animal fur, hair or feathers. Those of grasses are notable in this respect. Often these are not simple devices, but may be doubly or trebly hooked, or supplied with down-pointing barbs exactly like those on an Eskimo harpoon. One of the largest 'hookers' is the capsule of the American proboscideas, sometimes sold as a curiosity; about 15 cm long, they have two long, crook-like hooks capable of fastening themselves into sheep's wool. An aquatic version of this is the Chinese *Trapella*, but it is not certain whether its hooks are intended to fasten on to water animals or to anchor the seeds in the mud.

A seed distribution device which has gone a little too far is that of the tropical pisonias, which have narrow angular fruits covered with very sticky gum. Its sweetness attracts birds who are then forced to carry off the fruits which adhere to them; unfortunately the gum drips on to adjacent twigs so that sometimes small birds cannot get free, and *P. brunonianum* is hence known as the birdcatcher tree.

Nature's least pleasant refinement in this direction, monopolized largely by the two families *Pedaliaceae* and *Martyniaceae*, are the 'trample-burrs'. In their simplest form these are large, very sharp thorns which stick out from seeds and impale animals' feet. One group of fruits resembles the caltrop or 'crow's foot' used in warfare in past centuries – an iron device with four points so designed that one is always facing upwards, whichever way it lands, to impale the hooves of cavalry horses. One such, *Tribulus*, is called the puncture vine and often damages sheep's feet. A particularly vicious device occurs in Madagascan plants known as grapple trees. Their seed capsules are covered with springy interlacing hooks which allow a gazelle's hoof to slide between them, and then drive their sharp points ever deeper into the animal's hock as it tries to knock the painful object off.

The devices so far described are basically passive ones, however ingenious the mechanisms. Many plants have devised more positive apparatus and spread their seeds explosively. In most cases the explosion is the result of tension created in the seed-containing structures as they mature and become dry. Gorse, like many other members of the pea family which make hard pods, is a simple example: when dry the two sides of the pod suddenly split apart and curl up spirally, flinging out the seeds in various directions. Tropical pea relations sound like fusillades of gun-shots when their large pods explode.

Many of the balsams have similar habits – one is even called *noli-me-tangere* (touch-me-not) – the ripe fruits bursting by torsion of several special strands at the slightest touch. As in the fern *Dryopteris* already described, higher plants have sling-mechanisms: in the wood sorrel special tense layers of cells, like elastic, provide the power when activated. These are built into the seeds themselves, which rupture the capsule.

The acanthus adds to the force of the capsule exploding by using a sling – a rigid hook-shaped structure arising from the partition in the centre of the fruit. The capsule explodes with a loud report and the hooks ensure that the 3 cm seeds go in a definite direction; they are projected as far as 9 or 10 m.

One of the noisiest seed-distributors is *Hura crepitans*, a tropical tree of the spurge family, sometimes called the sandbox tree or monkey dinner-bell. The 2 cm long seeds can be thrown up to 14 m as the many-ribbed 8 cm seed capsule or sandbox (so called because it was once used as a receptacle for sand to blot ink) breaks into sharp sections.

Silent seed expulsion is the mode, however, of the Mediterranean squirting cucumber and some of its tropical gourd-like relatives. The squirting cucumber has foliage and small yellow flowers like those of a cucumber, and the 5 cm fruits look like a plump hairy gherkin, or perhaps a hairy green egg, hanging on a curved stalk. As the fruit ripens, a layer of cells within becomes tense, and the tissues at the stalk end weaken allowing the fruit-stalk's bung-like end to loosen. It becomes entirely detached when the weight of the fruit is sufficient, or if an animal touches it. As the bung becomes detached the tense layer of cells expands, suddenly compressing the fruit and thus violently ejecting the acrid mucilage containing the seeds in a stream through the bung-hole – reputedly to 7 or 8 m.

In this last case either ripeness or a touch sets off the mortar; a touch is also effective with balsams, but in most explosive seeds the final agent in setting off the device is hot sun, which carries out the last stages of drying and priming.

Seeds usually have a food reserve inside to start the seedling on its way. In fruits we often have also a food supply of a different kind, in which the seeds are embedded. The purpose of this pulp is to attract animals to eat it, and in so doing will either scatter the seeds or, more often, take them in also and eventually void them after they have passed through the digestive tract. The seeds will usually be combined with dung which provides them with a fertile start. This might be called sewage dispersal. Owls and bats regurgitate unwanted hard matter which may include seeds. Fruits have developed odours and tastes according to the creatures which eat them (thus bat-fruits are likely to be musty-smelling) in a way similar to the association between flowers and animal pollinators. The largest pulpy fruit, incidentally, is the jack fruit, which can weigh over 30 kg.

Bats are large eaters of many kinds of fruit; they may transport them considerable distances, sometimes between islands, and they are especially important as dispersers because the fruits in question are often despised by fruit-eating birds.

Some fruits are not immediately recognizable as edible to humans; thus many tropical pea relations have leathery but in fact nutritious pods, and often extremely hard seeds within to withstand mastication. Other fruits eaten by animals, even by monkeys, may seem to us distasteful, acrid or nauseous. Occasionally fruits are incredibly sweet, like the amusingly named serendipity berry, in which the sweetness protein is, weight for weight, 300 times sweeter than sugar. The active component of this and a similar shrub, katemfe, have been isolated, although little practical interest has been aroused so far, beyond that of local Africans who have long used the berries to sweeten bread and palm wine.

Some seeds have attachments called arils which are usually fleshy and edible, rather than an outer mass of edible matter as in plums or apples. Our native yew and tropical fruits such as lychees and mangosteens have arils, in the former immediately accessible, in the latter two surrounded by an outer rind. Sometimes, as in certain tropical lianas (e.g. the family *Connaraceae*), the seeds eventually project from the aril, or actually dangle from it, and are thus more likely to catch the eye of birds. Very often such fruits are brilliant orange or red, and the seeds may be red also, or vividly bicoloured black and red, or black and yellow. In getting at the pulp the animals concerned will easily detach the seeds from the parent plant.

In some areas where many species have arillate fruits other plants mimic them, producing seeds of similar colours which are in fact inedible and often very hard. Similar deceit can be noticed where there are many plants with edible berries.

There are two especially remarkable fruits with arils. One is nutmeg, where the aril is known as the mace, and is eaten by fruit-pigeons. When the East India Company wished to corner the nutmeg market they forbade anyone to grow it outside the specified areas that they controlled. But the fruit-pigeons spread it beyond these areas, with unfortunate results for those natives among whom the unsolicited seedlings germinated.

In the fabled durian the outer flesh of the rind smells like 'a mixture of onions, drains, and coal-gas' (Corner), but the aril actually around the seeds is without odour and has the most delicious taste, combining a fruit flavour with an indescribable creaminess. Dr Corner must forgive me for quoting him again here, but he puts this fruit brilliantly into its tropical context.

> Usually the fruits detach when ripe and crash to the ground, where the pulp turns rancid in a day or two. In Malaya the smell of fruiting trees in the forest attracts elephants, which congregate for first choice; then come tiger, pigs, deer, tapir, rhinoceros, and jungle men. Gibbons, monkeys, bears, and squirrels may eat the fruit in the trees; the orang-outang may dominate the repast in Sumatra and Borneo; ants and beetles scour the remains on the ground.

In fact all this edibility does not help the durian over much for, although some seeds are defecated later by large animals like elephants and rhinoceros, most are usually left immediately below the parent tree, where they start germinating without delay. This immediacy is typical of relatively primitive trees of the tropical forest, where conditions remain similar all the year round. Enough youngsters force their way up to take eventual advantage of the light dominated by their parent to ensure regeneration of the forest in natural conditions.

Specialized versions of the arillate fruit are found in seeds dispersed by ants. These have a fleshy, often sticky edible part called the elaiosome, usually oily and containing fatty acids especially attractive to ants. This the ants eat after transporting the seeds into their nests above or below ground. Cyclamens are an example, made more ingenious by the way the flower stem coils itself downwards like a spring until the seed capsule is brought right down to ground level.

Ants are very important seed distributors; some 1,500 plant species in Australia alone are dependent on them, always providing a reward in 'the form of a "food body"'.

Fruits like melons may create an animal-attracting display by splitting open when ripe. This is well shown in the widely grown tropical *Momordica charantia*, a large pumpkin with bright orange flesh and crimson seeds.

It will have been noticed what a wide range of creatures the durian fruit attracts. This is due to its oiliness, an attribute which brings carnivores as well as vegetarians to eat some fruits; wild cats and jaguars seek out fallen avocados, civet cats and the oil squirrel eat oil palm nuts. These are also prized by vultures, and there are South American oil-birds so called because of similar preferences. Wild cats and dogs as well as crows and magpies eat olives.

Such oily, carnivore-attracting fruits not only increase the range of unwitting distributors, but point the links with antique times when reptiles ate and spread seeds, as was the case with cycads, *Ginkgo* and other ancient gymnosperms. Fossil reptiles have been discovered with numbers of such seeds intact in their stomachs.

Fish certainly eat many seeds, again those which are typically oily; the tuna is reported to eat the olive-like fruits of *Posidonia*, one of the few marine flowering plants. However, it is not known for certain whether seeds eaten by fish are passed through unharmed, although it seems more than possible.

It is obviously desirable for a fruit not to be eaten until the seeds are ripe. The reds and purples of ripe fruit become imprinted on animals' minds; unripe fruits are green to camouflage them, bitter, and sometimes spiny. The bitterness, often created by alkaloids, acids and tannins, helps to keep out weevils and caterpillars as well as putting off over-inquisitive apes. The cylindrical fruits of *Monstera deliciosa* contain sharp crystal-like fragments before they ripen. The durian has a coat of thick spines. Dr Corner sets the scene again: 'Unless a young fruit be well camouflaged, armoured, or unpalatable, it cannot survive. Have you seen monkeys searching incessantly for food from sunrise to sunset and the splutter of objects that descend?' Even in this country we know what havoc squirrels can do, but they usually leave unripe fruits alone.

Occasionally nature slips up in this respect. The red berries of the garden shrub *Daphne mezereum* are normally eaten pulp and all by birds, which void the hard seeds later. Greenfinches were among the birds involved. Unfortunately, British greenfinches, at least, quite recently developed a taste for unripe fruits, which they crunch up seed and all, and it is now much less common to find a mature crop on one's garden plants.

It is easy enough to understand how a very hard seed, such as that of *Daphne mezereum* or of a yew, passes through a bird; cherries and plums may be ingested or sometimes discarded after the pulp is eaten. Some plants like hawthorn have lots of very small seeds in the pulp, relying on the typical gobbling of bird or animal to allow a good proportion to pass through unharmed; sometimes indeed this ensures successful germination. Unexpected seeds such as desert mesembryanthemums are thus eaten and spread by ostriches, even though they are quite inedible. Occasionally more than one animal will be involved; Darwin recorded how fish greedily eat seeds of riverside grasses like millet, the fish are eaten by storks, and the seeds germinate after being voided by the storks. In the same way worms, which certainly ingest small seeds with soil and carry them underground, may be eaten by birds and the seeds reappear when voided by the worm's consumer.

Nuts are not designed to be eaten: they have hard exteriors, difficult for many animals to crack, surrounding the flesh we find so tasty. But many animals do eat them, and as with all seeds nature accepts a proportion of losses. Fortunately many animal nut-eaters have a hoarding instinct and cache much of what they collect, but then forget most of the caches. Oaks, beeches, walnuts, chestnuts, hazels, pines and monkey puzzles are all spread by cacheing animals, which include squirrels, rats, kangaroo-rats, hamsters and porcupines. The chinchilla only exists in nature where the algorabilla fruit grows. Tapirs bury seeds of an Amazonian monkey puzzle which results in small groves of the trees appearing in unexpected places.

Numerous birds, notably the jay, also hoard nuts. German observations have established an average of 4,600 acorns cached by each jay in one season, the birds sometimes flying as far as 4 km. The Californian woodpecker pushes nuts such as pecans into fissured tree bark: this is of little use to the seeds concerned but a proportion are stolen and buried in the soil by rodents.

The most extraordinary nut-case is that of the Brazil nut and its relations: up to two dozen nuts are packed almost as neatly as orange segments into a round box with a lid which falls off when the fruit, weighing up to 2 kg, crashes to the ground. Here the hoarding instinct of the agouti plays the major part in dispersal. Any nut-box that does not crack on hitting the ground may remain intact for years before it decays and releases the nuts.

Some seeds avoid being eaten at all. Such is that of the tropical nettle relation *Sloetia streblus*. On the seed's ripening the flower lobes swell up, becoming sweet and juicy. If a bird or other creature bites these fleshy lobes the nut is released explosively, so that it both avoids any possible damage by the animal and is shot some distance from the tree.

Animal dispersal can produce astonishing results. H. N. Ridley, when Director of the Straits Settlements Botanic Garden, described how various trees in the garden were eventually surrounded by a host of other plants, mainly brought there as seeds by birds and also squirrels. One large fig eventually had fourteen adventitious plants around it.

It will be appreciated that seed dispersal mechanisms and associations are quite as specialized as those for pollination, and show marked evolutionary trends. The most primitive seeds are those of trees in tropical rain forests, about half of which are carried in edible fruits like the durian already described, and most of whose seeds are left below the parent tree when the animals that prize the flesh of the fruit have broken it up. In *Mora excelsa* the huge flat seeds cannot roll or be otherwise distributed. In such cases seedlings sprout but remain in a state of retarded development until age or accident allows light to reach them. Then they grow apace and the strongest and best-placed win the race. Evolution developed seeds which can resist cold and drought. Most of these are small, enabling trees to grow in areas away from the tropical forest. Some of the most specialized seed dispersal methods are those of parasitic plants, which are described in the appropriate place in Chapter 23.

It must also be emphasized that in many cases more than one mechanism may operate. Himalayan balsam seeds are flung out by an explosive mechanism and may then be transported further by water. Seeds of the sand couch grass, whose binding properties can create coastal sand-dunes, are first blown about by wind and then often moved along the coast by offshore currents. Cactus fruits may be eaten by birds, rodents and ants, and exposed seed may also be transported by wind and rain.

In some plants the same flower-head can produce quite different devices. In some daisy relations there are two or, as in the common field marigold, even three different kinds of floret, inner ones with a plumose pappus for airborne dispersal and outer ones without, which remain in the flower-head much longer and are shaken out around the parent. The small Mediterranean annual *Fedia cornucopiae* produces three distinct kinds of fruit, which can be spread by wind, water, large animals and ants.

In most cases, too, we can note nature's usual over-production and wastage. Immense hazards face seeds and spores, which cannot choose the place or the conditions in which they eventually lodge. But if a tiny proportion of the seeds – it may be one in a million for an orchid – germinates successfully, this is adequate for the perpetuation of the race.

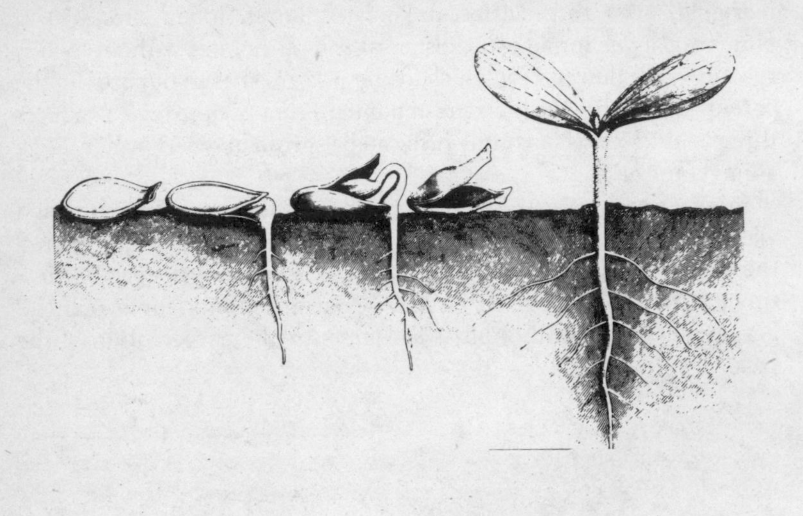

16 · *Birth*

It has been calculated that, if every spore from an average-sized giant puffball reached maturity, the resulting puffballs could form a five-fold ring round the globe, and if each of these produced totally successful spores the result would be a mass eight hundred times the earth's weight. But fortunately only a tiny fraction of spores or seeds reach maturity and ensure their species' continuation.

Although some might argue that the actual production of seeds or spores is in effect a birth, it is really an intermediate stage between one mature plant and another, a period in which these disseminules, as the botanist may call them, find a way, by one means or another, chancy or sophisticated, to a final lodging place. It is in this lodging that birth, or germination, will occur if conditions are suitable.

The period between seed production and germination is not just devoted to distribution and finding a suitable germination situation; it may be a very long one, and the seed is often dependent on certain external factors before it can germinate at all. In most cases these time-lapses and external 'triggers' are concerned with ensuring germination at a propitious moment, or spreading it over a long period of time.

The simple spores of alga, moss, fern or fungus, with little food reserve, send out a thin hollow growth with no special differentiation, which gradually develops into the next plant phase in the cycle – I use this phrase because, as we have seen, many of the more primitive plants have a sexual and a non-sexual phase.

In contrast, most seeds have some food reserve. In grains and many leguminous plants, this may make up as much as 90 per cent of the weight of the seed and allow it to be self-sufficient for several weeks, as well as letting animals, including ourselves, cash in for food. Even in fairly small seeds such as lettuce the food reserve allows the embryo to develop unassisted for a few days. At the other end of the scale the coconut has such an enormous reserve that seedlings grown in darkness, and hence unable to build up a new food supply, have only used half their built-in food after fifteen months. The coconut can thus establish itself in places with very little fresh water.

The tiny orchid seeds have no food reserve and are in fact barely more than a handful of undifferentiated cells; but their germination and development are dependent on association with a fungus, as we shall see later, so this is a special case. The maidenhair tree seed is well developed and has a big food reserve, but may have no embryo because it may not even be fertilized when it falls off the tree. But some kind of embryonic plant is usually present: typically one can make out the start of root and shoot, together with seed leaves, or cotyledons (in which indeed the food supply may be stored) often very different from the adult ones. The most developed seeds are those of grasses, where there are embryonic roots and leaves and stem already showing the pattern of nodes of the adult plant; these grains are comparable with a mammalian foetus in their state of development.

Seeds before they germinate have practically no internal activity. This is due to an extremely low water content – between 5 and 20 per cent compared with the 80 to 95 per cent in the tissues of actively growing plants. Moreover, water in a seed is tightly held by internal forces which are not overcome until the germination 'trigger' has operated: seeds are virtually unfreezable, and the water can normally only be extracted artificially by boiling them or subjecting them to intense vacuum, neither of which is likely to happen in nature. This is, of course, why advanced seedsmen market many seeds in sealed, impermeable packets filled in conditions of very low atmospheric humidity after careful drying. The water content of the packed seeds is around 4 per cent.

Some seeds must become virtually desiccated before they will germinate. Seeds vary a good deal, however, in their toleration of desiccation. Birch seeds have been stored for long periods with a water content as low as 0·01 per cent without loss of viability, although this is exceptional. But sugar maple seeds die if their water content goes below about 30 per cent. Wild rice and some citrus seeds are similar. Seeds of the coastal cordgrass lose their vitality in forty days if kept dry, but retain it for at least eight months if stored in seawater. Some seeds can be wetted and dried several times without harm, even when germination has begun, including wheat and beans.

The initial inert nature of seed and spore means that they can stand difficult conditions for a time. Most spores can and do germinate very quickly, although this is by no means essential; club-moss spores are known to survive for up to eight years, for instance. Seeds can last for very different periods. Some are only viable, that is capable of germination, for a few weeks. Many weed seeds can remain viable for decades, and there is a wide range of seeds capable of germinating after one or two centuries. Indian lotus seeds certainly not less than 1,000 years old have been germinated. Three viable lotus seeds were found in Japan in 1951, with the remains of a canoe at the bottom of a lake; the canoe wood has been carbon-dated to over 3,000 years and it seems likely that the seeds are the same age. But the longest viability known at present is of seeds of Arctic lupin discovered in permafrost conditions which can be dated to 10,000 or possibly 15,000 years.

For sheer endurance one must turn to bacteria: motile kinds discovered frozen in Antarctica, and estimated to be up to a million years old, have been successfully revived.

There are many reasons for a seed losing viability, but these need not concern us here, especially as the ageing and death of seeds is far from fully understood. In any batch of old seeds there is a proportion of dead ones, which increases with age. In some cases, such as lettuce, maize, cabbage and clover, the plants grown from old seed may be stunted or produce less yield, but in others, such as mung beans, the opposite is true. A great deal depends on the conditions in which old seed has been stored.

For a seed to germinate it needs water, oxygen and the right temperature. Some germinate as soon as they are ripe, especially tropical seeds like durian and mangrove. Many seeds do not necessarily perform as soon as the ideal conditions are available, and the reasons for this dormant period are quite as interesting as record viability. Sometimes it is merely to ensure that seeds germinate in the spring following their ripening, when growing prospects are good, rather than in similar conditions in autumn when winter might destroy tender seedlings, as in annuals and most alpines. This is typically controlled by a cold but fairly moist period. Many Mediterranean plants are inhibited from germinating by an initial high temperature period, to prevent seedlings being subject to scorching summer conditions. In many other cases delayed dormancy helps to ensure that a proportion of a batch of seeds germinates each year over a long period. In many plants, such as some clovers, a proportion of seeds is programmed to germinate at once and others have a delay mechanism. The familiar weed fat hen has one kind of seed which is large and germinates immediately, and three distinct types of small seed which germinate after varying periods.

A winter delay is usually due to a process called after-ripening, which may be controlled by temperature, by far the commonest factor, or by the gradual decay, washing out or volatization of built-in chemical inhibitors. There is nothing comparable in the animal kingdom, where once fertilization has taken place the production or offspring in time is inevitable, whether conditions are good or bad. As P. A. Thompson has written, 'many plants have developed mechanisms which enable them first of all to ignore the present and to forecast the future, and then to measure the passage of time and to verify the advent and passing of winter'. Though winter cold is the prime controller of after-ripening, the difference between day and night temperatures is also often of crucial importance.

Desert temperatures, which are linked with the season, control the germination of appropriate plants: thus if it rains in the Colorado desert and the temperature is 10°C, it is largely winter annuals that germinate; between 26° and 30°C it is summer annuals that emerge. Cacti in such hot regions germinate best between 30° and 40°C, and the operation is very rapid, to ensure establishment before the moisture runs out.

Rather surprisingly, light is also essential to the germination of many seeds such as tobacco, foxglove, many primulas and some lettuces. The dark-induced dormancy of the latter can be broken by quite low illumination for one minute. Some light-activated seeds must be subjected to long or short days before they germinate: thus birch seed must have sixteen hours of light a day before it will sprout.

The germination of foxgloves in woodland clearings is often spectacular, indicating that the seeds are capable of discerning variations in light strength, for it is no use a foxglove sprouting under a dense leaf canopy. This 'photocell capacity' in fact measures different qualities of light, ensuring dormancy in the red light that does filter through leaves.

Other seeds cannot bear light: onions are a homely example. In the bitter gourd or colocynth, which typically lives in gravelly wastes, germination is triggered by moisture but only in the complete absence of light, which ensures that the seedling is buried and has some chance of rooting down before it is exposed to the searing desert heat. Seeds which need to germinate in darkness can have their dormancy 're-set' by exposure to a short flash of light.

Aquatic plants also need adequate light to germinate. This means shallow water, in which seedlings are much more likely to grow and survive. Some aquatic seedlings float, which provides an additional dispersal method. This occurs in many temperate plants of river margins, some salt-marsh plants and in the yellow lotus of India.

'One year's seeds, seven years' weeds' is an old gardener's saying, and weeds are past-masters at spreading their germination over decades as and when they are turned up to the soil surface by cultivation, this matter of light being the main reason for their germination, as for the fantastic display of annual poppies in Flanders during the First World War, when shells churned up the soil. It has been estimated that there are up to a billion weed seeds per hectare of cultivated land, waiting in the moist soil for the light that will release them from their incarceration. The temperature and gas content at different soil levels also play quite an important part.

Hard coats on seeds, impermeable to water and sometimes also to oxygen, may be weakened by varying conditions of temperature and moisture, or they may have to be abraded by movement over hard surfaces. Others are broken down by decay caused by bacteria or fungi. Some seeds, like those of the lotus, actually contain antibiotic substances to delay even this activity. Many seeds must pass through an animal's gut before they will sprout readily, as with many bird-eaten kinds; this passage usually removes germination inhibitors. One of the oddest examples of this is a species of tomato unique to the Galapagos Islands which will germinate only after it has passed through one of the giant tortoises – but not if it is eaten and voided by any other animals! Though the tortoise may be slow-moving, its digestion is also slow, and the tomato is spread around quite effectively in this way.

Scorching by fire – caused naturally by lightning – is yet another trigger, exemplified by many plants, including the Australian glory pea which sprouts in great numbers after a forest fire, which its later brilliant flowers then seem to imitate. In the tropics certain acacias and albizzias dominate the landscape because it is regularly burnt and their germination is fire-stimulated. Fire is also one method by which the seeds of some very solid pine-cones, which do not open up in the more orthodox way, are released. After fire, light, air and plenty of nutrients are available for quick germination.

In Britain the willowherb earned its other name, fireweed, because it grew most readily where fire had been made. It therefore extended along the cuttings made for the new railways and, a little later, followed the first motor cars along the roads because their drivers made fires when they picnicked. I have heard of fireweed's appearance on burnt sites in the Himalayas where it had not been seen in living memory.

Some seeds demonstrate even more cunning adaptations. Thus in Death Valley and similar Californian desert places there are many annuals. Light showers happen not infrequently, but the seeds of these desert plants are able to distinguish between a succession of these and a real downpour, even though they may be equally wetted in both cases.

The reason behind this seems to be that water washed out a germination inhibitor, but that this substance is somehow replenished in between showers. A number of seeds can certainly pass through 'go' and 'stop' phases several times.

Even very considerable artificial rain will not trick desert seeds, because they are geared to what Bünning described as 'the time structure of the environment', as well as having chemical germination inhibitors which have to be completely leached out. In this way are created those sheets of flowers which appear after the rains in parts of North Africa; they are aptly known as ephemerals, for they may shrivel up within a few weeks, having flowered and seeded. Winter-flowering annuals are the typical plants, sometimes the only ones, of severe deserts which may receive rain only once every several years.

Desert plants seem to have more 'tricks' up their seeds than most, perhaps because they need them most. Many of them have dual types of seed. One saltwort, *Salsolo volkensii*, has one type which is green, with an embryo containing chlorophyll, and one which has none and is yellow. The first will germinate after rain, but the other has a built-in dormancy of up to five years, while retaining a full germination capacity. The dwarf composite *Gymnarrhena* has two different fruit types: one is aerial and one is subterranean. Most are of the first kind; they have a small 'parachute' or pappus which carries them off in wind, to germinate as and where they can, with a rather high mortality but a good capacity for dispersal. The subterranean fruits are produced from flowers just below the soil surface; they have no pappus and are larger, and remain in the tissues of the dead parent, which become woody, germinating *in situ*. They give rise to much larger and more drought-resistant seedlings than the aerial seeds, and are helped to develop because the dead parent's roots shrink, creating capillaries in the soil. Even in fairly light rain these quickly fill with water which is taken up by the dead roots which swell and hold the moisture. Thus the subterranean, non-dispersed seeds have water and are not subject to surface evaporation.

One other peculiar example, the winged thorny-spike, *Pteranthus dichotomus*, produces three kinds of fruits depending on the amount of growth it makes in its season. After bad years the only type of fruit is one which germinates readily in the next rainy season. After good years the three types of fruit not only have between them the power to germinate at once, at a medium interval, and much later, but require different combinations of light and temperature to do so. This marvellous permutation mechanism ensures that there are always viable seeds in the soil, ready for any set of suitable climatic circumstances. These might include the shelter and especially the soil compaction, resulting in better water-holding of the sand, of camels' footprints, which can often be seen massed with seedlings.

Germination starts with the uptake of water, and cannot occur without it. The first stage of this uptake is not necessarily controlled by the seed; even dead seeds can take up water, and the process can also occur when there is no oxygen available for the seed to develop. Where there is a very hard or impermeable seed coat this naturally holds up water intake, and the gardener may speed things up by removing or chipping the coat, as is typically done with sweet peas. In this phase very strong suction forces are at work: up to 2,000 atmospheres have been recorded. Once the seed has an adequate water content, however – 50 to 60 per cent of its weight – this 'colloidal imbibition', as it is known, is replaced by osmotic forces, and it is roughly at this time that life re-starts for the seed, and the tissues concerned are variously activated. The stages of activation caused by the gradual wetting of the tissues and the resultant stirrings of latent enzymes remind one in a way of the process necessary before a rocket can be fired and put into orbit.

Once all systems are 'go', cells start dividing and structures forming the initial root (radicle) push downwards, conditioned by gravitational forces whatever position the seed may be in, while a little later the first shoot (plumule), equally conditioned by gravity though in this case negatively, and also by light, pushes upwards. Very often the plumule's stem is bent in a crook shape, and it is this relatively tough crook which heaves its way up until it is clear of the soil and the much more delicate tip of the plumule can then erect itself. The large seeds of many tropical forest trees can erect a plumule to a metre or more (one Malayan tree to 3 m), taking the seedling well clear of the debris and giving it a springboard start towards whatever light may penetrate the dense leaf canopy. Their seeds have large food reserves to enable the seedlings to get through this difficult phase with minimum light.

It is when the shoot or seed leaves first see the light that chlorophyll usually begins to form. In a few flowering plants such as the spindle tree the embryo seed leaves already contain chlorophyll, and conifer seedlings are capable of producing it even in darkness. Some conifer seeds, incidentally, may germinate while still in the cone, not a very satisfactory occurrence.

In a few plants, notably citrus varieties, the sexual embryo derived from flower fertilization can be augmented or even suppressed by several non-sexual 'embryoids' derived from other parts of the ovule (nucellar cells). The seedlings that result are identical to the parent.

The success of the seed obviously depends on the radicle finding adequate moisture and food supplies, but it is sustained for a varying period, as already noted, by its built-in food supply. The reason why coconut and other palm seeds can survive so long is due to the fact that the cotyledon grows *into* the food supply and finally absorbs it completely, taking its time to send out exterior roots and meanwhile sending up a shoot and leaves.

Some seedlings, notably those which make subterranean tubers, bulbs or storage roots, may in their first year merely expand their seed leaves and concentrate on developing their underground parts. Some lilies may not show above ground at all. In most cases, however, adult leaves rapidly follow the seed leaves and growth into an adult plant proceeds steadily.

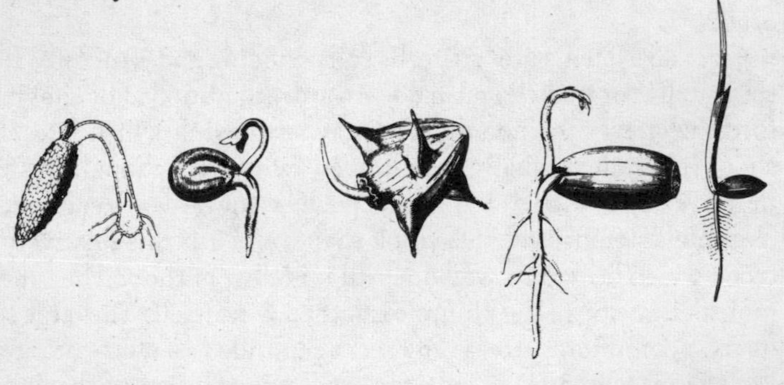

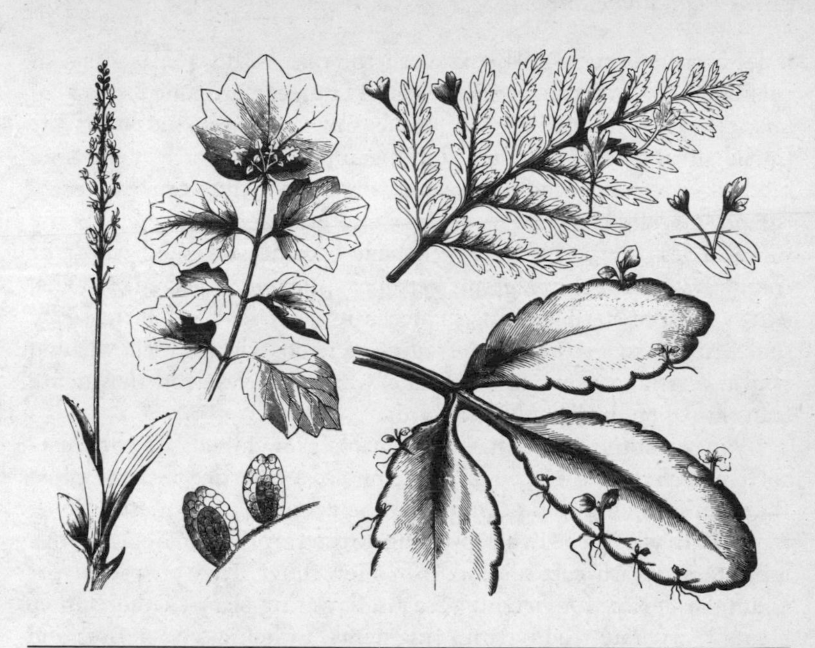

17 · *The Opportunists*

Opportunism is described by Webster as 'the policy or habit of adapting one's actions . . . to circumstances . . . in order to further one's immediate interests, without regard for basic principles or eventual consequences'. This is extremely apt of many plants which, yet again hedging their bets, have developed other means of increase, of the kind technically known as vegetative, which often supplement the seed. Almost half the perennials from Great Britain, for example, can increase vegetatively, or clonally as the method may be called. It is this kind of spread which makes many weeds so successful, or so vicious, according to one's viewpoint.

Such vegetative methods are often found in colonizing plants which, starting from seeds in a new environment, find it unsatisfactory for flower and seed production. Some plants which increase in this way, indeed, become sterile. This is often the case with water plants. An example is the sweet flag, which owes its European introduction to man and its wide spread to floating, broken-off rhizomes. Another is the Canadian pondweed, introduced to Britain in 1842, which at one time blocked waterways after increasing very rapidly from broken pieces; the plant brought in was of one sex and unable to reproduce. Some more or less sterile plants which have become successful weeds or colonizers include the creeping *Veronica filiformis*, a plague of lawns, which is quite sterile; mints, which rarely set seed and increase by underground roots; the fast-spreading bamboos which only flower and seed at very long intervals; while duckweeds, which seldom flower, are transported on the feet of water birds.

Many primitive plants have detachable growth buds. In the liverwort *Marchantia* these buds, or gemmae, are carried in little cups. Raindrops striking these can propel the gemmae a full metre. Some mosses carry gemmae in a cup-like leaf rosette, others in a cluster on a long stem. Splash-cups also occur in a few fungi. They are precursors of similar devices for ejecting seeds in flowering plants. Other simple plants regenerate readily from fragments, including bog-mosses and some seaweeds. Two species of the Sargasso weed are only able to multiply in this way.

Fungi, at any rate the larger ones which produce mushrooms and other fruiting bodies, tend to expand radially. They do this because they exhaust the soil and fresh food is thus only to be found outside their point of origin. Hence we get 'fairy rings', some of which can be shown to be several centuries old. There is typically a ring of lusher, darker grass just outside the fungus ring, created by the activity of the fungus threads or mycelium on the soil on which it is feeding; this activity releases ammonia, and salts formed as a result stimulate the grass. Sometimes there is a completely bare ring inside the darker one; this is the result of the soil entirely drying out when fully packed with the fungus strands, and gave rise to the idea of the fairies dancing in a ring. As the fungus strands die, they decompose and the results of their decay may produce yet another dark green, lush-growing ring.

This is the normal, irreversible spread of the soil-inhabiting fungus. Others increase in wood, both living and dead, equally inexorably. One or two develop special conduits. The honey fungus, a deadly tree parasite dreaded by gardeners and foresters, produces black underground 'roots' called rhizomorphs whose appearance provides the alternative name bootlace fungus. These can push through the soil for several metres in search of new tree roots. The bootlaces sometimes enter water-pipes in this search: there is a record of one blocking a 15 cm pipe 4 m below ground for some 2 m.

Another dreaded fungus with similar capacities is the dry rot fungus which devours dead wood and turns it to powder, as many a householder will know. Here again root-like strands are produced, which can penetrate inert material such as mortar for many metres in search of new sources of food in the form of joists or floorboards; they produce the characteristic drops of moisture which allow secreted enzymes to break down cellulose.

That curious fungus-alga association the lichen, growing on bare rocks, is often circular, but while a fairy ring's age is measured in metres a lichen's will be in centimetres. Other more three-dimensional lichens fragment, and here we can touch again on folk-lore. Crusty lichens of the genus *Lecanora* grow all over the steppes and deserts of Asia. As the crust develops it rolls up and then breaks. These small pieces are easily blown about, and if there is a rainstorm are carried by the water. In such conditions great masses become deposited in hollows as the water seeps away, and there are records of enormous heaps, enabling a man to collect up to 5 kg a time. This lichen can be ground and baked like corn and is the most plausible explanation of the manna of the Bible; botanists call it manna-lichen.

Higher plants have roots, and many of these have the power of budding and sprouting at intervals. Aspen, sea buckthorn, cherry are trees which do this readily; new trees are formed up to 30 m away from the parent trunk. More insidious are the herbaceous plants involved, for they move farther and faster. They provide a roll-call of inveterate weeds: bracken, couch grass, ground elder, bindweed, and the dreaded horsetails whose underground parts are more shoot than root, penetrating as far as 20 cm below ground. They carry small tubers at intervals, which become detached if the stems are dug up, and act as growth points if a selective weedkiller destroys the upper part of the shoot. It has been said that it would be impossible to measure a single old bracken plant, so extensive is its growth. If you destroy the fronds, there are growth buds of different ages ready to sprout new ones in turn. A plant will grow outwards at a metre a year, and the root network below the ground has been estimated as weighing 2,500 kg per hectare.

Some plants originally brought in as ornamentals have been found to have the same tendencies, such as the enormous Japanese knotweed or Mexican bamboo, various spurges, the acanthus and the tree poppy. Most often small pieces of root can regenerate, which is also true of tap-rooted weeds such as dandelions.

Some plants only move in one direction, like the fairy-ring fungi exhausting their food supplies; these include the raspberry which can grow forwards a couple of metres a year.

Others make such a mass of packed roots that nothing else can grow among them. Such are nettle, golden rod and giant reed. Butterbur combines this root-massing with the production of such luxuriant leaves that no seedling has a chance to appear. Many tropical arum relations have similar habits.

The mode of spread of some of these plants exhibits a remarkable geometry which is designed to make the most efficient use of space. The pattern involved is roughly hexagonal, and is created by parent stems putting out subsidiary upright branches, or ramets as they have been christened, at 120°. This allows spread into new territory while largely avoiding clashes between ramets of the same origin. Such strategy is found among gingers. Other plants, like bamboos and tufted grasses, produce tight-packed ramets on an ever-advancing front; botanists call this a phalanx, and it virtually ensures that no other plant can survive in the clone's territory.

Fleshy tubers and bulbs are only extensions of a root-bud, providing plenty of sustenance for the young plants. The Jerusalem artichoke is an example, paralleled in the tropics by the gourd-relation *Thladiantha*. In the latter strings of five to ten tubers are formed along underground shoots up to 50 cm long. Various bulbous plants similarly produce young bulbs at some distance from the parent. The spread of such underground bulbs away from the central parent is sometimes, as in the leek-like *Allium paterfamilias*, caused by the contraction of long horizontal roots beyond the bulblets, which are pulled outwards.

Most effective in opportunist terms, and deadly as weeds, are plants producing masses of loose bulbils (miniature bulbs, or bulblets). These are especially effective in cultivated fields, where ploughing will scatter the offspring widely. Onion relatives, celandine and especially various *Oxalis* are examples. It is ironic that most weed *Oxalis* were brought into various countries as ornamentals, like the pink *O. floribunda* in Britain and the yellow *O. pes-caprae* in the Mediterranean. Anyone who has seen this yellow goat's-foot, so named, of course, from one means of distribution, sheeting olive grove or vineyard will know how successful it is. The loose bulbils are scattered by any attempt to pull up the parent, or by the drying of the topsoil in summer, or they may be pulled downwards by the contraction of the fleshy, conical parent tuber at the end of its season. Defoliation simply encourages production of bulbils. Until recently, there was no effective chemical control for *Oxalis*, and even now it is very chancy. Before its introduction the Ministry of Agriculture recommendation for infested holdings was to run young pigs in them – young ones because their digestion destroyed the bulbils, which that of mature pigs does not. It was never a very happy solution for a gardener. It is ironic that the original plant introduced into the Mediterranean (for ornament in Malta) was a sterile form: none of its uncountable progeny bear seeds.

Above ground the easiest way for a plant to spread seems to be by runners and stolons. Runners are low-arching shoots which produce plantlets at intervals at their nodes, as in strawberries, and stolons are much longer ones which eventually bend down and root at the tip, as in brambles. The blackberry was introduced to the perfect growth conditions of New Zealand and became one of the country's worst weeds, giving rise to the story that there were only two blackberry plants in the country, one enmeshing the North Island and the other the South. It could almost be true. Plants of various kinds leap-frog in these ways, including the attractive striding or walking ferns, and the popular spider plant we grow indoors.

This method could be called infiltration – a guerrilla strategy which can, however, lose as much territory behind the advance, as older parts of the clone die among other plants, as it gains.

Offspring of 'guerrillas' are tied to their parent until the connections decay, and in fact the older part of the plants passes food to the young ones until they are established. Only after that do the new plants themselves contribute to the onward-moving runner or underground rhizome.

Although it builds up large colonies this method does not necessarily make for wide distribution. This only occurs when offsets are formed which soon detach from their parent. A number of primitive plants have this capacity, most elegantly displayed by water plants. The small floating frogbit sends out runners which initially produce new young adult plants; late in the season new runners carry tight buds. Once mature, these become loose and sink to the bottom, to rise again in spring. In the floating stages currents can of course carry them some distance. Pondweeds release similar winter buds when the annual stems start decaying in autumn.

Many land plants have detachable aerial growths. These may be bulbils, as in many of the onion family where they entirely supplant the flowers. Certain lilies produce bulbils; in the alpine orange lily this method often replaces seed entirely, because this handsome flower grows in meadows which are cut for hay long before seed can be produced. The nodding saxifrage lives in such cold places that it does not always succeed in setting seed. Bulbils can be very resistant: those of the Sinai blue grass can tolerate 80°C, and sprout roots within a few hours when the rains come. Occasionally bulbils are eaten by birds as if they were seeds, and are passed more or less undamaged: thus grouse eat those of the alpine bistort *Polygonum viviparum*. Another resemblance to seeds comes in the tropical *Remusatia* whose bulbils are hooked and form burrs, so that animals can transport them.

Actual vivipary also occurs, the carrying of fully formed young. Several grasses such as the Alpine poa make such plantlets, ready to grow with both leaves and partly developed roots. Some have picturesque names, like mother-of-thousands whose young are carried on long, very thin stems, and pickaback plant whose progeny appear on old leaves and root when they touch the soil. Several ferns have reproductive buds of this kind.

Some of the succulent kalanchoës bear youngsters round the edges or at the tips of their leaves. These are dislodged easily by touch, but *K. tubiflora* has a special projecting device. The youngsters are produced on short stalks near the tip of the cylindrical leaves. These tiny plants have rounded leaves which form a distinct cup. If a raindrop strikes one this pushes it down: the stalk acts like a springboard and rebounds upwards, jerking off the plantlet and casting it up to 1½ m away.

Not content with a 7 m tall flower-head, the Brazilian *Furcraea foetida* produces masses of plantlets among its flowers. The spectacular 2 m spike of its relative *Agave attenuata*, which bends over in a great arc, carries flowers in its lower half and plantlets in the recurving upper part. Agaves die after flowering, which may take many years to achieve, and usually also produce offsets round the parent rosette, like many plants of this formation.

Among successful water weeds are the notorious water hyacinth and one of the water ferns (*Salvinia*). The former has an inflated bladder at the base of each leaf in its rosette; the latter makes small flat fronds in pairs. Both multiply by simple division. Water hyacinth is another sexually sterile plant in its introduced form, which has choked innumerable tropical watercourses, while water fern prefers still waters like the Kariba Dam, where it forms dense, thick mats literally kilometres long, on which other plants grow; decaying fronds underneath the living give great depth to the carpet, so that dead buffalo have been seen supported on these vegetable islands.

Such water weeds not only make rivers impassable; they block all the light and use up all the oxygen in the water so that aquatic life is destroyed, which in turn can affect humans who depend on fishing. It has so far proved impossible to destroy these plants on a large scale by any means short of scooping them out, which only leaves mounds of stinking decay on the banks. Manatees may provide a solution, as described in Chapter 28.

Succulents have another trick, that of brittleness. Leaves of stonecrops, for instance, and the related crassulas and similar plants, break off very readily. Even when lying on the soil surface they will produce roots at the broken end which grope down to the soil, pull the leaf after them, and create a new young plant. Many stem succulents have equally amazing regenerative powers, as anyone knows who has tried growing them. Some species rely on brittle shoots, such as the candle plant, in which the constricted junction between the swollen shoots is so brittle that the shoots may break off when the upper one becomes too heavy, in a strong gust of wind, or if knocked by an animal. Once lodged on the ground, roots grow at any point of contact. Some houseleeks make miniature rosettes which detach equally readily, although in some cases they are initially tied to the parent by a runner, hence the name hen-and-chickens. The widespread Spanish moss of tropical America, in fact a moss-resembling bromeliad, is spread partly by birds which use the long flexible stems for nesting material.

Finally we should note the sheer root-forming capacity of broken twigs, as in willows and sea buckthorn, and of root fragments of reeds, sedges and so on. Knocked off in storms, these may be carried miles in a torrent and have the power of rooting and producing new plants if finally lodged in a suitable moist place. Such innate vitality is the force behind all aspects of vegetative opportunism.

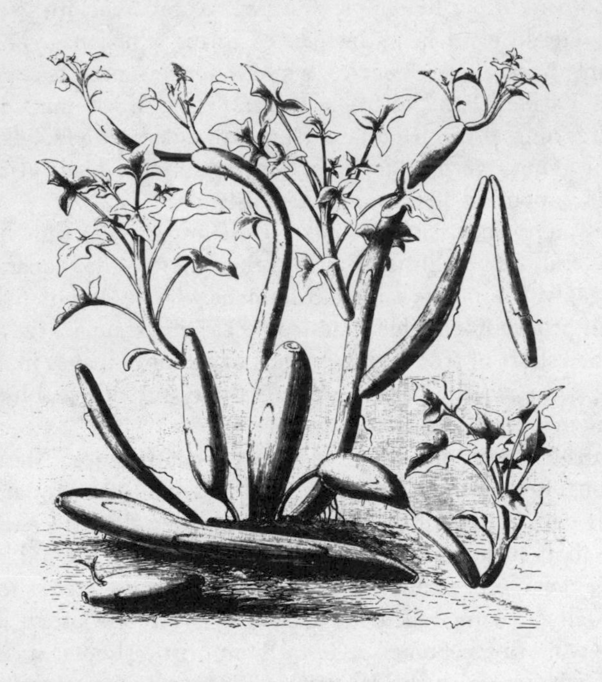

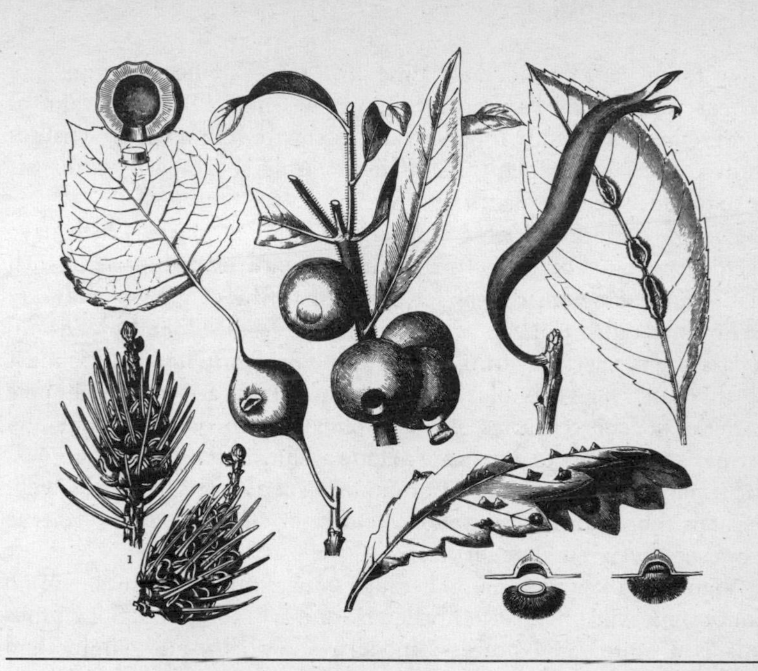

18 · The Allotted Span

Do we call a strawberry immortal because it can in theory increase without sexual activity for ever; or a bracken plant, spreading its underground stems indefinitely across the countryside; the suckering aspen forming large thickets; or the box huckleberry colony in Pennsylvania, 2 km across, thought to be linked with the original individual colonizer and reputed to be 13,000 years old? Where does an individual stop and start in these circumstances?

Plants are too varied to have an average lifespan; there is no question of three-score years and ten. Tiny plankton may have a life measured in days or even hours, although one can hardly talk of individuals when division, budding and other mechanisms are creating new lives from old so fast. But as soon as we come to multi-cellular plants the lifespans increase: seaweeds may be relatively primitive but an individual may grow for a very considerable time; the same applies to fungi and lichens, which may last centuries. And we are still in the realm of plants which can readily fragment and re-start, while as the last chapter showed many higher plants maintain a considerable capacity to do this or to produce buds or plantlets.

Certainly there are some very antique garden plants which have been perpetuated by man from pieces for centuries. All parts of such plants are biologically speaking identical to the original – they are technically called clones – whether the increase is done by division, separating runners, layering or grafting. Some cultivated olives must be thousands of years old, and in Britain there are some old apple varieties several centuries old. It is obviously to the gardener's and farmer's advantage to propagate a clone as long as it remains vigorous, for it may either be of hybrid origin or a special form from the wild, perhaps the result of a mutation, neither of which would breed true from seed even if it were fertile.

Vigour is a key word here, because some cultivated clones, and no doubt some wild ones, certainly deteriorate over the years. Sometimes this is a quite rapid process; modern show chrysanthemums and dahlias do not usually have a life of more than ten years before vigour starts declining. Sometimes this is due to diseases, notably viruses, as discussed later; sometimes probably to instability in the cells, which changes the effective number of chromosomes as the plants grow. It seems often to be associated with over-propagation. Sometimes we can only call it a kind of old age; and in such short-run cases it is noticeable that all the individuals of a clone fade away more or less together.

An annual's lifespan is known, although its season of growth may sometimes be from one year to the next. A biennial takes two seasons in principle, spending the first building up its vegetative part and the second producing flowers and seed. Other monocarpic plants build themselves up over a longer period, which may, as in giant puyas, agaves or the talipot palm, approach a century, before flowering, seeding and dying.

It is with plants which normally flower and seed regularly, becoming larger every year, that we can talk about lifespans in animal terms as something controlled by external as well as inherent circumstances. Even where such plants increase vegetatively there is often a basic parent plant whose life may be observed apart from its offspring. In many classes of plant, however, there is no internal demonstration of longevity. The age of some fungi can be roughly estimated from the diameter of their fairy rings, but with most 'lower' and a great number of 'higher' plants – herbaceous plants without trunks, and palms and other trees of the monocotyledon group which do not form annual growth rings – no evidence is provided. So that although we now believe that the record age of any individual plant on earth (as opposed to ever-expanding linked colonies such as box huckleberry, bracken or blackberry) is the 7,200-plus years of the Japanese *Cryptomeria japonica* on Yakusima Island (from carbon-14 tests), followed by the 5,000 years of the bristle-cone pine, which has the visual record of its rings, it is possible that some giant kelp in the Pacific, some dragon tree on Sumatra, some South American liana, or some double coconut tree on the Seychelles, is in fact older. There are suggestions that some cycads may approach 14,000 years, but this seems improbable.

Such very large plants undoubtedly die of sheer old age; they have an upper limit which is conditioned by accretions of waste products and slowing down of metabolism as in animals, for ageing, illness and death are facts among plants as among animals. Such circumstances may well be imposed by the limitations of external circumstances, notably the soil that the plant grows in, at which rate the seaweeds, fed literally by the waters around them, could have the highest chance of very long life.

Death certainly does occur from inability to get more out of the soil, to stretch roots further, from starvation in fact. Wear and tear from external elements play their part; trees are broken or overthrown in gales, lightning strikes them, fire destroys them, landslips bury them. Plants can perish from an unexpected drought, unless like succulents or bulbs they have special reserves. Flooding can suffocate the roots, or simply kill them if the water is salt or otherwise mineralized. Relatively tender species will go under if there is unprecedented cold (although sometimes, as in a recent Mediterranean frost disaster, trees such as olives, apparently totally killed above ground, were able to sprout again from their huge, antique roots).

There are a few internal ailments which can affect plants, although they are more curious than harmful, and seldom affect their vigour.

These are classified as teratological – monstrosities or malformations. Many gardeners will have seen flattened or multiple stems which are described as fasciated (literally, bundled); this can extend to flowers, whose numbers can be increased greatly beyond the normal. Other monstrosities are proliferation, in which extra small florets are produced around the normal one, and phyllody, in which the flowers are replaced by reduced leaves. Horticultural examples of the first are the hen-and-chickens daisy and Scotch marigold, and of the second the rose plantain. In peloria normally irregular flowers become regular, and sometimes much larger: the most notable example is the peloric foxglove, in which the upper flowers are replaced by one huge bell-shaped bloom. 'Doubling' of flowers, in which the centre of the bloom is filled with more petals than normal or in which one flower is produced within another giving a 'hose-in-hose' effect, occurs occasionally in nature, and the results are of course prized by gardeners, a great number of whose flowers are double since they give a bolder effect and last longer.

Fasciation is caused by some kind of damage to a growing point; it is often impossible to pin down the reasons why it occurs, though it may be mechanical, caused by frosting of the growing point, chemicals such as selective weedkillers, or possibly by radiation such as cosmic rays. It may only affect one shoot on an otherwise normal plant, and seldom recurs in herbaceous plants. Fasciated or 'cristate' cacti, often in weird cockscomb formation, and cristate ferns, occur in nature and are prized by fanciers. The other kinds of monstrosity mentioned are genetic mutations and can be propagated, usually vegetatively but sometimes by seed.

Besides such 'natural' hazards, plants have plenty of enemies. An incredible number of animals feed on plants, from minute mites, nematode worms and insects to herbivorous animals ranging from slugs to the elephant, a wasteful feeder if ever there was one, often knocking down small trees to get at their foliage, or gouging out trunks as in the baobab to get at its juices (which, extraordinarily enough, include a high proportion of tartaric acid and appear to be especially tasty to elephants). A red spider mite invisible to the naked eye multiplies so fast that in a short time the infested leaf has had all its sap sucked out and becomes yellow and useless. No gardener needs telling about such sap-suckers which include green, pink, black and woolly aphids, mealy bugs, capsid bugs, scale insects which settle in one place like minute limpets, all having powerful hollow stylets to penetrate the leaf tissues and suck their juices.

A less-known sucking pest is the minute nematode worm, or eelworm, which also has a piercing stylet. Eelworms often enter plant tissues and move about inside them. They may be present in astronomical numbers in the soil, and equally so if they enter a plant: 2 million have been found in a single onion head. When the plant dies the females produce eggs within their bodies; the females then die and their skin turns into a tough shell or cyst. These can remain inert in the soil for many years; but as soon as a suitable host's roots arrive in the vicinity, the chemicals released by the roots stimulate the eggs to hatch and start a new invasion. It has been shown that the kind which attacks potatoes causes the plant to modify its outer cells into 'transfer cells' which speed up the flow of sap to the nematode whenever it wants a meal.

Other all-too-familiar pests include leaf-miners, grubs which tunnel between the leaf surfaces; while caterpillars above ground and grubs such as wireworms below ground gnaw voraciously at plant tissues. In a caterpillar epidemic in a wood it is sometimes possible to hear the champing of thousands of tiny jaws, as well as the incessant patter of the creatures' droppings. Caterpillars can strip every leaf from a forest or orchard: this defoliation may kill trees outright, conifers being especially susceptible. The resulting situation sets off a chain reaction, notably increasing the risk of fire and erosion, and destroys the habitats of wild life.

The obliteration of vegetation by swarms of locusts has been a fact of Middle Eastern and African life since man began to grow crops – we have only to turn to Exodus: 'They covered the face of the whole earth, so that the land was darkened; and they did eat every herb of the land, and all the fruit of the trees . . . and there remained not any green thing in the trees, or in the herbs of the field, through all the land of Egypt.'

Snails and slugs also attack plants in great numbers with their ever-renewed rasping tongues. Tropical snails, which may attain 25 cm in length, are particularly voracious. Slugs exist in vast numbers, especially those that spend much of their life underground, emerging to feed at night; so far it has been virtually impossible to eliminate them.

Relatively minor problems are caused by insects such as leaf-cutting bees, which saw out neat portions of leaves to line the nests in which they lay their eggs. A great number of insects lay their eggs directly in plant tissues, to which the plants respond by creating galls. These are bodies of plant tissue, greatly varied in shape and size, each insect/plant combination resulting in a distinct, recognizable gall. Sometimes the galls do not start growing until the larva has hatched and started to feed, and the growth stops when it pupates.

Galls can be thickenings, folds, cups or spheres and the insect egg and subsequent larva can be either within the plant tissues or on their surface. They can arise on leaves, stems or roots, the latter typically caused in a similar way by nematode worms or mites. They are sometimes very elaborate, like those caused on sweet gum trees in western North America, on which the galls are hollow spheres with internal struts radiating from a central chambered capsule in which the larva develops. Very seldom are these galls seriously harmful to the plant host, although they can if very numerous make the host flower and fruit abnormally freely, as if the plant has an intimation of possible death. The insect gains shelter and a food supply for its offspring, for which the plant has no return. The gall-forming reaction seems largely to be the plant's method of isolating a parasite and any toxic substances it produces; the galls have a limited growth and are not tumour-like.

Some plants parasitize others with dire results, sometimes rapid, as with broomrape sucking the life from a crop of beans or clover, or protracted as with mistletoes and most tree parasites, where indeed a reasonable endurance of the host plant is important to the hanger-on. Others choke their hosts to death, like the strangler figs and, to a very much smaller extent, ivies, which in fact usually only kill trees already on the way out. More will be said about these plants later.

Fungi attack live trees; these are parasites too, but they do not just feed off the host like a mistletoe but actively destroy its internal structure, softening the wood so that insects and other secondary pests can penetrate more easily, and preventing its various conduits from carrying food and water as they should. These are fungi with large fruiting bodies – the 'brackets' which appear outside the trees, and are a sure sign of a usually incurable enemy within. Once they have finished the tree off – it will normally fall in a gale and reveal its rotten centre – other, saprophytic, fungi soon attack the dead wood. I have counted over thirty species of fungi, besides the original killer bracket species, on a beech tree a few months after its fall. In the tropics the ubiquitous termite is the first to take advantage of wood killed by fungus, and is soon found in every part of a diseased tree.

Lesser fungi also attack softer plants with a terrifying range of mildews, smuts, blights and so on, often fatal if remedial measures are not taken by cultivators. The reproductive powers of these are astronomical: one square centimetre of a mildewed rose leaf can produce half a million spores. Fungi ruin or destroy at least 10 per cent of cultivated crop yields annually over the world, and equally attack timber: in the states of Oregon and Washington alone it is estimated that up to 35 per cent of all the white firs are affected by blister rust and every year nearly three million cubic metres of timber are lost. The sweet chestnut has been entirely eradicated from eastern North America, where it once made great forests, by fungus attack; fire blight disease – so called because it blackens leaves as if scorched – has killed large numbers of apple and pear trees and other trees of the rose family on the European continent.

Fungi are the prime destroyers of dead wood, whether it is a fallen tree or its stump or cut timber; it produces volatile compounds which actually encourage the growth of fungus. In tropical forests decay is very swift; in colder, dry climates wood can persist for centuries.

Fungus spores are largely carried in the air, and their lightness ensures that they can travel vast distances. It has recently been shown that birds may also transport them very extensively. Thus it has been suggested that American cereal rusts are carried by blackbirds, which begin their migration when the rusts are infectious in Mexico; their northerly passage carries them right across the main American and Canadian wheatfields in seasonal turn.

Galls can be caused by fungi; some are serious diseases like black-wart potato disease and the two-host fungus puccinia which attacks cereals after making cup-galls on barberries. The aptly named witches' brooms and various tree-stem swellings are also caused by fungi.

Fungi may enter plants through wounds, and they can break into roots. Those affecting leaves appear often to enter through the breathing pores (stomata). It seems possible that the germination of fungus spores and their eventual entry into the tissues via the pores are partly controlled by the concentrations of gases around the openings. Some fungi have an 'open-sesame' power which actually makes leaf pores open for them. This is certainly the case with canker disease of almond and pear trees in which leaf wilting is a symptom; it has been shown to affect several unrelated plants in the same way, and fusicoccin, a substance isolated from the fungus, has been used to make alfalfa hay wilt artificially, which greatly speeds up its drying.

Bacteria likewise cause plant diseases which may be fatal; remedial measures against these are even more difficult than against fungi. Once again these very basic organisms can cause galls. Known as crown galls, these are irregular growths which look and behave like tumours, growing faster than the normal tissues; secondary galls which arise on infected plants appear to be caused by chemical substances produced in the original gall and circulated in the sap. Crown galls can cause the plant's death; their resemblance to some animal cancers is striking.

The most insidious enemy of plants is the virus. These minute entities, whose very 'life' has in the past been in question, have the capacity of invading cells and multiplying within them, sometimes apparently transmuting the nuclear matter so that cells behave abnormally. Although some plant viruses exist in soil, and can infect roots directly, they are normally only found in cells. Most are transmitted from one plant to another by sap-sucking insects, by underground nematode worms which likewise suck sap, sometimes by birds. Occasionally they are transmitted by human activity, often by grafting and pruning tools, but also by the fingers, as when a greenhouseful of tomatoes is being 'side-shooted' by hand; while the nicotiana virus, which affects tobacco, tomatoes and many other plants, can be transmitted to plants following finger contact with infected cigarettes. Russian research has suggested that cells communicate by emitting ultra-violet rays. It is believed that this can increase the number of cell divisions, and also transmit illness or infection, as with virus. Healthy cells have, it is reported, been infected with virus from neighbouring infected ones even when separated by quartz glass.

Viruses attack, or if you prefer inhabit, almost every kind of plant, often giving them a mottled or streaked appearance, sometimes distorting the foliage or stems (there is a curious virus disease of apple trees called 'rubbery wood'), frequently stunting plants seriously and almost invariably weakening them. Occasionally, if the already infected plant is externally wounded, they cause woody tumour-like galls. Sometimes there are no visible symptoms, but comparative tests with uninfected plants have shown that such masked virus infections do reduce growth and cropping capacity. It has been suggested that hardly an old fruit tree exists which is not virus-infected and hence is performing below par.

It is almost impossible to destroy viruses once a plant is infected, but certain means of outwitting them are described in Chapter 28. Virologists have, incidentally, recently shown that some plant and animal viruses have a strong superficial similarity, and that certain animal viruses can affect plants. It is already known that some plant viruses can multiply within their insect transmitters, and can sometimes even kill them.

One should perhaps place on record that certain variegated ornamental plants owe their extra decoration to a virus which does not weaken them unduly, and so do those tulips called Rembrandts, Bybloems and Bizarres by the fanciers, in which the flowers are streaked or feathered in another colour. These are some of the oldest cultivated tulips known, going back to the early seventeenth century, and they have survived despite the enemy within.

There are, of course, viruses of animals as of plants. Recently it has been discovered that other organisms previously known only from animals, mycoplasmas and spiroplasmas (one causes contagious pleuropneumonia of cattle) are responsible for a group of plant diseases known as 'yellows' which cause serious losses among important crops as varied as coconuts, citrus, cotton and rice, the causes of which were previously believed to have been viruses. They cause leaf yellowing, stunting and growth abnormalities, notably the production of leaves instead of flowers. Like viruses they are normally spread by insects and by grafting on to infected stocks. Mycoplasmas are the smallest self-replicating cells known; though similar to bacteria they have only a fifth of the gene count. In plants they normally inhabit the sieve tubes, growing within the plant's living matter or cytoplasm.

These insidious organisms can in some cases be controlled by tetracycline antibiotics, as they are in animals, which do not control viruses. This is not to say that antibiotics are the answer with affected crops: such treatment would be expensive and might create undesirable residues, encouraging organisms resistant to antibiotics in the consumer. As with viruses, the main line of attack against mycoplasmas at present must be against the insect transmitters, while in some cases heat treatment is successful.

Viruses, mycoplasmas and spiroplasmas are classic examples of virulent organisms transmitted almost exclusively by animals from plant to plant. We have already noted how fungus diseases may be carried by birds or other animals, such as the bark-beetle which carries Dutch elm disease at present decimating elms in Britain and Europe. This kills by blocking the sap conduits of the tree, causing starvation; its spores are carried in the mass of sensory hairs around the head and mouth of the young beetles whose larvae tunnel in the bark. A similar beetle/fungus association is responsible for the death of many British beech trees. With the elm disease, incidentally, more virulent strains than any previously recorded have appeared, which can infect elm varieties specially bred for resistance.

Viruses are sub-microscopic and there are vast numbers of organisms scarcely larger: microscopic fungi, yeasts, bacteria and single-celled algae, which live on leaf surfaces. They are not direct parasites nor disease carriers, and may feed on substances of external origin coating the leaf, sometimes even competing with plant parasites; but some can certainly degrade the waxes of the leaf skin, thus letting more moisture escape and many make use of materials which leach through the skin. Some exude growth-promoting substances which may conceivably affect plant growth, and it seems probable that this unsuspected army of minute squatters hastens the leaf's ageing processes.

One has to enter the tropical forest to appreciate the sheer weight of attack on living plants by antagonistic organisms. Anthony Smith has described such a forest in *Mato Grosso*:

> In many ways it is a forest of death more than the exuberant life one might have expected from above [e.g. when flying overhead]. There are great dead

trunks, either lying flat, or leaning on the living, or even standing vertically and surprisingly on their own. There are all the brittle lower branches, and there are the husks of former forest giants. Some tall trees, obviously flourishing at the top, are supported by half-rotten trunks, either soft and dusty on the outside or having a tough shell around a cavernous interior. One longs to find a whole tree, a tree whose bark is as pristine as a Norway spruce, whose leaves are uneaten, whose interior is entire. Even to find a whole leaf is a problem; the insects see to that.

The travel journalist Brian Jackman has described this forest as filled with a 'rank, greenhouse smell . . . of plants rotting and regenerating in a kind of endless leafy anarchy'.

Trees in tropical forests can be extremely weak. A friend has told me how he had seen surveying teams in Ecuador cut 'sightlines' through a forest, without which they could not use their instruments. Within twenty-four hours these narrow corridors were criss-crossed with fallen trees, for each tree leant against its neighbours and could not support itself. Even in natural tropical forest this friend remarked on the almost continuous sound of trees falling.

Man is of course a major enemy of plants, often assisted by his livestock; he uses them for food and an enormous variety of other useful products, he may practise 'cut-and-burn' cultural methods of making temporary plots in a forest, he bulldozes trees and plants remorselessly as he builds towns, roads, railways or airfields, they are targets in warfare, and at the second remove he may destroy them by the byproducts of his activities – pollution by smoke and effluent. All this I shall go into in later chapters.

It is largely due to man that plants become extinct, although species of limited range must be considered on the way out anyhow, due to all kinds of factors often dating back for millennia.

Plants are not without certain defences against natural enemies, however helpless they may be against a concerted attack by locusts or man. Their armour and poison are fully described in Chapter 25.

Some substances whose function is not always clear exist in plants. These are a whole range of oils, latex or milky sap, gums and resins. It has been suggested that these are byproducts, resin for instance being a carrier for surplus carbohydrates. It is equally possible that they form a food reserve. Plants manufacturing resin or latex have special channels for these substances; in both cases these ramify into every part of the plant, as in conifer leaves which have special resin channels. This would appear to be an unnecessary complication simply for waste products. On the other hand it can be shown that such products are actively lost by the plant in apparent excretory fashion: thus a common juniper may lose up to 30 g of ethereal oils in a day, while a mature rubber tree loses up to 50 g of sugar a day from nectaries outside the flowers.

Recent research on tropical gum-producing trees, including the vast *Acacia* group, suggests that gum is naturally produced at leaf-fall and conserves water within the plant tissues. Most gums 'bind' from 12 to 15 per cent water and this means that a tree with gum in its system never becomes completely desiccated, clearly an important factor with desert-inhabiting acacias for example. It might also explain why *Cryptomeria japonica* has seven times as much resin in its system when growing in the hot parts of Japan as when grown in cold areas. But it has also been suggested that resin increases a plant's resistance to cold, which would accord with the fact that so many conifers inhabit cold climates.

Some of these substances certainly make plants unpalatable to feeding animals, and equally certainly seal up wounds rapidly against fungus spores, bacteria and insects. Latex dries hard on exposure to air (in its original gritty state opium is dried latex which has exuded from scratches made in poppy seed-heads) while resins and gums form a sticky mass which hardens in time. The latter, with their water-holding capacity, thus not only seal off wounds but may protect damaged tissue against harmful drying out.

Wounds are in any case closed eventually, especially in woody plants, by the formation of callus tissue, a corky layer produced by the cambium cells directly below the bark. However, this takes time and is more in the nature of a long-term tidying-up operation than a quick protective seal.

Plants also have astonishing regenerative capacities if damaged. The sprouting of frosted olive trees from roots has been noted, and cut or broken tree stumps will very often sprout, or latent buds on roots will spring up. These latent buds exist in unexpected places, especially on woody plants. I remember during the last war seeing trees which had lost every leaf and most of their twigs and branches in bomb blasts; a few weeks later the entire trunk was covered with new leaves, and in due course fresh branches were made. One of Galileo's forebears noted this when he wrote in 1424, 'The wood of the tree is an image of hope: wounded, it bursts into leaf and is covered again.'

Once a plant is dead it usually vanishes swiftly and silently from the scene, for it is eminently biodisposable. There are tree trunks preserved in peat bogs, and others which have become fossilized, but these are the exceptions. The greatest forest tree steadily disintegrates, pulverized by insects, fungi and bacteria which feed upon it.

In its decay a tree may provide a nursery for youngsters, both its own and other species: in temperate forests one may find straight rows of young trees which appear inexplicable till one sees seedlings emerging from a half-disintegrated fallen trunk. One of the notable places to see this is the Olympic National Park in north-west America, where in the world's only temperate rain forest the debris at ground level is a severe handicap to seedling trees. On the very rainy Japanese island of Yakusima seeds often germinate on the upper branches of older, living trees, and the youngsters may send roots down over the bark of the 'host', so that when the latter eventually dies and decays the new tree appears to be growing on stilts.

The soft plant becomes earth in earth a great deal more quickly; frost is often enough to reduce its tissues to pulp, and even fibrous stems will decay quite soon, as anyone who has made a compost heap will know. Worms and other soil creatures large and small soon see to that.

And the compost heap reminds us how the remains of dead plants feed the next generation, for they are reduced to useful elements. More valuably, this decayed organic matter ends up as humus, a material whose chemical structure has defied research, but whose colloidal or jelly-like composition, swelling when wet and shrinking when dry, coats the soil particles and is essential in maintaining good soil texture, making the soil spongy and water-retaining yet crumbly and open and inviting to questing roots. Bacteria finally break down the humus into the elementary chemicals which plants absorb as food, and so the life cycle continues.

19 · Society

There is no kind of discrimination in the plant world. The rung that a plant has reached on the evolutionary ladder is of no importance. Socially, if not sexually, the different groups of plants mix freely, and interact with each other. Some live in groups of their own kind, others are individualists. Certain habitats are physiologically impossible for all but a few plants – thus only certain blue-green algae can stand temperatures over 55°C; but even spheres which might seem closed can be entered by apparently unsuitable plants. *Posidonia oceanica*, for instance, is a 'higher' plant which forms meadows in the Mediterranean at depths down to 60 or exceptionally 100 m. One or two other higher plants such as turtle grass and *Cymodocea* also form large communities where one would think only seaweeds could grow.

On the other hand no seaweed has found a fully dry-land niche, although there are species which exist with as little as 5 per cent daily immersion.

At the same time no plant is successful enough to occupy more than a very tiny fraction of its possible total habitat, although some are distributed extensively through certain areas, the most widely spread being very adaptable plants capable of migration or of seizing the opportunities offered by accidental transfer. External factors circumscribe a plant's potential distribution, and there is a vast array of possible factors which define any plant's position in the world.

The study of these factors, capable of far greater permutation than any football pool, is called ecology, which in essence means the relationship of living organisms with their total environment. For a rather more expanded definition we might take that of the great German biologist Ernst Haeckel, who wrote that ecology (or oecology as it was then spelt) was

> the knowledge of the sum of the relations of organisms to the surrounding outer world to organic and inorganic conditions of existence; the living together in one and the same locality, their adaptation to their surroundings, their modification in the struggle for existence . . .

The factors affecting plants fall into four main groups. These are climate; the physical and topographic features (physiographic); the soil (edaphic); and, finally the influence of living organisms including other plants (biotic).

The world's climatic variability means first of all an enormous temperature range, so that we have plants which thrive in the sweltering tropics and others which take on extreme cold conditions. Although some plants have a fairly wide temperature tolerance, the temperate plants will not by and large grow in tropical conditions nor vice versa. You *can* grow roses and apples in Africa, but only at an altitude which has tempered the heat; and this reminds us how the effects of latitude can be produced vertically. To ascend a mountain gives the approximate effect of moving polewards. This effect is most striking on tropical peaks such as Kilimanjaro and Kenya, where tropical flora at the base is finally exchanged for an alpine one near the summit, but it can equally be seen on Mediterranean and temperate mountains. Not for nothing do the advertisements for Cyprus or the Spanish Sierra Nevada suggest that one can ski in the morning and swim in the sea in the afternoon. A typical temperate progression on a European mountain sees deciduous trees tailing off at about 2,000 m. Hay meadows go a little higher; the coniferous tree line is about 2,500 m; small shrubs give out at 3,500 m; and the average maximum height for plants is 4,000 m. In the Himalayas this extends to 5,500 m, and the record is 6,180 m.

It is true that we grow certain warmth-loving plants in our gardens as annuals, but frost destroys them; what will grow in a given garden depends on the hardiness or tenderness of a given plant. Some succulents will flourish out of doors in the Scillies but cannot survive in the east or north of Britain. The United States has a much greater climatic range than Great Britain so that gardeners have to consider ten climatic zones.

Tropical versus temperate zones also means different hours of sunlight. On the equator the days stay the same length all the year round, and daylight lasts twelve hours. North and south the seasons provide short days in winter and long ones in summer, and indeed in the far north almost perpetual daylight in summer.

Another important climatic factor, especially in warm lands, is whether or not the rainfall is evenly spread or if there are alternate dry and wet conditions. Total rainfall is another very important factor, and so are cloudiness, evaporation, wind, and seasonal snow.

Fire is a surprisingly important external factor and observations have made it clear that lightning regularly caused many fires long before man entered the scene; they were also caused more locally by volcanic activity. An observed example of spontaneous lightning fires occurred in March 1965 when some 300,000 hectares of almost inaccessible forest in the Australian Gippsland mountains were burnt out following seventy-five separate lightning strikes during one major storm.

The regular occurrence of natural fire is intensified by what is called a 'fire climate', that is, a low annual rainfall (between 25 and 150 cm), and a completely dry period of not less than five months. Australia has such a climate: very hot, dry winds, in already high temperatures, desiccate vegetation, can quickly fan a spark into a blaze and propel it at an alarming rate. The characteristic gum trees are the most inflammable in the world, producing large quantities of tinder-dry leaf, bark and twig debris at ground level and giving off a highly inflammable gas which enables fire to leap 100 m or more between trees. Australian grass fires spread five to ten times faster than a bush fire; progress of 13 km per hour has been measured. A grass fire in central Australia in 1968, caused by lightning in drought conditions, burnt out over 2 million hectares. Africa is another continent with a large fire-prone zone, including savannah and dry forest.

However, many plants are adapted to regular fires. Gum trees can withstand an average blaze and some species actually seem to need the occasional searing to remain alive. Many forest patches in Africa are surrounded by *Erythrina* scrub which is very fire-resistant. Acacia is a successful colonizer for the same reason. Fire-resistant trees often have thick corky bark. The ohia lehua tree on Hawaii can be almost completely buried by red-hot cinders in eruptions and will still sprout; what is more, new roots rapidly form on a new level just below the fresh ash surface. Many warm-country perennial plants send up herbaceous or partly woody growth, not usually more than a metre high, immediately the rains start; this flowers within a few weeks and is burnt off every dry season. Sometimes such plants have huge woody underground trunks which sprout annual groves of young shoots, and may prove an unexpected hazard to cultivation.

In South Africa several species of the bulbous fireball lilies flower more profusely after fires; *Haemanthus canaliculatus* will not sprout without one. This species was only discovered in 1961 following the burning of a normally marshy area in a drought; the bulbs, which must have been dormant for decades, sent up their bright red flower buds four or five days later. American pitcher plants flower in greater numbers if the topmost layer of the marsh they are growing in is burnt over after the previous season's growth has dried up.

Scorched shrubs and trees often produce abnormal numbers of flowers and heavy crops of seeds, and as noted earlier many seeds germinate far better after fires, or may need fire to germinate at all.

Even certain fungi are stimulated by fire. The morel, a gourmet's fungus, is one. European peasants used to set fire to forest clearings in the past to obtain a crop (which was stopped when the fires tended to spread to the forests); and after the First World War abandoned trenches and burnt-out house sites became full of these curious fungi.

The overriding physiographic factor is whether the landmass is a vast continent or whether it is under oceanic influence. The presence of mountains is another, and the aspect of a slope can produce very striking variations – a ridge may have a forest on the windward and hence rainier side, dry grassland on the other. Deserts, salt-marshes, bogs, heath-lands, savannahs and coral atolls are other variants, and we must not forget the water habitat itself: sea or fresh, still or moving, the diminishing quantity and changing quality of light in increasing depths, the availability of oxygen due to currents or upsurges.

The soil provides another vast range of controlling conditions. Any gardener knows how variations of soil texture can affect his plantings; there are all the graduations from sand or gravelly soils, through that 'nice medium loam' which is described as perfection, to heavy, airless clays which remain sodden all winter and, often hardly giving one a chance to dig them over when the weather is right, rapidly become rock-hard on hot summer days. The presence or absence of decayed matter in the form of humus, with its enormous encouragement of the soil microfauna and flora, again affects plant life: microscopic soil animals break down matter into readily assimilated form, aided by soil fungi, yeasts and bacteria; larger animals, notably worms, can have an enormous effect on the soil, both in improving its texture and aeration and making minerals more readily available to plant roots. The degree of acidity, created by free lime or its absence, is again crucial; many plants are strictly lime-lovers or lime-haters.

The presence or absence of other minerals may again be most important; too much salt, for instance, drastically reduces the potential inhabitants. The tiniest quantities of some chemicals can affect plant communities, hence their name trace elements. Thus boron and molybdenum are vital to plants in minute amounts; other trace elements are iron, copper, manganese and zinc. Too much of such chemicals is normally as inhibiting as too little.

In the biotic sphere – the organisms affecting plants – we have the whole range from microscopic bacteria, fungi and animalcules, to larger creatures as diverse as greenfly and elephant, pest and predator respectively. Very often a plant's existence may be entirely due to the absence of serious pests and predators and it may be extinguished in a region to which such a creature is introduced. The rabbit's introduction to Australia resulted in steady eradication of plant life in many areas. The changes in rabbit population in Britain during myxomatosis and the reduction of the disease's effects have visibly altered many downland environments especially when coupled with a reduction in sheep grazing, and this has allowed many downlands to turn to scrub, because seedling bushes and trees are not eaten early. This in turn spoils the habitat for such plants as terrestrial orchids. Regular grazing has always been an important evolutionary factor, eventually changing forest into savannah.

Man is undoubtedly the most important and deleterious biotic factor with which plants have to contend today, but his relatively recent effects on plant life will be dealt with fully in later chapters.

Biotic factors are not, of course, all harmful. At the simple level some plants will benefit from the droppings of grazing animals, and earlier chapters have shown how many plants depend on animals of various kinds for cross-pollination and for dispersal of seeds.

These few examples must serve to illustrate the complexity of factors influencing the presence or absence of plant species. More important than the original species, however, are the specific communities adapted to given circumstances which arise in all the different habitats. A South American jungle community will not be the same as a Malaysian one, but it may share certain basic characters; the same will happen in associations of alpines in, say, North America and New Zealand. A huge ecological vocabulary has been developed to describe such communities and the types of association likely to be met. One authority (Nicholas Polunin) recognizes seventeen main types of vegetation, such as tropical rain forest, heath, desert, mangrove; and six classificatory units for them 'in descending order of ecological status'; formations, associations, faciations, consociations, societies and clans. The study of ecology in these classificatory terms reminds one of Debrett, or perhaps a vegetable Mayhew; but analogy with human society can quickly be stretched too far, except in so far as both the vegetable and human communities are making what they can of their particular niche.

One must also consider the development of a community from an elementary colonizing beginning; clear-cut *successions* of communities are recognizable, terminating in a *climax* which may eventually be overset by some disaster of interference and will then start again, usually for the worse. A typical succession may have ten stages from a bare-soil start, via successful initial establishment, competition and its reaction, to the stable climax.

Such a situation in temperate climates might begin with a freshwater lake, which becomes colonized by aquatic plants of various kinds including mosses. When silting of dead material plus lodged silt have raised the lake floor to a depth of not more than 3 m, floating-leaved aquatics such as water-lilies may enter, which being relatively bulky quite rapidly build up the silt as generations come and go. Next there is shallow water which can be invaded by swamp-lovers, of which the reeds are typical and swift-colonizing. But despite their efficiency they are continuously building up the soil level by decay and trapping of silt, and eventually make things impossible for themselves, when the reed-beds are taken over by fully land-living plants. At first these are low-growing and herbaceous, usually sedges and grasses, and once again, as their activities raise the soil, shrubs and finally trees find a footing. First these will be moisture-loving such as alders, poplars and willows, and then as they dry out the soil they will be overcome by the climax, a forest of trees preferring rather drier soil conditions.

A rocky environment is much more difficult to colonize: its first occupiers will be lichens and the tougher mosses, and as these decay, and soil collects around their colonies, larger plants can slowly find root-room. Here those plants will do best which can be thrust into minute rock crevices, like so many alpine and arctic species.

The seashore is often a perfect example of a graduated climax, with the great kelps at the lowest tide level, long, strong and slippery so as to move in the nutrifying water with no damage, but only exposed to air and sun for a small fraction of each day; then higher up the bladder-wracks and various other algae, with shorter fronds steadily more resistant to desiccation as they occupy levels nearer and nearer high tide mark. Finally salt-resistant lichens and tough mosses take over, and above these flowering plants once again able to resist salt and exposure, such as thrift, sea campion and marram grass.

Not for nothing is the business world sometimes described as a jungle. In the tropical rain forest, where there is little seasonal rhythm, growth is at its most lush. It can be very dense indeed, as evidenced by the 1972 discovery of a tribe called the Tasadays in the rain forests of Cotobato in the Philippines, whom just 25 km of forest had entirely isolated apparently since Neolithic times.

Tropical rainfall can incidentally be unbelievably heavy by temperate standards. The highest recorded is in the Khasi range in Assam, where the average is 1,500 cm per year and over 2,500 cm has been recorded. The botanist Hooker once recorded about 75 cm in a single night, as much as some parts of eastern England receive in a year!

In a tropical forest one can very often make out three distinct storeys or strata. The lowest is of small, tapering trees, trying to thrust their crowns between those of the middle layer. But they cannot reach high enough, for their 5 to 15 m is overtopped by the 15 to 30 m of this middle layer, whose crowns form a more or less continuous canopy. Above these may tower a few monster trees rather widely spaced. Sometimes there are a large number of the biggest trees, and these form a canopy below which crouch the lesser members of the hierarchy, including a ground stratum of tree ferns, palms and the like. Although there are sometimes groups of a single species it is much more common to find no specific dominance: any given species may be found only once or twice in a hectare, and sometimes quite extraordinarily rarely.

I have earlier mentioned how herbaceous flowering plants almost certainly descended from tropical forest trees, and this is reinforced by the number of plant families to be found in a tropical forest. As the explorer Richard Spruce noted in the Amazon forests, 'nearly every natural order of plants has trees among its representatives, grasses 20 m or more high, violets of the size of apple trees and daisies borne on trees the size of alders'. One should perhaps add that there are also a great number of families represented in the tropics which do not spread into temperate zones.

In some tropical forests, however, the opposite may occur. In south-eastern Asia, for instance, there are rain forests which have arisen from peaty swamps, where the water has few dissolved minerals and the water-table is very near the soil surface for much of the year, and may even be above it for several months. In these places a single tree species may be almost entirely dominant, although more often palms and screwpines are also found. This is true also of mangrove forest which develops on seashores and coral atolls, creek margins, and around brackish lagoons and swamps, growing in mud-flats which are usually covered with salty water at high tide. Both these forest types are virtually impenetrable because of the air-roots and stilt-roots of the typical trees, which are perfectly adapted to the situation.

Mangrove is often the first plant on a coral atoll or off-shore mudbank because of its drifting seeds and capacity to germinate in salty water, and if so it will usually monopolize the situation for many years, literally helping to build the island as debris and sediment collect among the stilts, until the centre becomes too high and dry for it. If other plants, such as *Cordia*, or the nipa palm, establish themselves first, the mangrove may not have a chance. Should a section of such a one-species forest be destroyed, it is often remarkable how some quite different plant, whose presence has been unsuspected, may spring up in the gap. Examples are known in Florida where a hurricane defoliated a mangrove swamp. When sunlight reached the long-shaded soil, seeds of various climbing plants such as the moon vine germinated and spread all over the broken trees.

After the atomic explosion on Bikini Atoll the coconut palms and breadfruit trees were largely obliterated; those that were left mostly lost their tops and apparently produced inferior fruit. But a dense growth of shrubs and vines took over – beach magnolia, morning glory and screwpine – so thickly that eighteen years later research workers found it extremely difficult to cut paths through them. During the last few years this vigorous growth has been bulldozed in certain areas to make room for the replanting of coconuts, bananas, breadfruit and other useful trees.

A brand-new environment can be colonized with great speed. Such virgin terrain might be exposed by the retreat of a glacier, by a landslide or by the changing course of a river. Perhaps the classic example is the volcanic island of Krakatoa, which literally exploded in 1883. Three smaller islands were left after the explosion, covered with ash up to 80 m deep. The first summer after the cataclysm a few blades of grass appeared. Three years later there were thirty-four plant species, including blue-green algae, mosses, ferns and some seed plants. In

1897 there were sixty-one species, a large number of the flowering plants being grasses; 1906 saw 108, and 1928, 276. Today the vegetation resembles both in denseness and number of species the typical climax forest found elsewhere in the Malay Archipelago.

Colonization of unpromising volcanic ash and lava can be seen regularly in Hawaii, where lower plants, blue-green and other algae, mosses, lichens and ferns, almost always precede seed plants. Ferns may germinate on the still hot surface within six weeks of an eruption.

In tropical rain forests many lesser plants take advantage of the trees. Even in temperate woods algae, mosses, lichens and ferns grow on or cling to trunks and branches. In the tropics these are augmented by more spectacular plants such as orchids and bromeliads, which perch on branches, taking advantage of the light and absorbing water either through their leaves, as in bromeliads, or through special air-roots which may hang in the moist atmosphere or, clinging to the host's bark, absorb some of the run-off. Others live in little pockets of humus created by decaying leaves in crotches. These are by no means parasites, although they might literally enough be called hangers-on, as the technical word *epiphyte* implies. There are also lesser epiphytes called epiphylls, mostly very thin leafy liverworts which cover the actual tree leaves with light-intercepting growths.

The plants which really avoid stratification are again the climbers. As described in Chapter 8 lianas fling themselves over the lower plants, cling between them and force themselves into the topmost crowns; some lianas are very much in the running for the longest plants in the world, with over 200 metres recorded. Sometimes they link tree crowns together like chains, so that a dead tree cannot fall down. Such climbers seldom harm their props, apart from constricting their growth or distorting their crowns on occasion. But they have found a way of cutting down expenditure on building stout woody trunks by this clambering through and over solidly formed trees.

In the main there is surprisingly little root competition in a jungle; and although I may have given the impression of an unhappy assemblage of plants at the lower levels, they are in fact fully adapted to the steadily reducing light; each has enough light, air, moisture, soil for its own needs, even if its subsistence level in these terms may be much less than that of the forest giants.

Water habitats can be as competitive as the jungle. The deepwater stage which is filled mainly with submerged plants rooting at the bottom can be almost impenetrable, and in many Indian and eastern lakes something very like a climax occurs with a few submerged plants coping with a considerable lack of light due to a very dense floating population, which may include pondweeds, duckweeds, water lettuce, water hyacinth, and floating ferns like *Salvinia* and *Azolla*. Above these the huge waxy leaves of the sacred lotus raise a canopy, so that there is a layer-system similar to that of the tropical rain forest. As with epiphylls, minute algae cover water plants with a semi-transparent sheath.

Sometimes plants are actively antipathetic to others. Thus certain plants of the daisy family undoubtedly reduce competition by toxic root exudations, sometimes in minute quantities. This family, the *Compositae*, has more than its fair share of weeds, as we all know, and this root poisoning character may well be an important element in their success. They even continue this assault when cut; flower arrangers will know not to place many members of the family – the shasta daisy is an example – among other flowers, which they would cause to wilt. Further examples of this root warfare are given in Chapter 25.

Plants are commonly in competition with each other or with their environment; the lushness of the tropical forest or the bleakness of an Arctic tundra are extremes but in both cases the fittest will survive. In the forest light is perhaps the key factor; in the tundra or on mountains it is often the terrain, the availability of soil or crevice, or the degree of root competition that a plant can stand. In both cases drying out of the soil may be critical. In temperate or cold climates, for instance, beech trees dry out and shade the ground so much that little grows below; conifers add to this a layer of dead needle-leaves which are largely uncongenial for growth of other higher plants. Water is usually the most important item in plant survival, and in very dry places one can often see plants of the same kind well spaced apart so that undue root competition does not occur.

I have used the word 'competition' rather frequently in this chapter, and perhaps implied constant struggle and antagonism. Certainly there are many habitats, especially in the tropics where plants grow fast, in which a degree of struggle does take place, and seedlings in particular find it hard to survive. At the same time one must emphasize that a plant community can only exist because its members are each adapted to a particular niche: some need much less light or water than others, epiphytes can exist on branches, roots penetrate to different depths, there may be bulbs deep in the soil and annuals at the surface, and so on. There is a large measure of integration.

There are also many examples of plants with apparently identical growth requirements successfully dividing a territory in which one would imagine they would be rivals. In such cases different aspects of life control the situation. Thus there are different kinds of American penstemons existing together at high altitudes; their growth and habitat are the same, but some have narrow-throated red flowers which are pollinated by humming-birds and others wide-throated blue ones visited by humble-bees. With American woodland columbines humming-birds again pollinate the red-flowered species and hawk-moths the white ones. It so happens in this case that humble-bees visit either and thus hybrids are created, whose pollination depends on their flower colour.

Not surprisingly the more specialized and difficult the habitat the more specialized the plants in it will be, and in some places there will be many fewer of them. Deserts, salt-marshes and alpine tundra are good examples.

The world, then, has an infinity of potential habitats, some very extensive, some very localized, for its plants. The resulting niches can be large or small, and in studying them we must not lose sight of the micro-habitats caused by some special local conjunction of features. These are likely to be most important in very severe situations, as in the Arctic. Here a boulder or tussock may have a temperature on its south side over 20°C higher than on its north one, so that growth occurs on the south while all around is frozen hard. In such places, too, boulders will be frequented by birds. Their excreta enriches the soil, and may introduce seeds, so that patches of flowering plants thrive and luxuriate.

Gardeners are well aware of micro-climates. Fanciers of less hardy plants will place them at the foot of a south-facing wall, where they will get all the sun available and usually rather dry conditions. Bulbs such as nerines and the Algerian iris will perform well here, and so will the more exotic kinds of fruit tree like apricots and peaches. A hollow will form a frost-pocket, because frost drains downhill just like a liquid; even a wall on a slope may create such a pocket if there is not a gate or other gap in it. Gardens on the west coast of Britain come under the influence of the Gulf Stream and are thus comparatively warm, as well as having a higher degree of humidity due to the prevailing winds. Sometimes gardens a few miles apart will provide astonishing contrasts – Kew's country extension at Wakehurst Place, Sussex, grows a number of quite tender species, but I know a garden barely five miles from Wakehurst Place which has to rely on absolutely hardy plants because of difficult winter conditions.

The study of the sociable plant and the communities that plants make is fascinating in itself. Behind it we constantly sense the delicacy of natural balance, which we ignore at our peril whenever we tamper with natural associations.

20 · *Exploiting the Environment*

A cactus desert looks sparsely populated, with many yards of sand between each plant. But if one could turn the land upside-down it would resemble a jungle, so widely do the roots spread in their search for water. This is a striking example of the external pressure posed by an environment. Every plant has its own solution and its own way to exploit the situation to best advantage. The essence of solving problems at this level lies of course in the long-term capacity to mutate, to produce the occasional individual able to stand a slightly different situation, which is at the heart of the theory of natural selection.

Ability to grow in a cold or warm climate is for instance a genetical characteristic conditioned by a series of mutations. We can imagine how some primitive plant gradually spread into a colder or warmer area, or was perhaps overtaken by a gradual climatic change, to which it had steadily to adapt or be doomed.

In a plant's natural habitat abnormal cold or heat is extremely unusual. It is only when we cultivate plants from other countries that we begin to see how differently they may react to superficially similar conditions. An alpine or Arctic plant is totally hardy in relation to cold; but if it is subjected to winter damp, or slightly fluctuating temperatures when it is accustomed to steady cold, it may very well die. Shrubs and trees introduced to a different country may be perfectly hardy when mature but late frosts after growth has started may damage buds and prevent flowering. In Britain the common holly is regarded as perfectly hardy; in cold areas of the United States a thriving plant will be treated with awe, because it nearly always succumbs to the scorching caused by reflected sunlight off the long-lasting snow.

Actual damage from cold is caused by ice crystals forming either within the cells or in the spaces between them. In the first case the crystals probably cause mechanical damage; in the second, water is withdrawn from the cells, which dries and deforms them. The hardiest plants are those whose protoplasm has a built-in resistance to the formation of ice crystals; they are also likely to come from areas where the temperature falls rapidly below freezing point, for it is around that point that ice forms most readily. Such plants probably also have a very highly concentrated cell sap, which lowers its freezing point just as the use of salt on roads makes snow freeze at a much lower temperature. Protein content is also high, although its effect on frost resistance can only be guessed at.

Something about the mechanisms involved has emerged from recent Japanese research on species of *Rhododendron*. The investigators measured the temperature and water content of the external bud scales and the unopened florets within. As temperature fell water froze at −5°C in the scales but not till −30° was reached in the florets. Apparently the water in the florets is transferred to the scales as temperature falls, so that the concentrated sap remaining in the florets resembles anti-freeze, and below −30°C the cells in the florets had practically no moisture in them at all. But what of the scales, receiving more and more water? Ice crystals do form in these, but *between* the cells. As temperature rises, this ice melts and the surplus water is transferred back to the florets, which are in due course able to open as unharmed flowers.

The toughest plants are some of the most primitive. The only plants that grow on snow are microscopic algae, yellow, green and red – the latter typical of the Antarctic where incidentally they subsist on guano dust carried by the wind from South American cliffs and islands. Flagellate unicellular algae can survive a temperature of −15°C. Fungi can grow at −9°C, and at the other extreme withstand 64°C – the latter is a 'toadstool' type found in the Wadi Rum, Jordan. In higher temperatures only blue-green algae and finally bacteria can exist, the latter up to 77°C. Some bacterial spores need literal boiling for subsequent germination. Very saline lakes are likely to encourage only specialized bacteria and one-celled plankton, but some larger green algae can live on salt crusts where the water has a concentration of over 30 per cent dissolved salts. A bacterium very recently discovered is able to live in highly alkaline solutions of sodium hydroxide, ten times stronger than any bacterium previously known will withstand; it is this kind of organism that may conceivably exist in such alien environments as that of Jupiter. And such bacteria must be very similar to the earliest life forms.

Phytoplankton and larger algae are abundant in Arctic waters, where the former provide a most important part of fish diet. Above ground blue-green algae form an important part of Arctic vegetation, where they are often the first colonizers of rocks and stones, their slime forming a home for other minute organisms; they appear to glide over such surfaces in an unexplained way. Lichens can also begin on bare rocks; they are found on high mountain peaks and in vast areas of both Arctic and Antarctic. The bushier ones, such as that silvery-grey one aptly called reindeer moss, are vital in the feeding of reindeer and other grazing animals, and one of them, the Iceland moss, is rich in a starch-like carbohydrate which makes it a passable human food. Mosses and club-mosses also inhabit these cold, arid areas.

The hardiest higher plants are Arctic and alpine species, and in some cases they spend a good part of the year under snow, refrigerated to a steady temperature and insulated against other elements. At such times the ground is likely to be frozen, so that the roots cannot take up water and the plant is dormant. Certain Andean alpines can even survive for several years under the snow, flowering freely when finally exposed.

The 'classical' alpine is dwarf, forming a rounded cushion or spreading mat which offers least resistance to wind and with adaptations to withstand dryness. Sometimes such cushions become very large like those called vegetable sheep found both in the Andes and in New Zealand, old specimens of which make literally sheep-sized mounds covered with tiny leaf rosettes.

An alpine has much reduced stems so that it is greatly compressed: there are a large number of virtually stemless alpine thistles. The 'ideal' alpine is very compact, often hard, and composed of small leaves which are frequently imbricated – one pair at right angles to the next, pushed close together – or in rosettes, both formations which pack leaves together and reduce evaporation. These leaves usually have protection and insulation from the elements in the form of a thick, waxy coat, secreted lime deposit or hairs. They are often white to reflect excessive sunlight, especially ultra-violet light. The ultimate in insulation is found in the Himalayan saussureas such as *S. sacra* in which every part of the plant is so densely furry that it is impossible to distinguish its separate parts. At the top of this rounded vegetable thermos flask, which is sometimes partly buried in snow, is an aperture through which bumble-bees can find their way to the flowers; they often stay the night in the shelter.

Alpines are likely to have deep and wide-spreading roots, or at least roots capable of forcing their way into the thinnest fissures in rocks. Long tap-roots are common, especially in plants which grow in screes where the stones continually move downwards and force the leafy part of the plant along with them if it is to keep its head in the air. With these the long thick root can often be seen stretching out uphill like an anchor cable.

The grassy tussock is another excellent alpine plant form. This is because the dead and decaying old leaves and leaf-bases form a felt-like mass at the base of the tussock, and it is here that the new shoots with tender buds are buried. Measurements made on grass tussocks at over 4,000 m in the central African mountains showed an internal temperature of 2·5°C when the external temperature was −5°C. The tussocks are also excellent buffers against variations in moisture, which the felty mass retains. The temperature insulation is especially important in this particular habitat because the conditions encourage 'frost-heaving' of the soil, which is obviously likely to damage or expose roots. The warmth of grass tussocks encourages a species of sunbird to build its nests inside them on Mount Kenya. Larger alpine plants may have shiny leaves which reflect excessive ultra-violet radiation, and these are usually combined with a thick exterior skin.

There are, however, many alpines which do not adopt such precautions, quite ordinary-looking plants such as primulas, columbines and *Adonis*. Admittedly they usually grow in more sheltered places than the cushions and mats, but not always: a photograph taken by Oleg Polunin at around 6,000 m – roughly the maximum altitude for alpines in West Nepal – shows an assortment of herbaceous plants, only two creeping or mat-forming, the rest low but upright. There are also some lichens.

Nature does not always carry out the obvious, however efficient it may seem. Indeed, a complete reversal of the obvious exists in the mountains of central Africa and in the Andes. Here quite extraordinary giant plants exist: grotesque tree groundsels and space-fiction lobelias up to 6 m tall, and even taller heathers and St John's-worts, the latter up to 12 m. The groundsels have leaves covered in wool for insulation, while some of the lobelias, like *L. telekii*, protect their blue, sunbird-pollinated flowers with long woolly leaf-bracts so that the fully developed plant looks like some weird brush or, as Patrick Synge once described it, like a 'gigantic woolly caterpillar petrified and stood on end'. In other cases the flower spikes are hard and shining. In the Andes similar plants occur, and in particular the fantastic bird-pollinated *Puya raimondii*, 10 m tall, with a palm-like trunk, a ragged skirt of elongated leaves, and a rocket-like head of flowers 5 m long or more: the plant perishes after flowering. There are also huge lupins (of the pea family) and espeletias (groundsel relations) of character similar to the African lobelias – striking examples of parallel evolution.

In these equatorial mountains the average temperature is much the same throughout the year. Between day and night, however, there may be a great variation: it has aptly been remarked that it is winter every night and summer every day. In the Andes the daily temperature range may be 40°C, with temperatures well below freezing at night. Moreover the intense heat and radiation from the sun makes the day temperatures rise very rapidly. Such conditions are testing indeed to the plant inhabitants, which exist up to 4,300 m.

Both the giant lobelias and tree groundsels have leaves arranged to form 'night buds', which fold up every night very tightly around the developing shoot or flower-head to resemble a monstrous globe artichoke. Mackinder recorded in 1900 that 'the lobelias closed their heads of leaves at night-time just like daisies'. In the morning the leaves can be seen unfolding on the sunward side as soon as the light catches them. In the tree groundsels the young leaves in the centre maintain a protective cone over the shoot tip at all times. Some of these leaves have thick felt on the undersides only, a fact which puzzled earlier observers until it was realized that this felt would be on the exterior of the night bud, and therefore in the right place for insulation. It also covers the breathing pores or stomata and, by curtailing air movement around them, reduces transpiration.

Almost the oddest device is the water reservoir of some of the lobelias. The rosetted leaves are so tightly meshed together that they can hold liquid as in a bowl. This liquid is undoubtedly produced by the plant, for it is present in dry weather. The reservoir, which can be 10 cm deep, surrounds the tip of the young shoot. In the cold of night the upper part of the liquid freezes, but below two or three layers of ice pieces up to a centimetre thick the lower part never does and the temperature remains at least 1°C and perhaps 2 to 3°C above freezing.

Apart from their growth buds the tree groundsels have the problem of protecting, as it were, the plumbing in their tall trunks against nightly rapid freezing. They do this by what one might call 'self-lagging'. The old leaves up the trunk die, but instead of falling off they form a tightly packed down-hanging layer of many overlapping surfaces, at least as thick on each side as the trunk they protect. A temperature of 3°C has been recorded in the pith when outside it was – 5°C. It is noticeable that tree groundsels whose lagging of dead leaves has been destroyed by fire are likely to succumb to the elements. The only species which does not retain its dead leaves has very thick corky bark, which provides insulation in a comparable way. Certain Andean trees have thick flaky bark for the same reason.

Why these equatorial alpines should have become gigantic is a mystery, but it is clear that they have solved the very hard climatic problems as efficiently as any dwarf cushion plant.

Before leaving the African equatorial mountains it is worth mentioning two other curious ways of solving the problems. As mentioned earlier, frost-heaving (a type of solifluction or soil movement) is a characteristic here because of the violent temperature changes, and on some soils this is accompanied by the formation of needle-ice crystals at least a centimetre long packed vertically together. Every night, this layer heaves the effective surface up. A small plant called *Subularia* has enough warmth of growth to prevent ice forming immediately around the upright leaf rosettes, and so survives in little vertical tubes.

This metabolic warmth can be noted in many European alpines, like soldanellas and crocuses, which melt holes as their flowers develop within receding snow-patches. While it has been impossible to estimate the actual temperature involved with these tiny flowers under the snow, measurements have been made in other alpine flowers. Thus a bloom of trumpet gentian had an internal temperature of 10·6°C when, just before sunrise, the external temperature was 8·4°. A campanula and a monkshood revealed temperatures of 16·6° and 14·6° respectively in an outside air temperature of 13·2°, while in a closed carline thistle no less than 20·4° was recorded in air of 13·2°.

Another African curiosity is the way in which several more primitive plants stand frost-heaving by going along with it – they are not attached to the soil. Lichens, the blue-green alga *Nostoc*, and several mosses do this in Africa, the latter forming almost spherical moss-balls 1 to 3 cm across whose nightly movements ensure that growth occurs all the way round. Many plants of the northern tundras also have to withstand frost-heaving.

Alpines, then, have to resist cold, rapid alterations of cold and warmth, the possibility of desiccation, excessive radiation, and often mechanical upheaval. In exposed places they may be affected by wind, although this is a problem to plants of many kinds: we have to look no further than our coasts to see the effect of a prevailing wind on trees, and such shaping is always noticeable at the upper margin of the alpine tree-line and at the northernmost extension of the Arctic forests, where cold also plays its part, creating bonsai-like, stunted trees. Where these are merely dwarfed the ecologists rather charmingy call them 'elfin wood', and when positively distorted 'krummholz'. The normal tree of cold regions is the conifer, with steeply down-sloped branches perfectly designed to shed excessive snow.

In temperate climates the typical problem is a relatively hard winter. Many perennials get through this by dying down to a root-clump and existing almost dormant and thus fairly resistant to bad conditions. Deciduous trees achieve dormancy by losing their leaves. Water plants may reduce themselves to submerged knobs of tissue, while annuals evade the hard winter as seeds. But it is the extreme problems which illustrate plant adaptability most vividly. Alpines and Arctic plants face one set of extremes, and the other most striking condition is extreme dryness. Plants adapted to dryness in whatever way are called xerophytes.

Dryness arises in many places; it may be intrinsically due to the habitat, as in seaside sand-dunes, but is at its most extreme in deserts. In very severe desert conditions nothing can grow, but certain plants can survive incredible aridity, as witness the grasses on which Bedouins' camels feed in parts of the Empty Quarter of Arabia. There are plants which exist in savage parts of the Sahara where the day temperature may be 60°C and it may nearly freeze at night, and where several years may pass without rain; similar conditions exist in south-west Africa.

The most difficult thing about these environments is the variation in conditions which occurs from year to year. Thus in the Negev desert there may be adequate moisture for four months in a 'good' year, but for as little as five to six weeks in a 'bad' one. Dew often occurs – in the Negev again it is recorded on 180 nights a year – but seems to be of very little value to larger plants, although it nourishes primitive ones. 'Subterranean dew' caused by condensation may also be beneficial. However, the main reservoir for water in a desert without a water table below is likely to be 'pockets' below stones, and roots are often observed wrapped around these sources.

Primitive plants certainly exist in implacable-seeming deserts. Patches of microscopic algae flush green after flooding, later leaving nothing but the finest dark crust among the soil particles. Stone fragments are often encrusted with algae. It has been found that these simple plants can lose almost all their internal water in dry conditions and become completely dormant, being resistant also to very high temperatures. They are capable of remaining in this desiccated state for at least two and a half years, probably much more. Any available water is quickly taken up so that they swell and pick up their normal metabolic activities – a resurrection which man's deep-freeze techniques have not yet achieved for himself. These algae certainly make use of dew or even damp air. Moreover they often live on the undersides of stones, and are capable of using the very low levels of light that penetrate there, some managing to operate in light intensities below that of moonlight.

Up the scale, lichens are widespread colonizers, just as they are in Arctic, alpine and other stony wastes. In his *Himalayan Journals* Sir Joseph Hooker recorded that he saw at about 6,500 m altitude a lichen he had last noted in Kerguelen Land in the Antarctic. Some live within rocks which often have a loose upper surface, and help to disintegrate it. Like their half-brothers the algae, the lichens can stand repeated drying out and wetting; this is particularly necessary with these because a 'wet' lichen may only stand about 30°C, while a desiccated one can survive 80°C or more. Within ten minutes of being wetted such a lichen will add 50 per cent to its initial weight by water-uptake. An example is the bushy *Ramalina*, which can photosynthesize even at −10°C, when some of its internal water will be frozen. Some mosses have similar capacities for drying up until they look totally dead, and when moistened coming to life as if nothing had happened.

Higher plants face two distinct problems in arid conditions. These plants, of course, take up water through the roots and lose it through the breathing pores in transpiration, and have to achieve a balance between these two; without transpiration they cannot carry out the photosynthesis that they need to grow and to produce flowers and seeds. Their first problem is actual soil drought, an insufficiency of moisture for the roots; some desert soils are 'permadry' just as Arctic ones can be 'permafrost'. Their second problem, air drought, may occur even when the soil is fairly damp, because the plant is losing more moisture by transpiration than its roots can supply. Photosynthesis can be completely disrupted if a plant becomes dehydrated, just as it is when the cells are damaged by frost.

It will be seen that, to quote from *The Negev*, 'each desert plant active during the dry season finds itself permanently between the Scylla of death by lack of water and the Charybdis of death by lack of organic material' and that transpiration could be regarded as a necessary evil. Apart from photosynthesis, however, some plants transpire to keep themselves cool, much as we sweat; the bitter gourd or colocynth (which caused those men served pottage in II Kings 4 in the Bible to cry out 'there is death in the pot') has broad, thin leaves which transpire so strongly that their temperature is lowered some 7°C below that of the air around, working on the evaporation principle like an old-fashioned earthenware milk-cooler.

Certainly photosynthesis can proceed at a considerable rate if there is enough water, so that plants grow rapidly in good conditions and prepare themselves for worse ones. They have a great number of solutions to the problem; indeed, the more extreme the conditions are the more numerous the answers plants come up with in a given habitat. And not only is there this general adaptability, but a great flexibility within a given plant. Each plant might be said to be continually balancing its water budget; they do this over each twenty-four hours, often achieving a credit at night and going into the red the next day. Equally they may have an annual water budget: survival is dependent on tolerating water levels which would bankrupt any normal plant, and these can be very low indeed after the dry season.

One oddity about many desert plants is that their capacity for potential growth is enormous. Normally they grow extremely slowly and may, as described later, actually shed much of each season's growth. But given continuous water they grow completely out of their desert character. Bean capers irrigated in this way grew from 2 cm tall to nearly 50 cm in ten months, whereas the normal growth in a 'good' desert year is around 5 mm. Bean capers are known to live up to 300 years, and a plant of that age is normally not more than 70 cm tall. One might well ask why they do not colonize better habitats if they are able to grow so well in them. But no, they stay among their habitual rigours like some Old Testament prophet.

Old plants survive whereas youngsters may not; hence there must be a good supply of seed and as many mechanisms as possible for spreading it in both space and germination period. Too many seedlings germinating together compete so hard for moisture that many die;

thus in a marked square metre only forty saltwort plants survived out of 850 germinating, even though newly germinated seedlings can stand twenty-five hours of desiccation. This saltwort is annual; its Negev neighbour, a perennial sagebrush, succeeds in germinating very few seedlings, but these have a high survival rate. With bean capers the Negev researchers found that the chances of successful establishment occur on average only once in five to seven years; but this is sufficient to maintain a population.

Desert plants, as already described in Chapter 16, have many devices to ensure seed germination at the most auspicious moment, or to spread it over a period. There are often two or three kinds of seed, one programmed to germinate at once, the other later, or in varying permutations of moisture, light and temperature. Further devices may ensure that desert seeds do not respond to light out-of-season showers.

Adult plants may cope with the desert by abandoning leaves altogether. Such are the 'switch plants', of which many brooms are examples, with green stems containing chlorophyll and bearing stomata, and hence able to photosynthesize. Often the stems are fluted in section; the main reason for this is that the stomata are placed at the base of the furrows, where they are sheltered from brilliant sun and from air movement, both of which would speed up transpiration.

Xerophytic grasses usually have leaves which roll up in dry conditions, concealing within the false tube thus formed the convoluted surface in which the stomata are located. In moist conditions they are likely to expand.

Many xerophytes do in fact have leaves. Frequently these are more or less fleshy, with thin-walled, elastic cells which can shrink or swell according to the amount of water available. Many desert plants exhibit some succulence, even if this is not obvious to the eye; I recall how specimens collected in the Jordan desert refused to dry in the press for several weeks, despite the heat and constant changing of paper. Desert acacia leaves contain nearly 50 per cent free water, and provide food and drink for gazelles.

Desert shrubs may cut their dry-season losses in the most literal way, by shedding part of their bodies. On desert goosefoots part of the green, photosynthesizing stem exterior sloughs off like a skin, once water in a reservoir tissue has been exhausted; as it dries out this tissue becomes corky, which not only cuts off the exterior green skin but insulates the central core that remains. Such plants, and many others, also shed branches or twigs, while leafy plants lose their foliage. Many switch plants have very small leaves which are immediately dropped; in the grey sagebrush the large, well-dissected winter leaves, perhaps a couple of centimetres long, are dropped, the size of new ones becoming progressively reduced during the dry season until they become nothing more than little projections of a millimetre or two. Thorny saltwort loses up to 90 per cent of its body weight by both branch and leaf shedding. The bean caper loses the main part of its leaves, only the stalk remaining; this carries stomata, and once the rest of the leaf has gone a thickening of the outer surface results in these pores being constricted below the skin so that they cannot open fully. In this way this species curtails its transpiration by no less than 96 per cent.

The bean caper has an even untidier trick up its sleeve. This is 'survival through partial death', and is shared with other desert shrubs. An unusual method of internal development allows the main stem to develop in several independent strips, each with its own root system and terminal cluster of twigs and leaves. After a time these strips curve apart and are barely held together by the soil at their base. Some of the strips may perish and others survive.

Yet other plants discard their delicate top growth and retire to underground safety in dry seasons. These are fleshy-rooted plants – bulbs, corms, tubers or rhizomes according to their structures. Often 30 or 40 cm below soil level, they shoot up at great speed when moistened by rain, with their flowers usually ready formed; they are some of the most striking and brilliant desert flowers, including tulips, irises and foxtail lilies. If the rains are inadequate they can remain dormant for at least one season, paralleling the Andean alpines which may pass years under snow. Several of the bulbous plants put up their flowers at the end of the dry summer, ripen seed and return to dormancy, sending up leaves when the main rains start. There are a few plants with bulbs more or less above ground, like nerines and *Bulbine*. The Mediterranean sea squill, with its huge bulbs, sometimes as big as a football, will survive out of the ground for many years, sprouting leaves annually. The Greeks hang the bulbs up in their houses as a fertility symbol.

Dormancy in these roots is assisted by exterior insulating layers of dead or corky tissue, and they can sometimes withstand very high temperatures: the Sinai blue grass has dormant bulbils which withstand 80°C. Within twelve hours of rain they are sending out roots, while within twenty-four hours of the first rains the soil around the sedge *Carex pachystilis* is entirely permeated with roots. This sedge never, in fact, loses its leaves: the upper parts shrivel up while the bases remain green but quiescent.

Perhaps the most extraordinary desert tuber is the smallest. This is the quantity named *Chamaegigas intrepidus*, a figwort relation known from one locality in south-west Africa, where the dry season may extend from May to January. Its tubers, the size of a pinhead, are found in dried-out ponds. Within *minutes* of the first rain falling, these tubers sprout cylindrical leaves a centimetre long. The second day the plant makes a rosette of four oval 1 cm leaves, which float on the pond's surface, in the centre of which the 8 mm violet flower shortly appears.

The true succulents turn every part of their body into water storage. There are stages of succulence from woody shrubs with fleshy leaves, like the bean caper already cited and shrubby members of the *Mesembryanthemum* family in South Africa, to leafy plants with fully succulent stems like crassulas. From these, notably in the *Mesembryanthemum* family, we can trace a complete range in which the leaves diminish in number and in size, ending up as a tightly compressed series of six or four, a single pair, or a roughly globular mass or plant-body whose origin as two leaves is shown by a groove or slit in the centre, from which the flower emerges. These include the well-known pebble plants of South Africa and their close relations. In the misshapen *Muiria*, an entirely green sub-spherical mass, the slit is so obscure that the flower appears to tear its way through the skin. When these super-succulents are resting in the dry period their skins often harden and become blanched, forming a thick and reflective shell within which the next season's plant-body can develop protected from excessive heat.

There are other exceedingly fleshy-leaved plants called window plants which live buried in the sand so that only their flattish leaf tops are exposed to the sun. In the well-named *Fenestraria* there is a 'window' composed of a layer of calcium oxalate crystals under the skin which reduces the power of the sun's rays while allowing sufficient to reach the chlorophyll layer deep down.

Such leaves, as in high alpines, often have anti-solar insulation such as waxy 'bloom', greatly thickened skin, extra layers of exterior cells to protect the chlorophyll, fur-like felt, white or silvery skin and sometimes a warty or tubercled surface. The American saltbush has an external layer of salt. In many cases the leaves shrivel a good deal in the dry season as their water reserve is used up.

Many succulents, including the hardy, often alpine houseleeks, adopt the rosette arrangement of leaves, which permits the maximum photosynthesizing surface in the minimum space. Often the leaves close up in dry conditions to conserve moisture and curtail transpiration; some of the houseleeks do this as do also their tropical relations the greenovias which look like rosebuds in the closed stage.

Rosette, leaf-compression, and finally 'plant-body' composed of joined-up leaves point the way to the ultimate in drought resistance: the sphere and cylinder, in which there is the minimum proportion of surface area for the maximum volume for water storage. Such are the typical forms of the cacti and of several unrelated families, spurges and stapeliads among them, which have come to the same end-point in response to the same set of conditions by parallel evolution. A handful of cacti grow normal leaves, like *Pereskia*, considered the most primitive of the family and perhaps a progenitor, which looks like a thorny camellia; others have ephemeral leaves which are lost as the plants mature, a reminder of primitive origins like the embryonic gills of mammals. But most have discarded them, and have green stems able to photosynthesize, heavily coated in wax, felt or hairs, and with relatively few stomata; they have made themselves into insulated, stoppered barrels or cylinders full of moisture.

A few cacti live in moist jungles where the problems are quite different: they remain leafless but their stems have become flattened and leaf-like, as in the popular orchid and Christmas cacti. Such plants possibly re-entered the primeval forest after an evolutionary spell outside. Most of the cactus family have become entirely adapted to their dry habitats, while in the spurge family, as in the daisy, lily, geranium, vine and other families, succulence is demonstrated only by a few genera which have, as it were, wandered over the xerophytic threshold, not necessarily representative of the family as a whole; some retain leaves, others lose them.

When leaves disappear the green stems take over photosynthesis and transpiration, as in 'switch plants'; the stems are likely to be protected in similar ways to the leaves just described, and also quite often by hair. Many cacti have ridged stems, and in any long dry period these ridges become more prominent and closer together as the body shrinks. The accordion-pleated stems are perfectly designed for this in-and-out movement. The flat, rounded leaf-like 'pads' of prickly pears simply become thinner till they finally drop, and many – notably the very prickly chollas – habitually lose pads or joints near the stem-ends in dry conditions, another aspect of shedding.

The stomata of cacti and succulents of similar structure are found at the bottom of the accordion furrows, where air movement will be at a minimum especially in dry conditions. Moreover, they open mainly at night. Since there is no light then, the carbon dioxide taken in cannot immediately be transformed into sugar as in normal photosynthesis. It is temporarily built into an organic acid, which is broken down to produce sugar during the next day. The acidity of the plant tissues is steadily increased by night owing to this 'crassulacean acid metabolism', as it is called, and reduced to normal by day.

The spines of cacti and succulent spurges almost certainly arose in the first place by the reduction of stems and leaves caused by dry conditions. Many other spiny plants can be shown to become most prickly in dry growth conditions: the common gorse which, if grown with plenty of moisture will have normal leaves, is an example. Some cactus spines have recently been shown actually to act as organs for absorbing water such as dew, although this is probably a secondary adaptation. Equally secondary, though often very effective, is the protection provided by the spines against many animals, even if some do manage to eat into the plants or, like woodpeckers, make holes in them for nesting.

Besides withstanding heat and drought, many cacti will stand considerable cold; many apparently hot deserts may have near-freezing temperatures at night, while a number of species grow in the Andes at over 3,000 m.

Some cacti are very large, like the 15 m, ten-tonne sahuaro of

Arizona, the Old Man of Mexico, and other tall cylindrical forms up to 12 m. Barrel cacti may reach 2 or 3 m tall. In Africa some spurges become trees with a stout trunk and much-branched rounded crown, like *Euphorbia ingens* which reaches over 10 m. It follows that these giant succulents are structurally rigid, and the large ones have woody skeletons.

Succulence is not confined to basically soft plants, as has already been stressed in Chapter 8, in describing some climbers with a large water-storing base from which the annual stems emerge, and the ungainly 'fat trees' and 'bottle trees' with swollen water-holding trunks.

The roots of xerophytes are naturally vitally important to them. In a cactus desert some roots spread horizontally not far below the surface, others go deeply down to a water table, and many plants depending on circumstances combine the two forms. The white broom of the Middle East may have radiating horizontal roots up to 2 m long and vertical ones a metre long. A bean caper covering 2 sq m above ground may have a root system tapping 35 sq m to a considerable depth. Sand-dune plants also have very extensive roots, as an attempt at extracting seedlings of, say, sea holly will demonstrate. An adult sea holly can have roots 3 m deep. Such plants also respond very quickly to being buried in sand by growing upwards.

The baobab, which lives in arid places, has horizontal roots which may spread out for 100 m around the plant, and many desert shrubs and trees have deep-penetrating roots; tamarisk roots, although normally not reaching more than 10 to 15 m, were recorded during the making of the Suez Canal no less than 50 m down.

Xerophytes' roots are also remarkable in that, unlike most normal roots which only produce water-absorbing root hairs near the growing tip, they often sprout root hairs all along the younger roots as soon as the ground is moistened after rain.

In many desert conditions the plants have to cope not only with aridity but high salinity in the soil; salt and other minerals accumulate in such conditions, as those attempting irrigation and soil-less cultivation in deserts have found. A salinity of over 0·5 per cent is harmful to most plants: salt is not only toxic but makes it more difficult for roots to extract water from the soil, because of unsuitable osmotic pressures with a higher concentration of salts outside the roots than within. The high salt concentration in the sap of some succulents, typically cacti, may possibly be created by this situation. The palm *Hyphaene* is characteristic of saline water in Africa, while the tamarisk is capable of thriving in places where the water table contains over 20 per cent of salts; it has glands through which surplus salt is apparently excreted, but this seems inadequate to explain how the shrubs exist in such basically toxic conditions.

The same of course is true of salt-marsh and sea-coast plants in general. Many kinds of plant are adapted to such life, including grasses, thrift, sea plantain, various sub-shrubs; there are also salt-marsh succulents, like sea blite and glasswort, which counter a low rate of assimilating water by storing it in their stems. The mangroves, possibly the only trees that actually grow in salt water, succeed by combining a high salt concentration in their sap with a salt-resistant skin to their roots.

Alkaline deposits and gypsum are two other fairly frequent circumstances which certain plants can overcome. Others can live happily where normally toxic quantities of heavy metals are concentrated in the soil. This can happen where natural ore-bodies are exposed, but it is to be observed largely on old mine workings, some of which are of course many centuries, even thousands of years, old. Grasses seem the most adaptable, putting up with lead, zinc, copper or nickel, and occasionally two together. Sorrels and plantains are other persistent colonizers, while there are strains of thrift and vernal sandwort in Britain and of monkey musk in the United States which are fully resistant to copper. Some species of *Astragalus* collect selenium, one species of catchfly enjoys cobalt, while the madwort *Alyssum bertolonii* and the New Zealand shrub *Hybanthus floribundus* are gluttons for nickel: they can absorb this metal to up to 10 per cent of their dry weight. The presence of known metal-tolerant plants is now being used by observant prospectors! Some plants accumulate gold or silver!

To adapt to a normally toxic metal in the soil is the result of mutations, and experiments have shown that perhaps three or four seedlings in a thousand have an initial tolerance. This quickly builds up to create fully resistant strains within a few years. Some strains actually become dependent for maximum growth on the metal concerned. But while a plant may, say, adapt to a high proportion of pure lead, the lead-resistant strain cannot necessarily face zinc or copper, nor is it likely to flourish in good lead-free soil.

It is interesting that mine workings are often very deficient in normal plant nutrients, notably nitrogen and phosphorus. The metal-tolerant strains are also able to cope with this problem, as well as the habitual dryness of such situations. Apart from colonizing naturally, such strains can of course be used to cover unsightly tips and form the first stage of a new vegetation cover as soil accumulates around them.

Other plants resistant to apparently toxic conditions exist on volcanic margins. An example is *Rhododendron kiusianum* on the Japanese island of Kyushu, which thrives among steam and sulphur fumes.

Algae have shown their resilience to pollution by successfully growing in rivers heavily contaminated with metals, including cadmium. Only some species succeed, and these are being used as indicators of metal pollution.

To live with roots in dry or toxic soil is a major problem. Living without soil at all is another. This is the state of the epiphytes, the plants that cling to tree branches, and sometimes to rocks, without their roots penetrating their host. In a way these can be likened to seaweeds, for the roots are sometimes no more than a holdfast, while the leaves, bathed in a humid atmosphere, absorb water directly. In fact many of the epiphytes, notably orchids, develop fleshy aerial roots which combine holding with absorption from the air.

The typical orchid aerial root is a shining silvery green, the silveriness coming from an exterior layer of air-containing cells, the velamen, which acts as insulation both by air against heat and by reflection against radiation. If the root touches a damp surface it develops white root hairs on the contact side. These are at once anchors and absorbers; in some cases the weight of a developed root system is at least equal to that of the leaves and remainder of the plant. These root hairs absorb whatever humidity is going, almost like a wick, as direct rain, as moisture trickling down the bark, and occasionally as mist. Where water is absorbed from the bark it is likely to contain various minerals derived from decaying leaves and animal excreta. The tree's bark is likely to dry out rapidly after rain, but silt from debris often collects under the centre of the plant and, in some, around the mesh of roots. Moreover, epiphytic mosses on the trunk very often provide a reservoir of moisture.

The stagshorn ferns have two entirely different forms of leaf, one flat, which spreads over the tree's bark like a plate, within which moist debris collects to provide for the plant's root system; the other, shaped like a stag's horns, spread outwards, functioning normally and carrying the spores. In the birds-nest fern the leaves form a conical vase, tightly overlapped at the base, in which rain and debris gather. Absorbing roots from the leaf bases penetrate this material and make use of it.

Many orchids live in areas with a dry season, and when this is upon them the roots harden up, the absorbing tips becoming covered in velamen so that they are protected and insulated.

In many cases roots eventually reach the ground and penetrate it; the same thing happens with climbing plants such as philodendrons and monsteras. At that point the whole root becomes enveloped in a fur of root-hairs, except for the smooth growing tip.

In some orchids the roots can photosynthesize, and are green; in these there is likely to be a much thinner velamen above the chlorophyll-containing cells. There are also some extreme cases where the roots have taken over the normal function of the leaves entirely, the plant being composed only of a radiating array of roots and sometimes a minute woody stem; the flowers arise either from the centre of the root-star or from the stem. In *Campylocentrum* the flowers are tiny and carried on masses of slender spikes; in *Polyrhiza* they are up to 12 cm long and of very extraordinary shape, the somewhat boat-shaped lip ending in a pair of long, curving lobes so that the whole bloom looks rather like a leaping frog. The effect of either is remarkable, and when not in flower almost unbelievable. There is an intermediate form, *Taeniophyllum*, in which the minute stem has scale-like leaves; the 6 mm flowers are carried between these. In this weird plant the aerial roots can be several metres long.

These orchids employ the crassulacean acid metabolism mentioned earlier. The gaseous exchange necessary for photosynthesis in the roots takes place at aeration points called pneumatodes, composed of air-filled cells. It depends not on night and day, as in normal plants with stomata, but on high moisture content. To achieve this, the water-absorbing cells in the velamen are extremely large and sponge-like. The special metabolism allows carbon to be fixed at any time and be stored, in the form of malate, until it can be processed into carbohydrates in daylight.

Another vast group of mainly epiphytic plants are the bromeliads. These bizarre but highly ornamental plants, widely cultivated, with very striking flowers, start off with tough, leathery leaves which resist drought, and their roots are mainly for anchorage, apart from a number of terrestrial kinds which have ordinary feeding roots. In the aerial species water almost always collects in a 'leaf-vase' – indeed, they are often called vase or urn plants, as well as air-pines ('pines' from pineapple, which belongs to the family). The vase holds water effectively, and is often provided with special absorbent cells near the inside leaf-bases, and through these minerals derived from debris accumulating in the vase are also taken into the system.

The most extraordinary bromeliad is perhaps the so-called Spanish moss, *Tillandsia usneoides*. *Usnea* is in fact a group of lichens, and this bromeliad looks exactly like a silvery branched lichen until it produces its small but distinctive flowers. It is, however, very much larger, making festoons of its threadlike growths up to 7 or 8 m long, which drape the Florida swamp cypresses, for example, and can even thrive on telephone wires. The growths – one can hardly call them leaves – are covered in small scales which retain any air moisture going; there are no roots at all.

I have taken a lot of space to describe plants which contend with dryness, for this is one of the hardest conditions for the average plant to challenge. Let me finish with a few comments on water habitats. Water might be considered ideal for plants, but it has its problems. Movement is one; in water rigidity is a mistake, while tensile strength and flexibility are essential. The waves on a rocky coast demand the greatest strength in any plants living there, as well as immensely strong anchorage; this the seaweeds possess in good measure, from the enormous lissom kelps to the flimsiest sea-lettuce. The large palm-like American *Postelsia* is unusual in having very stout but flexible stems from the tops of which hang a cluster of fronds. In fresh water plants tend either to develop very long, ribbon-like leaves or to have finely dissected ones, which not only avoid damage but allow the maximum possible absorption of minerals and oxygen.

Oxygen is as vital to water plants as to terrestrial ones, and most such plants have large spaces for the movement of air within the tissues; there are positive conduits in water-lily stems, for instance. *Jussiaea repens* is a water plant which creates its own miniature submerged jungle. It has long stout stems which grow transversely; they bear leaves above water but below it drop vertical roots and send up vertical shoots which end in a bulbous tip just above water level. These are a kind of reversed root specifically to absorb air and to transmit it via large air spaces to different parts of the plant.

Perhaps the most remarkable adaptation to water is shown by the family *Podostemaceae*. Looking very much like seaweeds, these flowering plants live in tropical and sub-tropical countries, and there appears to be a different species in almost every cataract and rapid they are found in. They attach themselves to stones and rocks by special root-hairs which secrete a glue-like substance. The leaves are very slender and flexible, with great regenerative powers following damage.

During rainy seasons podostemons have leaves only. As the water level falls, they develop flowers, and once the plants are out of the water they flower and fruit within twenty-four hours, pollination being usually by wind or insects. In the heat of the sun the plants soon wither; they slough off the outer layers of cells to leave the vascular bundles only, which become woody. The seeds, shed over the rocks, adhere to them by a mucilaginous seed-coat which quickly dries, leaving the seeds firmly fixed to await the rising of the water in the next rainy season, when they develop into young plants.

Seaweeds are most remarkable in their response to the lessening of light intensity as they grow further and further below the surface. The problem is confused because as the depth increases so the different component colours of white light are absorbed and effectively cut off. Near the surface there is the maximum of all colours, and red is very strong; hence there is a majority of green algae which can absorb the red. At a deeper level red algae take over from the green because green and blue light still filters through, although in ever-decreasing intensity. Diatoms also flourish relatively far down, preferring these colours, while green algal plankton is largely near the surface. Brown algae have somehow overcome the colour problem, for they are to be found all the way down, adapting merely to lessening light intensity. Seaweeds also vary considerably in their tolerance to different concentrations of salt in the water; those that live in pools near the high tide mark may be in high-salt concentrations when seawater evaporates, or virtually fresh water after heavy rain.

In such ways are the marginal areas of the sea colonized, with plants making the most of each set of factors; in the other ways described earlier more difficult areas of the globe are also filled with plants adapted to the circumstances. There are pure salt pans, completely arid deserts, where no plant life is possible; but they are small in area compared with the apparently inhospitable places where some plants have learnt to grow, preferring these conditions because they have become fitted to the purpose.

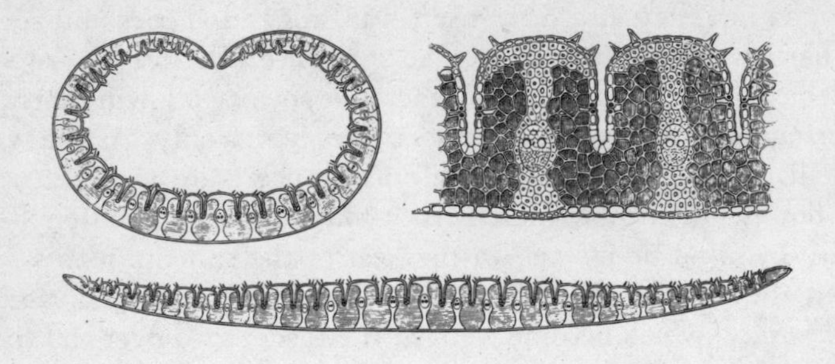

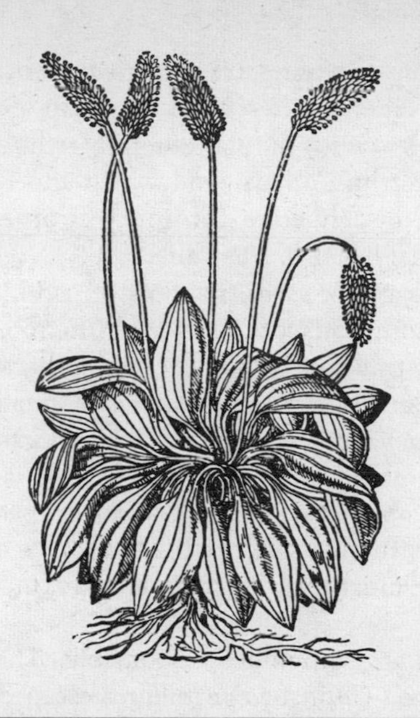

21 · *The Carpet-baggers*

Like carpet-baggers who, as Webster defines them, 'take advantage of unsettled conditions', the plant world includes aggressive, opportunist cosmopolitans – what we call weeds. These vegetable free-booters have more or less escaped the limitations of a distinct set of conditions and are able to live almost anywhere, overall climate being one of the few controlling influences.

Of course a plant is a weed largely by man's definition and is often created by his activities. Weeds frequently develop in the unnatural habitats created by man – by cultivation, burning, road-making, quarrying and so on. Here annuals are characteristically found. Some weeds, such as ground elder or gout weed, which was introduced to Britain as a herb by the Romans, were first used by man and then abandoned; fat hen was once cooked as a vegetable, as have been nettles, docks, thistles, chickweed and many another garden weed. Yet others were transported by man and transformed their character in new surroundings and among different competition. Such introductions might be accidental, as with the plantain in North America which was brought by the Pilgrim Fathers, and which the Red Indians called Englishman's foot; or deliberate, for utility or ornament, as with the blackberry and prickly pear in Australia.

Weeds came in sacks of seeds, in roots, in clothing, in kitbags, on animal fur. The more transport developed, the more the weeds were carried about; they travelled on boots, carts, settlers' wagons, loggers' trucks, railway trains, car tyres, boats, and finally in comfort by aircraft. Their initial centres are likely to be docks, railway sidings, mills, tanneries, timber yards, as well as rubbish and sewage farms; while in countries able to afford such luxuries, people who feed our feathered friends with bird-seed are likely to have a number of weird plants appearing, sometimes to their dismay including the illegal Indian hemp.

Botanic gardens are sometimes responsible. That little composite from South America, *Galinsoga parviflora*, escaped from Kew and in the past century has spread widely in London and southern England. It was called the Kew weed, Joey Hooker, after the Kew Director of the time (who presumably answered inquiries about it), and also, since people could not pronounce its Latin name, gallant soldier.

The pepper relation *Piper aduncum*, a tree growing to 6 m tall, was introduced to a Javanese botanic garden. Its fruit-catkins proved irresistible to the local fruit-eating bats, with the result that it has spread all over western Java, often crowding out the original flora.

The speed at which areas can be colonized is remarkable. Thus the perennial buttonweed, an example of many tropical American weeds, reached the Old World some time after 1900, having been first recorded as a common weed in Java in 1924. Between 1925 and 1929 its existence was confirmed in New Britain, New Guinea, the Philippines, Samoa, Singapore and Sumatra, and by 1945 in Fiji. A glance at a map will show how far apart some of these places are.

To be an effective weed a plant must have both aggressiveness and a marked lack of specialization, so that varied environments and soils can be quickly colonized. Drought, day-length, summer heat and winter cold should not affect its growth and flowering.

If the weed is an annual it will have powers of seeding profusely and over a long period, often starting to do so when it is only two or three weeks old. The seeds will have mechanisms allowing for germination over many years as and when circumstances are suitable. Seed distribution devices, such as the parachutes of composites or the explosive capsules of balsams, are often sophisticated.

There is an apparent contradiction in the fact that most successful annual weeds are self-pollinated and self-fertile. Male and female flowers on separate plants creates variability in the offspring without which they cannot adapt readily to new or difficult circumstances. To self-pollinate, or to produce seeds without fertilization as dandelions do (apomixy), is to deny such possibilities. Weeds, however, must often be independent of pollinating insects, which may not be available in poor weather conditions, and in fact most weeds have innumerable strains, so that a single apparently pure species contains individuals capable of meeting very different challenges. Such strains have arisen in the past as the species became adapted to less specialization, when presumably cross-pollination was the rule whereas now it is only occasional.

Annuals which self-fertilize are capable of recovering much more quickly from a disastrous year which has greatly reduced their numbers. Moreover, self-pollination or at least self-compatibility aids plants in dispersal, for in this way a single specimen can seed and give rise to a new population.

Perennials often dispense with fruiting and rely entirely on vigorous methods of vegetative spread, so that all the individuals in the species may be of the same stock. Even small temperate plants can grow quite fast: creeping buttercup is able to cover 4 sq m in a year, and cinquefoil covers 12 sq m a year with its thin radiating runners a couple of metres long, with a node capable of making roots and shoots every few centimetres. The record for spread must be held by tropical bamboos, with their shoots capable of growing 50 cm a day starting up continuously from the radiating underground roots, an armoury of thrusting spears.

Spread below ground is usually rather slower than above, but it is perhaps more effective. Though ground elder only colonizes about 3 sq m a year, its roots can regenerate from small fragments, so that any assault on the weed is unlikely to succeed. Some weed roots go very deep: field bindweed is recorded at a depth of 7 m and horsetails in light soils two or three times as deep again; the latter may carry regenerative tubers every few centimetres. Creeping sowthistle averages 1 to 2 m radial growth a year, although over 5 m has been recorded; this plant also has efficient parachute-spread seeds.

Weeds with tap roots also regenerate, making several new shoots if decapitated. Docks, plantains, dandelions and the like are like the legendary Hydra, sprouting two heads for every one cut off. The dandelions may indeed be cited as highly developed weeds. Besides their competence at springing up after beheading, they are efficient at seeding. We think of dandelions with their gossamer 'clocks' of parachute seeds, but in many species there are in fact two kinds of seed, the second, which is without a parachute, produced around the rim of the flower-head where a band of red attracts birds to them. The seeds can germinate within three days of wetting, and the plants flower within six months.

One must remember that weeds are not all herbaceous plants: there are many weed trees, like the sycamore, *Prunus serotina*, and a host of tropical species which spring up if primary forest is cut down.

Other methods of weed spread have been described in Chapter 17: the bulbils of *Oxalis*, the ready-rooting, easily detached leaves of stonecrops, the continuous division of water hyacinth, water lettuce, water ferns and duckweeds – the latter can produce 84,000 plants per square metre. These water weeds must indeed be considered as among the most successful of carpet-baggers, now creating a serious threat to waterways, irrigation canals, drainage ditches and paddy fields all over tropical Africa and Asia. Even the lovely Indian lotus and tropical water-lilies are often serious pests, together with a number of plants that live submerged, like tape grass, pondweeds and stoneworts. Such weeds choke waterways, impede navigation, block hydro-electric plants, reduce water flow by up to 80 per cent, deoxygenate and foul water when they die and rot, and compete severely with rice in paddyfields, reducing yields by up to 50 per cent.

Weeds not only fill disturbed places, they compete actively with other plants. Part of this competition comes from a basic struggle for water and nutrients in the soil, where the weed with its wide-ranging root system will often be at an advantage; it also comes from root secretions which actively inhibit root growth, and even germination, in other species. Early research work on this showed that a single oat plant could develop a root system over 100 km long if grown with no weed competition. Surrounded by weeds, an oat could only produce 500 m of roots in a season.

Finally, weeds may compete strongly for light. The large leaves of bindweeds keep light from the plants they scramble over; in the Mediterranean the acanthus prevents anything growing under its luxuriant spreading leaves, and nearer home the butterbur spreads its huge leaves so thickly that they effectively choke any seedling that dares to sprout among them.

Gardeners sometimes define weeds as 'plants in the wrong place', and this is true up to a point. A flower prized in one garden can be a serious pest in another, depending on the growing conditions and especially on the soil, for even weeds have a much harder time in clay soils than in sandy ones. Most people would prize the candelabra primulas introduced from Asia, but I knew one garden where they seeded so profusely that the owner killed them off with sodium chlorate. But in any case weeds are more than just 'in the wrong place'. To the cultivator they can represent great reductions in crops; it is common practice, for instance, to grass down orchards in order to stop fruit trees from growing too vigorously.

To the conservationist weeds often appear as ugly, markedly second-rate plants, reducing the impressiveness and beauty of landscape, reminders of the ever-accelerating scarring of the natural world. Among the few exceptions to this are the annual weeds which make the fields of Morocco or Turkey so lovely as they flower among the crops: larkspurs, cornflowers, poppies, Venus-mirror, gladioli and so on, which people off the breadline are able to cultivate and breed for their gardens, having eradicated them from their own fields. But in principle, if we do not keep a careful watch, weeds could be the inheritors of the earth.

22 · *Living Together*

There are among plants all shades of union and of sharing, ranging from virtually perfect and continuous cohabitation to love–hate relationships and ending up on the borderlines of parasitism when one partner goes too far. (True parasitism is discussed in the next chapter.)

Plants very often form associations with other living things, from related organisms such as bacteria and fungi to a whole range of animals from protozoa to mammals. I mean by this various degrees of literally sharing life rather than the partnerships involved in pollination and seed dispersal, which have been described earlier. Some of these are, however, very close to life-sharing, as in the yucca and fig pollination associations described in Chapter 13.

There are three terms covering life-sharing. The loosest association is *commensalism*, where one organism lives on, in or with another, sharing its food to some degree; in *mutualism* there is mutual advantage to both organisms concerned; finally there is *symbiosis*, in which two or sometimes more organisms are more or less fully integrated in a balanced manner, ideally beneficial to both. In each context the organisms involved are likely to be very different, and in most cases each partner can exist separately.

Some associations are very one-sided, and hardly come into the category of living together. Such are those of epiphytes which simply perch upon trees without detriment to themselves but equally without advantage to the host, collecting decaying plant material around their roots as a substitute for soil. Lianas and rattans use the physical support of erect trees; they and more direct climbers such as ivies and philodendrons with aerial roots may eventually choke the host with their own foliage, or drag it down.

Epiphytes exist in the sea too, and here we can see relationships where either host or epiphyte may find protection from rough seas or from excessive light, although, just as terrestrial climbers can choke or pull down their hosts, marine epiphytes can sometimes grow so profusely that the overloaded host is torn from its attachment.

Certain animals, too, have taken to living in the specialized shelter of plants. There is, for instance, a kind of frog which inhabits the leaf-vases of bromeliads. There it finds enough water for its needs as well as various lesser animals it can eat; the frog's head is shaped like a spiny helmet, so that if danger threatens it clamps itself down in the vase, the skull-helmet giving both protection and camouflage. On a much smaller scale the liverwort *Fraullania* makes small pitcher-shaped structures to store water, which a tiny rotifer often inhabits.

Ants frequently cohabit with plants, notably trees, in some part of which they usually live, although they occasionally fasten external nests to it. Ant nests almost always include soil or vegetable debris which is fashioned to provide a structure. Internal homes include hollow stems, or those with easily scooped-out pith, or large hollow nodes or thorns, which may initially be entered by the ants biting a hole. Sometimes the ants make a complete communications system within such tubular twigs, in other cases exterior covered ways are constructed between hollow ant-houses.

The whistling thorns, of which there are species in Africa and South

America, are acacias with round, hollow structures (stipules) at the base of their thorns. The ants bore into these, empty them of surplus tissue, and inhabit them. When the wind blows across the resultant apertures it makes an eerie whistling sound.

Other kinds of ant make use of leaves to create a nest, sometimes forcing the two surfaces of a leaf apart as if it were a paper bag; inhabiting the ready-made inflated sac to be found in some plants between ribs at the leaf-base; or glueing leaf-edges together to form a tubular housing site. Others make simple nests between the large spathes enveloping the flower-heads of rattans. Some species of wax vine grow with stems and leaves very closely pressed to tree-trunks, and in the pockets thus created colonies of ants are frequently found. Yet others inhabit the galleries formed within the paired, overlapping rings of spines of certain palms. One orchid, *Schomburgkia*, has long hollow pseudo-bulbs as storage organs which ants enter by a hole at the base. Ants also excavate the swollen, tubular, water-storing rhizomes of one or two tropical epiphytic ferns.

Although it is quite clear that the ants are very largely opportunists in finding suitable structures, the trees do frequently benefit, especially when it comes to defence. In effect an ant-inhabited tree maintains a standing army. Any attack on it results in myriads of the insects rushing forth, as many a botanist ruefully records. I quote from E. J. H. Corner another of his superb passages of observation:

> We pause in the excitement of a first expedition through the sweating glades . . . and a faint rustle fades into the trees. It comes again and vanishes as widely. We think of snakes, wasps, bees, and look around, but can discern no obvious source of the mystery. Yet now it is repeated right, left, and all sides as a series of tiny rattles through the bushes into the canopy. An uncanny feeling shivers over our clammy bodies that we are watched by a host of invisible onlookers. But is this snake underfoot any more than the trailing stem of an old rattan? We bend down with relief to examine it and, just as the sound commences, a small ant at a hole in an ocrea [leaf-sheath] beats its mandibles on the dry cover. Instantly a rattle comes from the box, to be taken up by the next and the next along the stem into the trees . . . The sentry has alerted the soldiers and the alarm is transmitted from post to post as a sort of 'action stations' to repel boarders up to the crown in the tree-tops. We lift the stem and out rush the black ants with ferocious bites.

Ants may keep off many harmful insects, caterpillars, snails, and other animals, perhaps even grazing mammals, and in particular their leaf-cutting relations which can so rapidly denude a plant.

Apart from providing a home for the ants safely above the possible flood-level in many jungles, the tree may also provide pastures for aphids and scale insects which the ants herd and 'milk'. In some extraordinary cases the tree actually provides ant-food itself, which one might liken to the protection money handed out to a gang in some human situations. In the trumpet tree these gift 'food-bodies' grow on the lower side of the leaf-stalk. These are effectively insect-mimics for the carnivorous ants which live in the tree's hollow, compartmented stems. Besides eating these bodies, which are constantly replaced, the ants pasture sap-sucking mealybugs within the stems, a further drain on the tree's resources. But the ants keep the trees free of climbing plants to which for some reason they are exceptionally prone, and probably protect them from herbivores such as aphids and caterpillars; so that this form of cohabitation provides a good example of mutualism.

In the whistling thorns the ends of the leaves grow a small sausage-shaped food-body, and there are also nectaries on the leaf-stalks. Similar nectaries are produced by species of *Macaranga*; these are brought by the adult ants to the larvae deposited in the hollow stems. This is again good mutualism: the hollow stems give ant-shelter, food for the occupiers, and useful protection by the ants against caterpillar attacks to which these trees are liable.

These associations must be long standing because trumpet trees and whistling thorns differ from their closest non-myrmecophilous relatives, the former in having no latex and the thorns in having less chemical anti-insect protection in their sap.

One odd ant-association is with epiphytic plants of the genus *Dischidia*. This has two kinds of leaf, one small, rounded and succulent, the other larger, forming a closed hollow 'bag', round or elongated, within which roots develop, thriving in condensed moisture. Ants frequently inhabit these ready-made shelters; they bring in debris which, together with their excreta, forms humus from which the plants derive nourishment. The ants feed on the plant's seeds, and in taking these some are dropped and can germinate. But they also sometimes pasture scale insects within the pitchers, so there is a lot of give and take in the situation.

There are at least 200 species of plants which rely heavily on ant excreta in various ways, and can be said to be 'ant-fed'. Like the insectivores of Chapter 24, they mostly live in environments where nutrients are low or difficult to absorb; most ant-fed plants grow as epiphytes in open scrub or forest where water is not readily available.

Ants are in any case frequently responsible for collecting plant debris into which epiphytes can push their roots, the latter providing a rigid structure to the 'flower-pot nest'. Apart from the provision of a root-run some plants seem actively to need the company of ants. The bucket orchid is very difficult to cultivate successfully in the absence of its ants, presumably because the latter contribute some essential element. If such an orchid is placed in a greenhouse any ants around will colonize its roots, so possibly they obtain something from the plant also. Such orchids grow naturally in ants' nests in the tops of trees, 'a conspicuous element', to quote P. H. Allen in *Flora of Panama*, 'in the unique arboreal myrmecophilous gardens . . . the association often including a purple or orange-flowered Epidendrum and several apparently specialized, succulent-leaved non-orchidaceous plants, among the most frequent being Peperomias and members of the Gesneriaceae'. The orchids 'are well protected by the belligerent ants and are painful subjects to collect and still more painful to transport'. The ants appear to start their nests by cementing debris together; they bring in the seeds of the plants concerned to eat the edible 'gift' attached to them, as described in Chapter 15; the seeds then germinate and more food-rich, moisture-retaining material rapidly accumulates around the roots, so that formic hanging gardens are soon created.

Gesneriads themselves often live in solo association with ants, species of *Codonanthe* being the commonest. The seedlings begin life in some small pocket of debris in a bark fissure. Ants immediately start building a nest around the seedling, and this provides it with additional humus. In due course the plant's seeds fall into the ant-nest or garden and germinate there; sometimes again the ants appear to carry the seeds into the debris. This they do also with a variety of other plants, so that they clearly have a very positive interest in creating ant-gardens.

There may even be two or three species of ant living in the same garden.

Another collector, G. T. Prance, records like P. H. Allen how these ants emerge and attack the instant one of their plants is touched: 'We have to snatch our specimens rapidly before the ants have time to crawl on to them.' He also says that he has never seen *Codonanthe calcarata* outside an ant-garden.

We can see in our own gardens ants acting in defence when fed. The cherry laurel and the Banksian rose, for example, have nectaries on the stems; these attract ants which actively keep away leaf-eaters such as caterpillars. Several European knapweeds secrete honey from special glands on the scales outside the unopened flower. A dozen ants at a time may be seen on such heads, and these direct formic acid sprays at certain large beetles which would otherwise eat into the flower-heads.

Rather similar activity occurs with the stagshorn ferns, in which large, flat fronds clasp the tree-trunk, while more normal fronds are also produced. Debris accumulates within the clasping frond-basket and the fern's roots make use of it. Much of this debris and soil may be brought there by ants, which then nest in the 'basket' in great numbers. Similar material collects and is used as ant-nests in the birds-nest ferns, whose fronds form an upright vase-like receptable.

There are a number of other 'air plants' on the upper branches of tropical trees which provide ant-shelter, in particular the genera *Myrmecodia* and *Hydnophytum*, both belonging to the coffee family. These plants have large often spiny tubers containing a labyrinth of cavities, apparently for aeration, which are positive ant-palaces. *Myrmecodia*, which looks rather like a hedgehog incongruously sprouting short, leafy stems, has ant-transported seeds and also provides nectar for its inhabitants.

A tropical tree called *Goniothalamus ridleyi* produces most of its flowers at the base of the trunk. Small ants usually pile fine soil over these to cover them entirely while they are still in bud; they obviously extract the nectar within the flowers, for any which open too high up the tree to be covered up are always swarming with the ants. Since the basal flowers are inaccessible to any other insect, it is assumed that the ants are responsible for their pollination.

Apart from the ferns already mentioned myrmecophily and its

various forms of mutualism are entirely restricted to flowering plants.

As well as finding shelter within plants, either by 'squatting' or by abetting plants to make 'flower-pot nests', and perhaps even by carrying seeds to increase the size of these, ants also cultivate plants for food. These are the leaf-cutting ants of tropical America, which can defoliate large trees overnight, and may travel hundreds of metres for the desired species. The leaf fragments, up to 2 cm across, are brought into special chambers in the nests: one observer described how the ants marched in 'like Sunday-school children carrying banners'. The fungus chambers, up to a metre long and 30 cm wide and high, may be as much as 5 m below ground, the total ants' nest spreading over 100 sq m or more. In the chambers the leaf fragments are chewed up and, with other vegetable debris and ant excrement, a hot-bed is created upon which a special fungus grows, in principle exactly like a modern mushroom-grower's beds. The worker ants actively tend these fungi, weeding out any alien growths that may appear, and transplant the fungi to new chambers as these are prepared. Something in the ants' saliva inhibits the growth of unwanted fungi and may also promote the wanted fungus's growth. This fungus, which forms the major food of these ants, is a small subterranean species which has never been seen outside the ants' nests, although this may well be simply because of the problem of knowing where to look. Neither is it known whether the fungus lives naturally in the soil and was first deliberately brought in by the ants, or grew in their nests by accident and was then cultivated.

Termites also have fungus gardens but the evidence for deliberate cultivation is less clear. The fungus concerned is of mushroom type, with a fruit-body having stem and cap. After developing as thread-like mycelium in the chambers of the termite nest the stem eventually appears, to push its way forcibly right up through the nest, often a metre or two tall, with the aid of a very hard and sometimes nipple-shaped top. Once it emerges, the top deploys into a cap which can be up to 60 cm across, looking for all the world like an umbrella over the termite mound.

The termites swarm on to this cap and bring fragments of it into the nest where it forms a major part of their food. Naturally spores are dropped within the nest where many germinate, produce mycelium and restart the cycle.

Periodically the termites throw out masses of the mycelium, presumably because they become too large and block passages. One of the mysteries of nature is that the fruit-bodies these rejected mycelial masses produce outside the termite nest are always very small, on quite a different scale from the giant 'umbrellas'. This may be due to specially favourable conditions for the fungus within the nests, to material provided by the termites or possibly to active cultivation procedures.

Apart from ant-relationships, and of course those concerned with pollination, plants and insects are not normally very cordial. In Chapter 18 I outlined how insects may feed upon plants and how they may also lay their eggs in plant tissues, to which the plants respond by creating galls.

Plants do, however, have successful relationships of various kinds with many other animals. There are several hundred known associations of aquatic invertebrates with one-celled algae, some of which are perfect examples of symbiosis. These include a marine worm (*Convoluta roscoffensis*), a rhizopod (*Paulinella chromatophora*), and a flagellate (*Cyanophora paradoxa*), for each of which the alga provides carbon and probably nitrogen. In the first two there is indeed no feeding mechanism, while in *Cyanophora* the algal cells cannot exist independently, having lost their walls, and moreover the two partners reproduce simultaneously. The algae thus resemble large chloroplasts, and are to be found in many protozoa; but whether these are symbiotic associations or a half-way stage between the animal and vegetable kingdoms is uncertain.

Besides food production, the algal or chloroplast component usually makes these organisms light-sensitive. This continues in algal associations with coral polyps, sea anemones, sponges and jellyfish where the algae utilize the animals' waste products in exchange for carbohydrates. Light sensitivity is used by these organisms to balance the feeding needs of each partner, although it appears that the alga's use of the animal's excretions has merely replaced its normal utilization of seawater, and the animal is by no means dependent on the alga.

One of the more extraordinary symbiotic developments is between the giant clam *Tridacna* and various kinds of microscopic algae, which colour its mantle in bright patches of orange and purple. The mantle has a row of blue 'eyes' which puzzled biologists until it was discovered that these were not literally eyes but lenses to focus light on to colonies of algae within the tissues, so that they could photosynthesize to best advantage! Excess populations of the algae are digested by the clam.

Certain fish carry algae on their gills, where they receive maximum aeration; another odd association is between a large sloth and an alga which coats its hairs. The alga is provided with an intriguing mobile habitation; perhaps the sloth obtains some small value from camouflage.

There are a number of linkings between lower and higher plants, in which typically the blue-green alga *Nostoc* is the lowly partner. This may inhabit pockets on the higher plant, or its filamentous growth may actually penetrate the roots. The alga fixes atmospheric nitrogen, which benefits the other plant because the *Nostoc* retains its colour and its fixing capacity even underground. Plants associated with it include liverworts, mosses, floating ferns, duckweeds, aroids, the giant-leaved *Gunnera* and many of the antique cycads, and it is economically important in rice-growing. Nitrogen fixation is more typically associated with a bacterium called *Rhizobium* which forms nodules on the roots of members of the pea family (*Leguminosae*) in particular, but also some others. Nitrogen fixation is also carried out by certain fungi associating with plants such as alder, sea buckthorn and sweet gale. Nodules formed on such permanent woody plants are often very large and clustered, although they do not last indefinitely.

It appears that the bacterial or fungal cells enter in a superficially parasitic way which stimulates the roots to make nodules, just as various organisms make galls appear. The bacteria are apparently unable to fix nitrogen when free-living; they have to be within a nodule, in which the plant contributes something towards their existence. There are, however, other bacteria (*Spirillum*) which do not form nodules, but live in close association with plant roots, notably of tropical grasses.

Nitrogen is very inert and reluctant to form compounds in nature; a good deal of energy is needed to tear apart the two atoms in its molecule. In nitrogen fixation the chemistry involves the combination of free nitrogen to create ammonia. The hydrogen may be produced by the micro-organism or be available in readily reducible form. The trace element molybdenum is involved, possibly as a catalyst. The red pigment within leguminous nodules is, incidentally, a protein closely related to the haemoglobin in red blood corpuscles.

Bacterial leaf nodules occur mainly in the madder family, *Rubiaceae*, where they exert a remarkable influence on the plant concerned, for without them the latter becomes dwarfed. Here the plant can only grow 'normally' when involved in the symbiosis; alone, it appears quite abnormal.

The total amount of atmospheric nitrogen fixed by these nodule-making bacteria, mainly those on leguminous plants, has been assessed at 100 million tonnes a year. To quote George D. Scott:

> It is . . . a sobering thought that, but for the advent of the leguminous plants and the symbiotic association between them and *Rhizobium*, the present-day status of soil nitrogen would be very low – much too low to support anything like the present world cover of natural vegetation, let alone the vast areas under intensive cultivation.

A good crop of suitable legumes can fix 250 kg of nitrogen per hectare during a season, which is roughly the equivalent of 2½ tonnes of fertilizer.

Fungi are especially important in partnerships. In the soil they and bacteria form all kinds of associations; and although these are casual, both organisms involved benefit greatly; the life of the plant is enhanced since many bacteria lack important growth substances which the fungi can provide. This is the beginning of symbiosis.

Perhaps the most remarkable fungus link-up is that with algae which produces lichens. In the more elementary kinds the algal cells or filaments are simply mingled with fungal strands, but in more advanced examples the arrangement is remarkably like that of a leaf in a higher plant, with a layer of photosynthetic algal cells near the upper surface, above a mass of fungal filaments. Sometimes the thallus of the lichen, as its structure is known, resembles a higher leaf in outline as well, as in *Sticta filicina* which adds to the resemblance by having air pores which can be likened to stomata. Besides being leaf- or scale-like, lichens can form thin crusts, resemble mosses, grow upright, or branch in complex ways.

Although both partners can be found independently, it seems likely that the alga concerned is usually unable to exist comfortably on its own. Fungus and alga can meet accidentally, and often do, but lichens also produce soredia, little pustules containing tissues of both partners which are readily disconnected from the main thallus. In either case the new lichen which starts growing is a mixture of cells, but in most lichens the arrangement already described soon develops. The fungus continues to reproduce as it would on its own, the combined plant forming fruiting points for this purpose, but the alga does not reproduce sexually when living with its colourless associate.

In this symbiosis the alga receives protection and basic food substances, the fungus carbohydrates. Lichens appear also to be able to extract minerals from the rocks on which they often live, as some fungi can on their own. As we have already seen, lichens are remarkably resistant to very difficult conditions, standing extreme heat, cold and desiccation. The bushy *Ramalina* from the Negev desert can tolerate over 80°C when dry and continues to photosynthesize when partly frozen at −10°C.

The complete integration of the two different partners is most clearly shown by the lichen's reaction to light and moisture. In dry conditions their surface layer contracts, and so do the algal and fungal cells beneath, so that less light reaches each individual cell. If the plant is moistened, the whole expands, exposing more of each cell, and in particular the photosynthetic algal cells, to the light. If one wets a dry grey lichen it can be seen how the surface rapidly becomes green due to this process. Lichens can take up free water at an amazing rate, absorbing over 50 per cent of their initial dry weight in ten minutes.

Many lichens are strongly coloured in yellow, orange, black and so on. These tints, long used by man for dyes, are due to a heavy deposit of coloured lichen acids in the upper surface cells. This occurs in lichens continuously exposed to sunlight: in shade, the growth is green. In this way the harmful effects of too much light are filtered out. It can also be observed that if a lichen becomes shaded after exposure to light it will change colour from orange or yellow to green, because the pigments are re-absorbed. This offsets the lichen's lack of light-sensitivity to control actual growth, or of movement to or from light.

The ultimate in lichens is *Sterocaulon*, in which a blue-green alga associates with the basic alga–fungus combination, producing curious surface growths which, it has been suggested, are capable of nitrogen fixation. Some authorities, indeed, consider that most lichens are associated with nitrogen-fixing bacteria, and cannot perhaps begin to exist without their presence.

These lichens are not only examples of a very remarkable and successful liaison between two very different life-forms which can live for several centuries, but are important in the setting of world vegetation. I have already noted how they are often the pioneers in very hard climatic conditions, carrying out the disintegration of rocks which is the first step towards soil. They are also important as food in the northern hemisphere, where reindeer and caribou subsist very largely on so-called reindeer moss – in severe conditions such lichens provide 95 per cent of their total diet. The Laplanders carry out controlled grazing of reindeer moss with their herds, and sometimes harvest it as a food reserve for their animals.

Fungi not only associate with algae, but combine with the roots of an astonishing range of higher plants, from ferns and liverworts to forest trees.

They do this in two ways, outside and inside the roots. The word mycorrhiza, meaning simply fungus-root, is used for the mainly external bonding, in which a sheath of fungus tissue surrounds the tips of the tree roots and also forms a network between the cells of the outer root layers. This mantle is only about a centimetre long, but its felty texture enables it to absorb water, and it effectively replaces the root-hairs which would otherwise do this. Mycorrhizal roots are always shorter and more branched than 'normal' ones; sometimes they look like masses of coral. In cycads these coral-like mycorrhiza occur on abnormal roots which grow upwards, developing at the soil surface; they are inhabited both by fungi and blue-green algae, and exist in parallel with ordinary uninfected roots.

Many fungi of 'toadstool' type associate with trees, notably with conifers, and are important, especially in the early stages, in aiding their growth; seedlings grown in sterile soil remain quite stunted compared with others allowed to form links with fungi. Although a few tree/fungus combinations are unique, a great many different fungi can combine with a great many different trees – the Scots pine, for instance, has had 119 fungus partners recorded – the results being essentially identical. Sometimes one tree can combine with six or seven fungi simultaneously, and it is common for one fungus to replace another in mycorrhizal linkage during the life of the tree.

In these combinations the fungus stimulates root production and enables the roots to absorb more food from the soil, while receiving carbohydrates and sugar compounds from the tree. Some have been shown to repel other, harmful fungi and prevent them infecting the trees, acting in fact as antibiotics. Some fungi are unable to produce spore-carrying fruit-bodies (i.e. 'toadstools') if severed from their tree partner. It is especially in poor soils that mycorrhiza are so important, and in the light of the great forestry schemes, mainly of conifers, on which part of our modern economy depends – largely in the production of ephemeral newsprint and perhaps less ephemeral literature – one can see just how important mycorrhizal fungi are to modern man.

Saprophytic plants are entirely dependent on mycorrhiza. These plants have no chlorophyll, and in that respect resemble their fungus partners, although otherwise they are higher plants. One such is the birds-nest or Indian pipe (names referring respectively to the knotted roots and the form of the above-ground growth). These plants rely entirely on a fungus for their carbon uptake. Indeed, the fungus seems to obtain little from this relationship. Birds-nests are plants of woods and forests, and the most remarkable thing about them is that their fungus relationship is shared with trees. The fungus establishes its two-way food communications with tree roots, and the carbohydrates obtained by the birds-nest seem largely to be obtained from the tree. 'Labelled' glucose and phosphates have been injected into trees and recognized in birds-nest plants in the vicinity.

Several saprophytic orchids such as the birds-nest orchid and the coral-root have similar coral-like roots and entire dependence on fungus associations. The coral-root has no true roots at all. It seems likely, though it is not certain, that they may also share their fungus with neighbouring trees.

One of the most remarkable orchids known is certainly saprophytic;

it has to be, because it is entirely subterranean, never seeing the light of day. This is the Australian *Rhizanthella gardneri*, whose flowers develop, mature and produce seed about 2 cm below ground; tiny flies which burrow in soil are the probable pollinators. Only after fertilization are the seedheads pushed up with the soil, presumably to ensure adequate seed distribution. The orchid has associations, which may possibly be semi-parasitic, on roots of various trees and shrubs.

It may be mentioned that our native birds-nest orchid can on occasion develop flowers and produce mature seeds underground, usually because of obstruction, but this is unusual.

Among the most spectacular saprophytic orchids are species of *Galeola*. These make climbing stems up to 30 m long, and their huge flower-heads may contain several thousand smallish blooms. Such examples of saprophytism in higher plants help to vindicate the theory that fungi are equally plants which have lost their chlorophyll.

But it is the internal or endogenous fungi associated with most orchids which give us one of the more curious combinations: a veritable love–hate relationship. A fungus called *Rhizoctonia* is the usual partner. I have already mentioned how simple the orchid seed is, a seed which can neither germinate nor grow unless provided with external food and vitamins. It only germinates successfully in nature, in the presence of the fungus which provides these essentials. (It is possible to germinate orchid seeds in artificial conditions by placing them on a jelly containing sugar and mineral nutrients.)

The process that follows is mutually aggressive. The fungus penetrates the orchid root and its filaments grow alongside the orchid cells, which provide a limited amount of basic foods in exchange for carbohydrates. In concentric zones, however, the fungus filaments are digested by the orchid's cells, and, in those orchids which can finally exist alone when their photosynthesis provides all the energy they need, the fungus is finally completely repelled by this process. Before this occurs, it sometimes happens that the orchid actually reverses the food flow and provides the fungus with carbohydrates. Some orchids, notably the saprophytic ones, continue the relationship for their lifetime.

Other fungi sometimes link with orchids, and one of the odder partnerships is between the honey fungus, a dreaded tree-killer, and the orchid *Gastrodia elata.* The tubers of the orchid cannot start growing unless the fungus attacks it in a parasitic way. The tables are then turned, the plant absorbing energy from the fungus to make its flower-shoot.

Just as the honey fungus leads a double life, on the one hand stimulating orchids and on the other killing trees, the race of *Rhizoctonia* which inhabits the ground orchid *Dactylorchis purpurella* and provides it with carbohydrates turns, under the name *Corticium solani*, in its adult or sexual stage, into a serious disease of crop plants such as sugar beet and potato by actively removing carbohydrates from the host. On the whole fungus relationships with orchids are charitable, for the fungus gets little out of it; one might say that the fungus gets its own back on occasion by turning on other plants – a Jekyll and Hyde approach to life.

One can see from these two examples how delicately balanced a symbiotic association can be, and how near to parasitism (which has indeed been called 'the most exquisite example of symbiosis'). One might regard invasion of root tissues by fungus filaments in mycorrhiza as attempted parasitism which has been balanced by who knows how many millennia of attack and repulse and final treaty.

Another aspect of this is shown by a fungus called *Endogene* which inhabits the root systems of perhaps every type of green plant, or certainly a very high proportion of them, and lives entirely within the plant tissues. Its filaments spread throughout the tissues, making reproductive bodies, little bush-like groups of 'roots' for absorption as well as little sacs in which oil is stored. The fungus undoubtedly assists the plant to take up certain nutrients from the soil, notably phosphates, which the roots cannot always get at, and it may well be extremely important in obtaining high crop yields and good growth. However, the balance is not well understood; clearly the fungus obtains nourishment from the plant as well. The final oddity of this relationship is that the fungus is almost always in time digested by its host – a form of symbiosis that has not entirely come off.

Finally one must mention that leaf surfaces carry often vast populations of microscopic fungi, yeasts and bacteria, apparently existing non-parasitically, some of which fix atmospheric nitrogen (there are such nitrogen-fixing bacteria on the leaves of Douglas firs) while others compete with parasites to the plant's advantage. Also, on occasion, they attack the waxy skin of the leaf which, by increasing loss of water vapour, may have a harmful effect or cause leaves to age. One can hardly call this superficial micro-flora symbiotic, but it is an association which probably exists everywhere and may have more to do with plants' well-being than we appreciate. There is one particular case in which it is certainly beneficial. This has been discovered in a grass from the Ivory Coast, where it was found that, contrary to expectations, the pasture land accumulated nitrogen. Carbohydrates appear to exude from the leaves, which are wet most of the time owing to the climatic conditions, and run down to form a mucilaginous concentrate in the leaf sheaths. In this mucilage live colonies of bacteria and fungi which between them fix nitrogen in useful quantities.

In some ways, perhaps, considering the interdependence of organisms throughout nature, symbiosis is not unexpected. It starts by casual rubbing together, like people in a pub who may find something in common and perhaps some mutual advantage arising from it. Specific friendships whose pattern is constant then follow. Such a perfect combination as a lichen, however, is in any terms a remarkable achievement, especially when the result imitates so well the form and activity of the leaf of a higher plant from which it is so far removed in evolutionary terms. The simpler plant–animal symbionts with protozoa and coelenterates are again, in their own way, perfect solutions to a problem which normal evolution has not solved. Such achievements, to quote George D. Scott, 'clearly vindicate the assertion that biological advantage accrues from the sharing of life with a companion organism'. It might be a text for marriage.

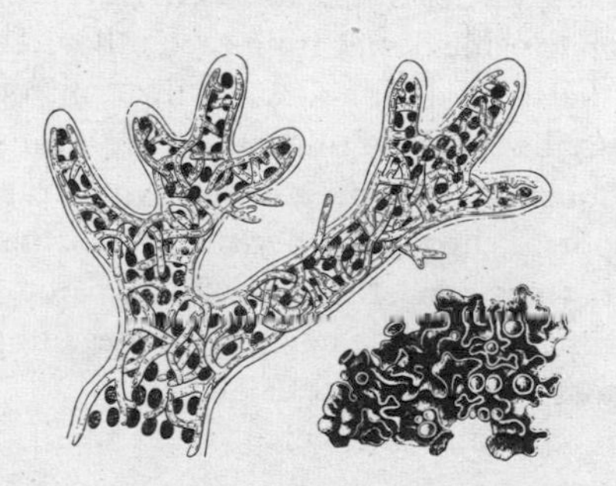

23 · *Stranglers, Parasites and Spongers*

In living together, one of a pair all too often dominates the other. Battening on to another organism is one of the standard solutions in both plant and animal kingdoms, a habitual way of life which raises mixed feelings in us, its observers and sometimes sufferers.

Plants are often capable of harming each other. This may simply be by clumsiness; the struggle for survival in terms of light causes, as I have already described, plants to grow epiphytically on others, or to thrust thorny shoots or entwining lianas, which can eventually bring down their supports, through other plants. Lichens on the bark of shrubs and even trees can stunt and finally stop growth, as many gardeners have discovered; the hanging drapes of Spanish moss must slow down the host's photosynthesis.

Ivy is often regarded as a killer of trees in Britain, although the evidence shows that it only overgrows and chokes elderly trees which are in any case on their way out. Many tropical climbers are equally vigorous and can have similar choking effects on the crown of a tree. The one group which really takes advantage of its position is the strangler figs. Their seeds are dropped into tree crowns by bats, birds, monkeys or squirrels. At first the resultant young fig grows purely as an epiphyte in debris lodged in a crevice. Unlike the small trees one may see in similar situations in Britain, however, the figs produce roots which cling to the branches, joining together to form a basketwork. As it gains size, the young fig sends down vertical roots which eventually reach the soil. Once this has happened the host is doomed, for the fig rapidly gains in strength, sending down further roots and forming an interlacing network around the tree. The roots increase in size and eventually prevent the host from expanding at all, thus effectively crushing it to death like a vegetable boa constrictor. Strangled at the trunk, choked at the crown, the host gradually fades out and dies. When finally its trunk decays the strangler stands alone, its roots forming a cylinder up to 30 m high. It has been estimated that, with a big forest tree, the strangulation process might take a century. One of the typical stranglers is *Ficus benjamina,* which, with its delicate drooping growth and willow-like leaves, is popular as a house plant.

It is interesting to note that if a strangler fig germinates on the ground and not in the crown of a tree it never forms a tall trunk nor has any strangling tendencies.

Instead of growing as relatively harmless epiphytes, the strangler figs – and a few unrelated trees like clusias – have become killers. We can presume that there was some such development from harmless to harmful, from independence to dependence, in the parasites, because they are plants which derive some or all of their nourishment from another. There are at least eight distinct groups of plant families which have become parasites, and this means that this solution to the living problem has been successfully adopted at least eight times (and tried how many more?) in the plant kingdom. Many families and genera are involved, and sometimes great numbers of species – 700 mistletoes alone – adapted to all kinds of circumstances.

Within all these families there are not surprisingly a great many methods and degrees of parasitism. There are semi-parasites, the spongers of the plant world, with their own root systems and green leaves, and others with leaves but a root system fully transformed to sucking sap from their host. Some parasites have scale-like leaves and even no green chlorophyll, so that they are entirely dependent on the host. Finally comes the stage when the parasite lives exclusively within the tissues of its host, only emerging to flower and fruit.

Parasitic plants can be trees or herbaceous perennials or annuals; those that live inside the host may transform it in various ways, notably in the form of witches' brooms, in which the growth becomes twiggy and bunched; or they may be invisible except when flowering, when their blooms appear weirdly through the bark or pop up through the soil from the roots. The flowers, sometimes very beautiful or striking, are frequently enormous compared with the host's roots, as in most broomrapes and in *Rafflesia*, which has the biggest flower in the world; one might compare them with the cuckoo which usually grows far larger than its foster-parents.

The pollination of parasitic plants is not markedly related to their predatory habits. Their seed dispersal, however, is highly specialized, and I have chosen to deal with it in this chapter for that reason.

Most of the mistletoes are dependent on birds. The dwarf *Arceuthobium* has, however, emancipated itself from this dependence, and has an explosive mechanism not unlike that of the squirting cucumber, its vertically positioned fruits shooting 3 mm seeds up to 15 m away at a speed of 24 m per second. Since the fruit is only 4 mm long the internal pressure built up before ejection is enormous. Like most mistletoe seeds they are coated with an extremely sticky substance, and if they touch anything in their flight, very often a leaf or twig of the host tree, they adhere to it. The seeds can slip down a leaf until they lodge between leaf and twig because at first they are slippery. They soon dry, however, and the sticky substance sets hard like glue, making sure that the seeds remain in position ready to germinate in moist weather.

The stickiness of the seeds just mentioned is the important factor with the bird-dependent species. So sticky are they that bird-lime is still made from them. To quote Job Kuijt, a major authority on parasitism, 'the irony of a bird being caught by the fruits of a plant whose dispersal he carries out did not escape the Ancients, who already used birdlime. The saying *Turdus ipse sibi cacat malum* ("The thrush prepares his own misfortune") is ascribed to the Roman Plautus.' Thrushes, waxwings and many other fruit-eating birds like mistletoe berries for their exterior pulp (so do martens and fruit-bats in certain places). In some cases the birds place the seeds on twigs in wiping their beaks free of the sticky pulp in which some seeds may be caught up. But in most cases the seeds pass through the birds, and are placed on branches in the droppings. The German word for manure, *mist*, is so similar to that for mistletoe, *mistel*, that the latter must be based on early observation of this fact.

The speed at which seeds pass through the birds' alimentary canal is remarkable, suggesting that too long a period under digestion may harm the seed (as it does indeed in fowls and pigeons where the action of the gizzard, which so often contains grit, usually grinds the seeds to nothing). In a thrush the seed passes through in about thirty minutes; in species of *Dicaeum* the periods observed vary from twelve minutes to as little as four! The dependence of species of *Dicaeum* on mistletoe flower nectar has already been described. When the berries appear these birds eat nothing else, and there are species of *Euphonia* and of silky flycatcher which do the same. It is clear that the berries have a very pronounced effect on the bird, for it usually sits quietly after feeding and then, at least in the case of *Dicaeum*, presses its anus to the branch it is perched on and rocks itself along for two or three feet, literally pasting the excreted seeds on to the bark.

Euphonia has a digestive tract quite different from that of its relations, clearly designed without internal constrictions to allow the passage of mistletoe berries, on which it feeds exclusively. *Dicaeum* has developed a dual digestive mechanism enabling it to eat both mistletoe berries and insects. Animal food is forced into a gizzard through an aperture too small for mistletoe seeds, and the hard unwanted parts of the insects are rejected in pellet form as with owls. Only then does a sphincter muscle below the gizzard allow the now liquid remnants to pass into the digestive tract proper. But when mistletoe berries are swallowed, the sphincter muscle does not close up at all, and the berries pass straight into the digestive system. Once again one can only marvel at the evolutionary process involved; of the processes involving seed dispersal this one is the most specialized.

Other parasitic plant seeds are eaten by all manner of animals, but are voided like other edible-fruited seeds and take their chance of finding a suitable place to germinate which of course involves the presence of a host. Many are brightly coloured to attract birds, and some of these, in the sandalwood family, have fibres or bristles attached to the seeds themselves which are almost certainly means of attachment to host-plant branches comparable to the 'glue' of mistletoes. The weird *Hydnoraceae* produce fleshy fruits containing starch in which the seeds are embedded, and these are eaten by all kinds of animals: porcupines, armadillos, foxes, jackals, baboons and many 'primitive' humans. Rats may eat the fruits of *Pilostyles*, a small-flowered relation of *Rafflesia*, and the seeds of this latter giant flower itself are probably mainly dispersed by ants, although it has been suggested that they may also be transported on the feet of wild pigs and even of elephants investigating the rotting pulp of the flower; although nothing is certain in this case, these are the only seeds where such a size range in dispersal agent is conceivable. Ants indeed transport many parasite seeds, usually because they contain starch and oils in an elaiosome as described earlier, although one naturalist (Lundströn, in 1887) suggested that ants carried the seeds of cowwheats because they resembled their pupae.

One ingenious way of ensuring that the parasite seed has a host on hand is that adopted by the North American broomrape relation *Orthocarpus*. The seeds of the cats-ear host have spreading bristles at one end and those of the parasite have a net-like outer coat well separated from the seed. The bristles of the host seed frequently grow through this net so that the two seeds are dispersed together and can therefore germinate simultaneously.

Many relations of *Orthocarpus* have pitted or sculptured seed coats, very elegant when viewed under a microscope; these pits appear to trap air when the rest of the seed is wet in rain, thus keeping the seeds on the surface of water runnels which sweep them into cracks in the soil – and into such cracks they certainly go, as witness the finding of *Striga* seeds 150 cm down in sandy soil. This family tends to produce a very large number of small seeds, rather as orchids do; *Aeginetia* has up to 70,000 seeds per capsule and a single plant of *Boschniakia* has been calculated to produce about a third of a million seeds.

Some though not all seeds of root parasites not surprisingly respond to substances exuded naturally by the roots of their hosts, like the seventeen amino-acids known to be released by tomato roots. In one remarkable experiment the immersion for only thirty seconds of broomrape seeds in an extract from a potential host's roots resulted in 60 per cent germination. It is likely that in most cases of this sort germination depends on a suitable period of exposure to root products and their concentrations.

Having caused the parasite seeds to germinate, the root exudation also frequently acts as a guide-line to the developing rootlet of the parasite, due presumably to its increasing concentration nearer the root. Some parasites, however, have such large food reserves in their seeds that they can germinate without this chemical trigger and hope to encounter a suitable host by basic spreading of roots, as in toothworts, or by wide spiralling of the seedling in the air, as in dodders. A dodder can live up to seven weeks without a host, by which time it has grown 35 cm. Dodders do not just spiral on chance; they are positively attracted by chemicals exuding from the host – in an experiment even severed leaves have been found to attract. It is also possible that even a source of moisture helps to attract the stem. Dodders have two kinds of spiralling which in fact continue once they are fixed to a host: a tight vertical coiling and a much more open, wide-angled type. In this, as in some other parasites that germinate in soil, the plant loses its baby-stage roots almost as soon as the stem has successfully fixed itself into a host stem; sometimes it appears literally to pull its roots out of the ground.

Parasite seedlings can remain alive without a host for very varying periods. Dodder and *Cassytha* have an endurance of seven or eight weeks; the little alpine *Tozzia*, and also the toothwort, can live for two or three months. The mistletoe *Gaiadendron*, which has a tuberous storage organ, can probably survive for six months without a host, and *Nuytsia* for at least a year.

Some parasites spend many years underground, such as *Tozzia*, which looks like any normal plant when its flowering shoot appears, and dies entirely when seed is set; and the toothwort, which lives on tree roots and stays underground for up to ten years. The above-ground appearance of this plant is restricted to its flowers; there are no green leaves. In this respect it resembles parasites which spend their entire existence inside their hosts except when flowering, like the giant *Rafflesia*. And both these show a curious resemblance to fungi in their entirely non-green existence.

Mistletoe seeds will not germinate at all in the dark and will eventually die if kept in it. On germination they are controlled by light and gravity, but these act negatively, so that the radicle of a seed on the underside of a branch will grow upwards (against gravity) or if on top of the branch downwards (against light). The seeds have enough built-in water reserve to germinate on dry wood or inanimate substances.

The basic method by which the parasite enters and retains contact with its host is through an organ called a haustorium, which was probably originally derived from an ordinary root. This is a rather general word covering a variety of forms of 'bridges' between host and parasite: to quote Job Kuijt again, 'the haustorium is the specialized channel through which nutrients flow from one partner to the other. It is the organ which, in a certain sense, embodies the very idea of parasitism.'

This organ begins as one of penetration, a tight mass of tissue forming a wedge which probes the skin or bark of the host. Penetration follows the production of a disc- or cone-shaped growth arising from the root or stem of the parasite in contact with the host, or made by the parasite seedling, which clamps on to the host with a cement-like hold. Such a clamp is essential to counter the force of the wedge. Its presence, possibly through enzyme action, helps to dissolve or distort the host skin so that the wedge can penetrate. Leaves are readily penetrated, but stems can withstand entry for a time, sometimes until a split has occurred in their skin or bark. It seems probable that after a long build-up some parasites make a sudden thrust by rapid enlargement of the cells of the core, or wedge, which takes advantage of any split; haustoria can be found with successive layers of wedge tissue which have obviously failed to enter on earlier attempts.

Once within, the conductive strands of the parasite force their way among the host cells, forming a linkage into the xylem, the host's central conductive cells. As discussed earlier, the xylem's flow contains minerals and at times other food materials including sugars and carbohydrates. The parasite is therefore tapping, as if by a drip-feed, a very rich source of nutrients and does not need to bother with the other nutrient flow, that in the outer phloem, coming down from the leaves.

Once the connection has become established the haustorium (which is basically similar whether branch or root) may encircle the host with a tight-fitting mantle- or saddle-like growth, or may become very massive and create a positive fusion of tissues, as in the spectacular 'wood-roses' from Mexico and Central America. When cut and dried the wood-rose is a fluted, cone-shaped structure on the host branch, which is in fact made of the host's tissue; the parasitic complement of the mistletoes involved, the absorptive organ, is an equally large, dome-shaped structure which falls away on preservation.

Some mistletoes produce a series of what are aptly called 'sinkers', small narrow cones of core tissue connected at bark level which, if separated from the host, resemble a piece of wooden rake.

Another remarkable conjunction of tissue occurs in a quite different group of parasites, the *Balanophoraceae*: these are weird plants at best, producing a flowering head often looking more like a fungus than anything else. These make swollen tubers, sometimes rounded, sometimes branching and knobbly like coral, vastly larger than the roots they finally entirely engulf, and within which the tissues of both partners appear to combine both intricately and harmoniously, the conductive tissue of the host penetrating right into the tuber. At first sight this might make one think of symbiosis as in a lichen; but the parasite has nothing to offer, it has simply stimulated the host to produce extra tissue, just as others make the host produce a wood-rose or at least an enlarged area of tissue around the entry-point of the sap-sucking connection.

In the case of those parasites or semi-parasites with roots below ground and shoots with leaves above, which for some reason often have an enormous thirst for water, the roots have to search actively for contacts with host roots. Turner in *The Grete Herball* of 1526 wrote of secondary root contacts that 'I know that the freshe and young Orobanche [broomrape] hath commynge out of the great roote, many lytle strings . . . wherewith it taketh holde of the rootes of the herbes that grow next unto it.' Annual parasites inevitably have a rather desperate period doing this, because if the seedling fails to make a contact it perishes; but a surprising number of perennial parasites make an annual search for host roots, some if not all of their old contacts decaying.

Broomrapes not only tap the host's nutrients but have a remarkable way of sucking water out of it. The nutrient taken from the host, say a bean, is sucrose, but this the broomrape converts into glucose and fructose. This has the effect of raising the osmotic pressure in the broomrape and thus of making water pass very rapidly from the host's cells into the parasite.

Broomrapes make a network of fine roots with small connections. The desert parasite *Pholisma* makes a wide-spaced root system about 60 cm below the surface; these produce smaller roots which establish haustorial contacts if they can. Some species produce extraordinary mats of parallel roots whose purpose, unless they provide a storage capacity, is obscure.

Sandalwoods are trees, although by no means all their relations are. If the parent plant suffers injury or death, the roots can often regenerate shoots; some have been recorded 6 or 7 m away. This is however unusual, for most parasites do not have such extensive root systems.

Mistletoes have abandoned the soil except in one or two cases, and grow exclusively upon trees. This can be a drawback, because if the tree, or its neighbours, overgrows or excessively shades the parasite, the latter may die or at least be unable to flower. Mistletoes often grow on isolated trees or those at the edges of forests, and although this is presumably due to the habits of the seed-depositing birds, it is also a favourable habitat for the mistletoe. Some mistletoes overcome the problem by producing long trailing shoots which can anchor and make haustorial contacts all over the host; others 'walk', by sprouting new shoots from creeping secondary roots (as distinct from the original basic sinkers). Another creeping species twists its leaf-stems around anything they touch, exactly in the manner of a clematis; but where a twist is made, a haustorium is produced. Sometimes this occurs on another growth of the same mistletoe, and indeed also happens with secondary roots, but it does not matter for the result either way is an interwoven mass of the parasite capable of spreading further and further.

A few unusual mistletoes still root into the soil. One of them is the well-known Australian 'Christmas tree', *Nuytsia floribunda*, which makes a tree up to 10 m tall and bursts into orange-yellow flower in December. It parasitizes small plants such as grasses, and is capable of sending up youngsters from its roots many metres from the original parent. Two other species are known whose seeds germinate on the ground in an orthodox way and then climb any tree nearby and send sinkers into its branches.

One final mistletoe curiosity is a Brazilian *Phrygilanthus* which may make a network of subterranean roots up to 20 m long which usually exist without parasitic contact on the host's roots. However, they may encircle the bases of adjacent trees and establish haustorial contacts on *their* roots; one specimen has been recorded which had parasitized five neighbouring trees, all of different species. Whether this species originates on a tree and sends roots down into the soil or vice versa has not been established.

Apart from mistletoes, which are so important in magic and folk-lore and which form the central theme of *The Golden Bough*, the most familiar parasite group is that of the dodders and their relations. These make incredibly extensive networks of very thin stems, enmeshing the wretched host so that it sometimes entirely vanishes under the thread-like growths. They are very widespread and, since they grow so intimately with their hosts, are very easily distributed as seed along with crop seed, especially as the dodder seeds are often very similar to those of certain crops, and therefore difficult to separate from them. Although we usually see dodders on herbaceous or annual plants, they can infest trees, and there are records of tropical trees entirely covered with their yellow mesh. They can even attack submerged aquatic plants. One calculation of the total length of stem produced by a single 'representative' dodder plant (whatever that may mean) runs to nearly a kilometre!

In *The Botanic Garden* Erasmus Darwin has a charming verse about the dodders:

> Two Harlot-Nymphs, the fair Cuscutas, please
> With labour'd negligence, and studied ease;
> In the meek garb of modest worth disguised,

The eye averted, and the smile chástised,
With sly approach they spread their dangerous charms,
And round their victim wind their wiry arms.

He then compares them to the serpents which crushed Laocoön and his sons:

Round sire and sons the scaly monsters roll'd,
Ring above ring, in many a tangled fold,
Close and more close their writhing limbs surround,
And fix with foamy teeth the envenom'd wound.

The leaves of dodders are reduced to scales and the yellowish stem contains very little chlorophyll once the seedling has found a host. The stems form haustoria on the host at every possible opportunity, and even if the growth of the host, or some other accident, breaks the stems, their growth continues from the isolated haustoria. The plants are usually considered to be annuals, but it seems likely that in many cases haustoria persist on the host through the winter and make new growth in the spring.

Dodders belong to the *Convolvulus* family, but in Australia we find the superficially very similar but perennial *Cassytha* which belongs to the laurel family – a good example of parallel evolution.

The most extraordinary group of parasites is that which lives entirely within its hosts except when flowering. These are *Rafflesia* and its kind. Not only does *R. arnoldii* (or *R. tuanmudae*) have the largest flower in the world, a great fleshy cup over a metre across, but the flowers burst straight from the roots or sometimes the stems of the host. Just as no one knows exactly how the seeds are dispersed, the way in which seedlings penetrate their hosts is also a mystery. What we do know is that the parasite exists within the host's body in the form of thin filaments, remarkably like those of a parasitic fungus, which penetrate virtually every part and organ of the host. Rafflesias do not have haustoria in the ordinary sense, although perhaps the seedlings do. Even the flower buds originate within the host tissue, not as happens with other root-infesting parasites on its surface, and then burst through it.

Rafflesia usually grows on roots, so that its enormous flowers appear incongruously on the jungle floor while its photosynthesis is carried out for it 50 m above, but it also inhabits lianas, with the result that its monstrous blossoms may sometimes be seen hanging high in the air.

Some of its relations have more delicate flowers, as with the Japanese *Mitrastemon*, which actually produces a witches' broom on roots below the surface, from which the flowers appear like a fairy ring. In other genera the flowers emerge on the host's stems, as with the tiny flowers of *Pilostyles*. In one species, *P. haussnechtii*, the parasite's flowers are distributed very regularly along the host branches, a pair just below each leaf-stalk, showing what Kuijt describes as 'an uncanny synchronization of growth rhythms of host and parasite'. Having no chlorophyll, this group, like the toothworts and balanophoras, can grow perfectly well in dense forest. Complete parasitism is clearly the ultimate goal.

If the basic types of parasite are not odd enough some further curiosities may be mentioned. There is one mistletoe, *Gaiadendron punctatum*, which grows only on epiphytic plants and not on the trees the latter cling to. Sometimes parasites grow on others, recalling Swift's lines about the flea:

> Hobbes clearly proves that every creature
> Lives in a state of war, by nature:
> So, naturalists observe, a flea
> Hath smaller fleas that on him prey;
> And these have smaller fleas to bite 'em,
> And so proceed *ad infinitum*.

At least one case is known of four mistletoe relations each growing on another in a chain, the basic one parasitizing a tree, the Indonesia *Macrosolen*. To cap it all the final parasite was attacked by a fungus which itself was parasitized by another! Another case, from El Savador, combined a palm tree which was being attacked by a strangler fig, and in this case the strangler bore one parasitic plant which a second had attacked. Here again, the last parasite was spotted by a fungus and this disease organism was under attack by another. Quite often more than one parasite inhabits the same tree; there is in the Kew herbarium a probably unique branchlet of *Nothofagus antarctica*, one of the southern beeches, carrying the branching flower-heads of three distinct species of *Myzodendron*.

I have already mentioned how a parasite may often attack its own roots or branches, sinking haustoria into them from young growths; they seem to have no discrimination in this respect, but appear to do no harm and indeed simply augment the mesh of growth. This lack of discrimination extends to inanimate objects: many cases have been observed of haustoria having been formed on dead roots (not so surprising since they are organic), pebbles, sand grains, and in one fantastic case the Australian Christmas tree already mentioned made ring-like haustoria on a polymer-insulated electric cable about 2 cm across, which penetrated sufficiently to cause short-circuiting.

Many parasites seem not to be choosy about their hosts; the mistletoe *Dendrophthoë* has been observed on 343 host species. However, other kinds are virtually specific: there are species only growing on olive, pine, juniper, cacti and so on. It seems likely that this happens only when the host species is very common in a reasonably large area. The root parasites are especially catholic in their tastes; they will grow on almost anything, although some are curiously restricted as to the class of plant they choose – some only grow on monocotyledons, others avoid these, ferns and conifers.

Some plants seem to evade parasites, either by having an especially thick or hard layer under the bark, which repels attack mechanically, or what appears to be a chemical, or more correctly biochemical, incompatibility. *Begonia* and *Oxalis* have very acid tissues and seem to resist dodders entirely.

Distortion, debility and death are the typical outcomes of a plant parasite. Some cause enormous damage and economic loss. Wild trees can be seen weighed down under blanket-like growths of mistletoe, which are forest and plantation pests in many places, attacking such crops as conifers, pears, pecans, walnuts, citrus and cocoa. The damage by the dwarf mistletoe genus *Arceuthobium* alone, on the much-cultivated ponderosa pine, is greater than any estimate for the combined loss caused by all other diseases of this tree. Dodders are regarded as noxious weeds in forty-seven states of the United States, notably in leguminous crops such as alfalfa and clover, and also on flax, which it can ruin in a very short time. In Puerto Rico the similar *Cassytha* is called Public Enemy No. 1. Broomrapes and parasitic figworts can often be seen decimating crops such as beans and clover; they also attack tobacco, tomato and hemp. The most destructive broomrape is the tropical *Aeginetia* which attacks rice, maize and sugar cane in Asia; it does not merely reduce yield but, in the cane, converts the desirable sucrose to worthless reducing sugars. Such plants are especially lavish with their seed.

It is especially under rather primitive agricultural conditions, notably where rotation is not practised and modern methods of seed preparation are not available, that such parasites flourish in crops. Control is difficult at best: herbicides give some hope, but sophisticated seed separation is necessary with dodders. Another method is to use 'trap crops' – plants which stimulate the parasite's seed to germinate but are resistant to attack. It may also be possible to breed resistant crops. However, as we have seen, there are a great many parasites and they have many ways of life, which makes them very difficult to deal with.

Finally one may ask whether any plants parasitize animals. The answer is no, with the exception of the fungi. These cause many skin diseases, including ring-worm, athlete's foot and some unpleasant tropical ailments in humans; they are also responsible for various allergies. Fungi also cause the notorious 'vegetable flies' or 'vegetable caterpillars' by parasitizing insects and growing their branched spore-carrying bodies out of them. The spores germinate on any moist caterpillar skin, and can at once penetrate into the body, their strands branching and passing in all parts of it. A dense mass of filaments finally fills the body entirely, and needless to say this process causes the insect to die. The fruit-bodies then appear: the biggest European species, found on beetle larvae, is up to 17 cm long. They may sprout from the caterpillar or, if it has pupated, from the chrysalis. Since this is often buried, the fungi may give the impression of emerging from the soil.

One species used to be an important drug in China where specimens were said to be worth four times their weight in silver. An Australian vegetable caterpillar with black fruit-bodies up to 40 cm long was used by the natives in tattooing.

There are, then, very many plants which have become parasitic on others. While lower plants can achieve mutually helpful associations with each other, and with higher plants, the latter cannot apparently do so among themselves. One recalls the phrase quoted in the previous chapter: 'parasitism is the most exquisite example of symbiosis'.

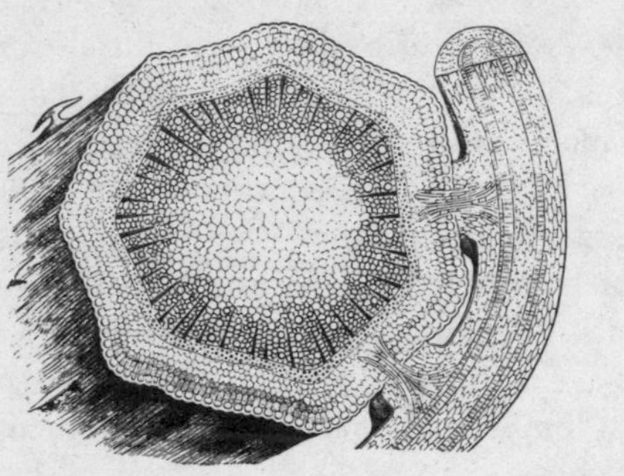

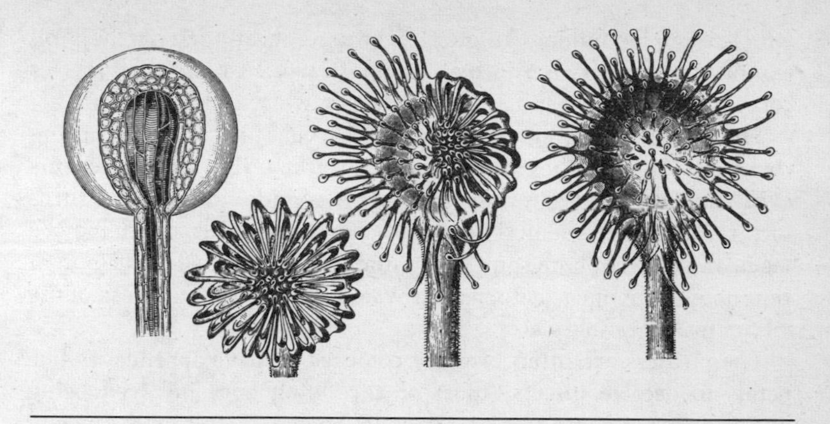

24 · *Killers*

It was the late John Collier who wrote a story about a man-eating orchid which eventually produced the heads of its victims as flowers. There are travellers' tales of plants that ate dogs, if not men. We must be grateful that man-eating plants exist only in fiction; the small-scale reality is macabre enough.

Several distinct plant families have achieved this flesh-eating habit; the one thing they have in common is a tendency to live in nitrogen-deficient habitats, so that the animals consumed help them to balance their diet. There are well over 500 species of positively insectivorous plants, which can be divided into three distinct groups, in each of which a trap of some sort is prepared. They often live in swamps or bogs, which allows them to use traps which are water-filled or glands which remain moist. A fourth group, discussed at the end of the chapter, provides some wider but not entirely proven aspects of carnivorousness.

To say that the first group is inactive belies the macabre ingenuity with which their traps are constructed and baited. These are the pitcher plants, in which some or all the leaves have been curiously modified. Some carry clusters of pitchers at soil level, erect or almost lying on the ground, sometimes among orthodox leaves. Others – climbing plants which scramble among jungle vegetation – bear their pitchers on extensions of otherwise normal-looking leaves. A number of species carry both terrestrial and arboreal pitchers, which differ in shape.

The pitchers can be likened to jars, urns, cups or funnels of varying curvature. Usually they have a flap above the orifice which prevents excessive rain from entering. Each of these pitfall devices has three factors in common – a method of attracting the animals, which may be visual or edible, or both; an almost infallible way, like a lobster-pot, of retaining the trapped visitor; and a system for reducing the flesh of the victim to a digestible state.

The pitchers are often brightly coloured or have appendages like petals to deceive insects (most of the flying ones involved being nectar-collectors) into thinking they are flowers. Closer to, the insects may find a pleasant scent and often a secretion of sweet honey-like juice round the rim of the pitcher; there is also often a vertical pathway on its exterior, sometimes with carefully provided footholds, so that wingless insects, especially ants, may also be enticed.

In many of the pitchers the top is constructed of rounded, very slippery ribs which terminate in down-pointing spines. An insect probing over the rim for more nectar finds it impossible to prevent itself from slipping downwards. Once within, the way out is barred by various horribly ingenious means. In the terrestrial sarracenias and darlingtonias the inside of the pitcher is lined with spikes or hairs projecting downwards at a slight angle, which either prevent the insect getting a grip or actively stop it climbing up. *Sarracenia psittacina* is like a fish-trap: its pitchers lie on the soil to attract ants, centipedes and the like, for which, once within, long bristles in the narrowing funnel ensure a one-way trip. In the tropical *Nepenthes*, those with pitchers at the leaf ends, the edges of the trap, surrounded by barbs, lead to ice-smooth waxy walls which afford the insects no grip at all.

Victorian botanists in particular described these devices with marked relish, as this quotation from *The Natural History of Plants* will show:

Every animal that wishes to save itself from a Cephalotus pitcher has three obstacles to overcome: first, a circular ridge projecting inside the pitcher; secondly, a bit of wall thickly covered with little papillae, sharp, ridged, and pointed downward, the whole being comparable to a flax-comb; and, lastly, on the involute rim round the mouth of the pitcher, another fringe composed of hooked, decurved spines which bristle like an impenetrable row of bayonets in front of such animals as may have surmounted the other difficulties.

In the sarracenias and darlingtonias the pitchers partly fill with rainwater and the trapped insects drown and then simply decay. Special cells at the bottom of the pitcher absorb the resulting fluid as a kind of liquid manure. In others such as *Nepenthes* and *Cephalotus* a liquid comparable to mammalian gastric juices is secreted which actively digests the prey, the fluid remains of which are once again absorbed by special cells.

Enormous quantities of animals can be trapped. Pitchers can be seen filled to half their length with the indigestible chitinous remains. The biggest pitcher is that of *Nepenthes rajah*, 50 cm long, up to 16 cm wide, with an opening 10 cm across, and capable of holding 2 litres of liquid – big enough to conceal a pigeon! It is not recorded, however, that a bird has ever been found in a pitcher, but they are reputed to trap large rats.

One should note in passing some resilience of the insect world in relation to pitcher plants. There is a blow-fly which, to quote Kerner and Oliver again, has a special appendage on each foot which 'may be likened to the grapple-like climbing irons of Tyrolese mountaineers', enabling it to clamber readily among the pitcher's thin, down-curved hairs. Moreover, its maggots inhabit the charnel-house of decaying insects, and when ready to pupate cut holes in the pitcher to escape. There is a small moth with a similar leg device, whose caterpillars cover the hairs with a web. Finally we may note spiders which spin webs in the upper parts of *Nepenthes* pitchers in order to have first choice of the insect visitors; one of them at least has a special skin which allows it to shelter within the digestive fluid if it is threatened.

The next group of insectivorous plants relies partly on stickiness and also on movement to entrap insects. One might call them 'flypaper' traps. These are the butterworts (so called because of the buttery colour of the flat, rosetted leaves and their use in curdling milk). The leaves are covered with glands which secrete mucilage, some of which is always present on the surface. Any large body of organic origin causes the glands to produce a much larger amount of the sticky juice, and also an acid fluid, like that in advanced pitcher plants, which can digest animal matter. Insects, especially small ones, are readily caught by the initial stickiness and quickly become hopelessly bogged down, soon to be digested. The presence of an insect also causes the leaf margins to curl inwards. This seems to be partly in order to push any creature which may be stuck near the edge into the centre, where there are more glands, and partly to hold the prey fast. Some time later the edges curl back again to expose the flat leaf.

The sundews have leaves which end in a round or long, narrow pad, decked in red hairs, each of which glistens with a drop of clear liquid at the end. These tentacle-like hairs are longest at the leaf edges and shortest in the centre. Insects are attracted to these glinting tentacles, possibly because the drops resemble nectar, possibly because they mimic the moulds which grow on decaying meat, hence the red colour of the hairs and leaf. At any rate, once a tentacle is touched the insect is doomed, for the drops are extremely sticky, and an insect of any size will have touched several tentacles, or even be involved with more than one leaf.

This is the first phase. Once the prey is ensnared the tentacles bend over it. It is a fairly slow process; a hair can bend to 45° in about three minutes, to 90° in ten. It is, however, inexorable, and its result is that within an hour the prey is enmeshed by tentacles from all around, one circle of tentacles following another about every ten minutes. Should two tiny insects land on the same leaf, the tentacles divide into two groups so that each insect is well held down. The purpose is to bring the prey into contact with acid-secreting glands on the leaf surface, which are most numerous in the centre. One might say that the bending inwards of sundew tentacles and the inrolling of butterwort leaves are to provide 'temporary stomachs'.

Once the insect has been digested, which may take from two to several days, the tentacles straighten out and remain dry-tipped. This allows the hard parts of the prey to fall off, after which the liquid is produced again.

There are two remarkable things about the sundew, apart from the fact that the hairs can move at all; one is the way in which the stimulus to bend over is passed from one hair outwards, and the other is the extreme sensitivity of the hairs. Kerner describes an experiment in which a fragment of human hair, a fifth of a millimetre long and weighing a thousandth of a gram, caused a tentacle to bend; yet rain does not affect them.

Orchid growers sometimes make use of one of the large-leaved sub-tropical sundews to trap insects in their greenhouses. The biggest species, the South African *Drosera regina*, has leaves 60 cm long and can hold small animals. An Australian relation of similar size, *Byblis*, certainly catches frogs and lizards and is reputed to ensnare rabbits and squirrels when plants grow close enough to form a large clump.

There are plants called *Drosophyllum* which resemble sundews in that the leaves are entirely covered with red, sticky hairs. These hairs do not move, but they are so effectively adhesive that Portuguese peasants, in the area where the plant is common, used to hang up the leaves for a natural fly-paper. Other plants also have sticky, glandular hairs on the stems, and apparently digest the insects which get stuck on them; indeed, the silenes are often called catchflies because of this. In some cases sticky stems certainly exist in order to prevent small unwanted creeping insects from reaching the flowers. This occurs in the spectacular silversword of Hawaii.

But the real miracle among the insect-catchers is the Venus flytrap, familiar to many because it is widely sold for cultivation. It is a small, low-growing plant; its mature leaves, in rosette formation, each end in a pair of rounded, bristle-edged lobes which lie apart, almost flat. On the surface of each there are glands, those on the outside producing a sweet-smelling liquid attractive to insects, the others within a very acid liquid. There are also three hairs on each leaf.

These hairs are triggers. If only one is moved, which happens if a raindrop strikes it, nothing happens; but if two hairs are touched, or one hair is touched more than once, the leaf lobes move swiftly and silently together (in about a quarter of a second), the bristles on the edge interlocking like fingers of a hand, fatally clamping in any fly which has landed on the lobes. At first the lobes are concave on closing, but once they have locked together they flatten, crushing any soft body within to pulp. Then the acid-secreting glands start to work; eventually they digest the fluid results, and dry up when the trap re-opens and resets itself. This may take from eight to twenty days depending on the size of the prey.

Not only can this plant count the necessary stimuli before its trap operates, but it has a built-in timing device: not less than one and a half and not more than twenty seconds must elapse between the stimuli. Several light stimuli will also make the leaf respond, but more slowly. The number and timing of stimuli required for closing depend on exterior temperature. The trap can be sprung artificially, and also by proteinaceous material, whether in the form of a live fly or a piece of meat or cheese, so that the flytrap responds to both mechanical and chemical stimulation. The secretion of digestive acid will not take place in the absence of protein material in contact with the glands. The operation of the flytrap results initially, as described in Chapter 11, from electrical impulses created on stimulation, which cause cells in the hinge to expand very rapidly.

The Venus flytrap has a few similar relations, one of which, *Aldrovanda*, lives in fresh water. It carries its lobed traps at the length of a long slender stem; like the bladderworts it floats without roots. Instead of the interlocking bristles, the inner edges of the lobes carry a series of small conical structures, which act as locks when the lobes meet together after small swimming creatures set off the numerous trigger-hairs of the trap.

It is intriguing that individual families have devised quite different modes of trapping insects. Thus *Aldrovanda* and Venus flytrap belong to the same family as sundews. The sticky butterworts are relations of the bladderworts, mainly water plants with fine feathery foliage and odd little yellow flowers projecting from the water. They have no roots but float just below the surface. Small bladders are carried on fine stems among the leaves, which are in fact traps; they range from about 2 to 5 mm across. A small opening is guarded by a lip-like, air-tight valve, which hinges to allow easy entry but, like a door, cannot be forced open from within. When 'tripped' by contact with an external bristle-trigger the trap springs open and the organism that unwittingly knocked at the door is sucked inside in a jet of water, for the bladder creates a partial vacuum within.

Creatures that enter die of suffocation or starvation; they decay, and the liquid result is absorbed by star-shaped cells on the inner wall of the bladder. Most of the prey are minute crustacea such as daphnia which, perhaps in seeking refuge, are deceived into thinking that because of its similar shape the bladder is another crustacean. This resemblance extends to the antennae and swimmerets of the animal, mimicked by the bristles radiating around the orifice. Up to twenty-four crustacea have been found together in one bladder.

Some bladderworts live in damp moss and debris on rocks and bark, where very minute bladders provide a deadly refuge for many small creatures.

One of the more interesting species lives in the 'urns' of an epiphytic bromeliad (a *Tillandsia*) such as have been described on page 256. As already mentioned, the water-filled urn not only provides moisture for the bromeliad but home for many tiny animals, and there to prey upon these lives the bladderwort. In passing I should mention that this plant produces runners to reproduce itself non-sexually; these arch in the air and, with luck, find lodging in a neighbouring *Tillandsia*.

It is not only the higher plants which have this carnivorous capacity, which is sometimes combined with movement and unusual speed of reaction. There are at least fifty species of soil-inhabiting fungi which trap microscopic eelworms or nematodes with a lassoo-like contrivance. These fungi produce little rings on their main strands, in each of which there are three sensitive pads. If an eelworm burrows through the ring, these pads become suddenly inflated, pinching and trapping the animalcule. In some cases a poisonous substance is produced at the same time, and it has been shown that this eelworm-killer is effective at a dilution of one part in 5 million. The fungus ring then develops strands which penetrate the eelworm and suck out its life. Yet other fungi have sticky strands which immobilize eelworms which touch them. Some make a mesh of loops that act simply like a fishing net, and some appear to lure their prey with chemical secretions.

Although the mechanism of the active ring traps is far from understood, it is fascinating that trap closure is stimulated by a substance called acetylcholine which is the basic neurotransmitter in humans – a common character presumably of the greatest antiquity.

These primitive traps probably pre-date the insectivorous higher plants by millions of years, and show once again how basically similar 'inventions' can occur in quite unrelated organisms.

Beyond these three main groups of positively carnivorous plants – those with pitchers, fly-papers and active traps – there is a grey area of plants with sticky hairs on which insects become trapped. Many botanists have considered these as methods mainly to prevent insects climbing up to flowers where they are not welcome as pollinators, but there is a growing belief that the hairs contain digestive substances and that the plants absorb nourishment from the decaying insects. There is evidence too that some plants, including the sun pitcher *Heliamphora* and the myrmecophile *Dischidia*, can absorb decaying animal matter through unspecialized tissues.

Plants with sticky hairs include London pride, *Pelargonium zonale*, petunias, and some wild tomatoes and potatoes. Some have the hairs on the flower stems, others on leaves. Teasels have paired leaves which form water-holding cups up their stems, and it is thought possible that the remains of drowned insects provide nourishment absorbed through the cuticle in these cups.

Even the clear-cut carnivorous plants with specialized apparatus do not rely on food of animal origin, any more than cannibals relied upon that; it is a bonus to them in difficult growing conditions deficient in certain nutrients. Their sophisticated equipment is remarkable, and it now seems probable that less specialized opportunism is much more widespread than was previously thought.

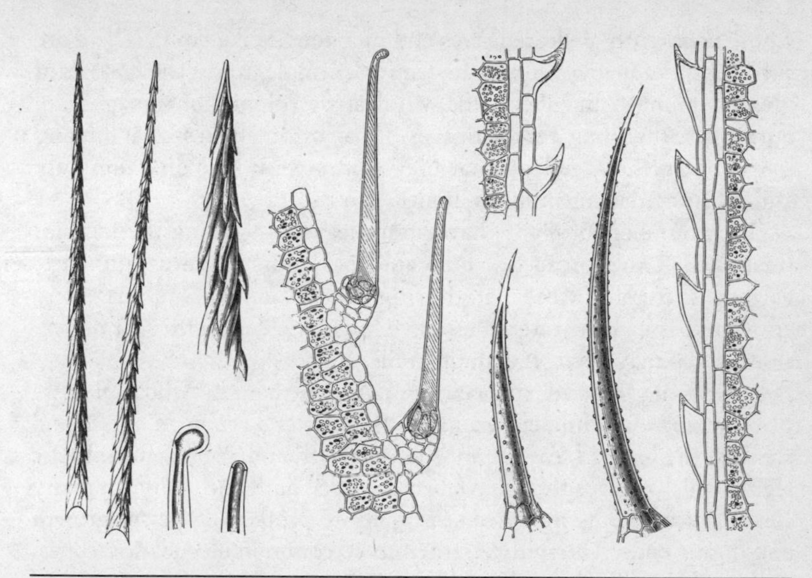

25 · *Armour and Poison*

During the last century hill villagers in northern Nigeria were harried by Moslem invaders. To defend themselves they surrounded their villages with a succulent spurge called *Euphorbia desmondii*. Its angular stems, as thick as a man's arm, and covered with stout spines, form a tall, impenetrable hedge. To attempt to chop into this plant stockade would be folly, since the stems contain poisonous latex which blisters the skin and would blind an eye.

Many plants have passive devices such as thorns and poison against predators of various kinds; not man initially, but primarily grazing animals. The most obvious defence is to brandish weapons. Yuccas, agaves, aloes and many bromeliads have stiff rosetted leaves with sharp points, radiating in all directions. One of the last, the giant puya, is becoming rare in its native Peru largely because local shepherds destroy young plants, since sheep often blunder into them and impale their eyes. On a smaller scale there are plants such as some grasses whose apparently delicate leaves end in a needle-like point. These are sufficient, in the lowly alpine fescue for example, to cut the nostrils of grazing animals, and shepherds will destroy colonies of this grass by burning. Other long narrow leaves, those of other grasses and many sedges, carry saw-teeth so that they can move in one direction only and, if jammed in an animal's mouth, can cut it readily.

Great numbers of plants have spiny leaves, including the familiar thistles, the widespread clot-burs and cockleburs of warm countries, various successful weeds of the nightshade family and the spiny acanthus. The giant water-lilies have huge prickles on the leaf undersides which may keep off aquatic snails.

Leaves may become reduced to spines so that the whole plant is prickly: gorse is a homely example. In butcher's broom, *Colletia* and the like the leaves have been dispensed with and the photosynthetic stems which replace them have become very hard and spiny. In many cases leaves become modified seasonally to protect the newer growth and flower buds. The spiny *Astragalus* so common on Asiatic steppes, making low hedgehog-like clumps, is an example. In the Mediterranean shrubs the previous year's leaf-shoots become hard and spiny; Theophrastus christened these phrygana, a name still used for these plants as a group. Spiny burnet, Spanish gorse, and the dwarf spiny spurges are typical. Reduction of leaves to spines or their disappearance in favour of spiny stems is typical of plants living in arid conditions.

Spines on stems are common too, often on woody plants where, apart from deterring grazing animals, they help to prevent attack by soft-bellied snails and suchlike from ground level. Frequently stem-spines cluster round soft leaves to protect them against browsing. Palms often have fierce spines at the leaf bases, or all along the stems, often down-pointing. Spines are often barbed, at the top or all the way up, so that they make nasty wounds. Enlarged, they could serve as models for some of the vicious weapons devised by man for hand-to-hand fighting.

Cacti and some succulents are of course supreme examples of spininess, and the beautiful flowers of cacti emerging from their armoury remind one of Beauty and the Beast. However, cactus spines often have a beauty of their own, both in their colour and curvature, and in the geometry of the arrangement of the clusters on the plant, almost always a stem without leaves, as we have noted earlier. Spininess is of extra importance to such plants since they usually exist where green food and juicy sap are very much at a premium. Larger animals usually keep their distance, and any traveller will know that prickly pears, agaves, columnar cacti and the erect cactus-like spurges already mentioned are used for hedges to surround penned animals and indeed entire villages.

Prickly pears, incidentally, not only have orthodox spines but masses of unspeakable miniature barbed hairs called glochids, which become readily detached and penetrate skin with alarming ease, where they produce intense and persistent irritation.

On a smaller scale sharp hairs called trichomes can tear the flesh of grazing caterpillars, as for example on *Passiflora adenopoda*.

Many plants have a covering of small, visually innocuous bristles, which however produce an unpleasant itching and often inflammation if touched, especially by tender mouths. Comfreys and buglosses are good examples. The apparently juicy mulleins so common in Mediterranean lands escape being eaten because they are usually coated in a grey or silvery felt, which comes away quite readily and sticks to the mucous membrane in the mouth, where it irritates violently.

A refinement of the bristle is the stinging hair. No one needs to be told about nettles, but there are other plants which sting, notably in the families *Loasaceae* and *Euphorbiaceae*. Sometimes the hairs concerned are pointed, when they seem to be direct developments of ordinary protective hairs; in nettles and loasas, however, the top end is a small round 'bead', set obliquely on the hair, which is a single hollow cell. Its upper part is stiffened to a glass-like hardness, either by silica or lime compounds. Any pressure breaks off this hard tip, leaving sharp edges which penetrate the skin, while the soft base of the hair pressurizes the contents, so that the liquid within is forced into any wound which has been made, just like a hypodermic needle. The liquid combines corrosive formic acid and histamine, one of the main causes of human allergies; and as we all know, a nettle sting can cause severe irritation. Some of the tropical nettles, indeed, are so virulent that they can produce symptoms similar to snake-bite. There are even nettle trees, such as *Laportea gigas* from Polynesia, which grow over 13 m tall and whose large heart-shaped leaves deliver long-lasting stings.

The Central American climber *Loasa canarinoides* adds poison to its sting, while the spurge relation *Cnidoscolus urens*, from the same area, is virulent enough to bring up pustules on the skin.

It is amusing to recall that Roman soldiers are reputed to have deliberately stung themselves with nettles to give themselves the impression of warmth in cold Britain, while Greek girls used to beat themselves with nettles on Good Friday as a penance in remembrance of Christ's sufferings. It is not surprising that animals with tender mouths and nostrils avoid these otherwise succulent and nutritious plants.

In some cases thorns exist for other purposes besides protection. Brambles, roses, and the tropical rattans or climbing palms up to 200 m long have thorns to help them hook themselves over bushes and into trees. In these the thorns are a device parallel to suckers or tendrils in other climbers.

One can also point to animals which consume spiny plants: cacti are attacked by birds and rodents; donkeys eat thistles and camels eat camel-thorn, an acacia with 8 cm iron-hard spines; giraffes browse upon other thorny acacias; birds and small rodents tunnel into cacti. But these are really examples of animal resilience in the face of a specific problem; no one can deny that spines, thorns and barbs do in general provide considerable protection against the lesser kinds of 'herbivore' – sap-sucking insects – and it is salutory to recall that the cochineal bug was the first line of attack against the prickly pear after its disastrous introduction into Australia.

I have earlier mentioned that many plants have sticky hairs on their stems, either for active digestion of trapped insects (for instance, *Drosophyllum*) or more often to prevent unwanted animals from reaching the flowers and taking nectar or pollen. Some plants appear to create physical barriers for the same purpose, as in the teasel, where pools of rainwater form in the cups made at intervals up the stem by the joined bases of each pair of leaves. This appears to be largely against ants. Other plants employ rings of spines or prickles, as in many palms.

Yet other plants contain an internal deterrent in varying degree. Thus alpine rhododendrons, cowberries and the like have very thick leaf-skins reinforced with silica, which though primarily to resist climatic rigours are distasteful or uncomfortable to animals. Barks may contain unpalatable substances, like the tannin which saturates redwood bark and helps it to resist fungus diseases and parasites. Fruits may be bitter when immature to prevent them being eaten at the wrong stage. Many plants have extremely acrid sap, like *Peganum harmala*, widespread in Turkey. The Mediterranean oleander, which survives in very dry conditions which few other shrubs can tolerate, has bitter and very toxic leaves which even goats will not sample. In Africa the succulent carallumas have poisonous latex which makes them distasteful to stock, even goats.

The tall-growing alpine meadow gentians, whose juicy, bitter roots are prized medicinally, are avoided equally by cattle above ground and rodents below, whatever their hunger pangs. Spurges with their acrid latex are equally widespread and are not eaten by grazing animals, although caterpillars will eat them. Such latex can blister the human skin or an animal's mucous membrane, and the watery sap of some other plants is also remarkably corrosive: thus giant hogweed will raise an enormous blister on human skin and mark it for months, a poison associated also with tiny stem-hairs which often create allergic inflammation without release of sap. Similar intense and long-lasting pain is created at a touch by the American poison ivy.

Many plants combine such an unpleasant taste or acridity with actively poisonous substances, as in the oleander just cited. Here we have to be careful, however, for what will kill a human being does not necessarily kill other creatures, even mammals. Whereas the poison in most members of the buttercup family, such as the yellow monkshood aptly called wolfsbane, acts against man and herbivores alike, the very deadly fungus *Amanita phalloides*, the death cap, can safely be eaten by rabbits, and equally by slugs. The berries of lords-and-ladies, fatal to many animals, are avidly taken by thrushes which distribute the seeds. Fruits of *Strychnos*, the source of strychnine, are eaten by monkeys though they take care to discard the deadly seeds.

However, many grazing animals are killed by poisonous plants; in this country they may die from eating yew, for instance; in South Africa the low-growing shrub *Dichapetalum cymosum* has been responsible for heavy losses of cattle, while in New Zealand the shrub

Coriaria ruscifolia earned its place in history by decimating the flocks of the early European settlers. In North America species of *Kalmia*, attractive enough to be prized in gardens, are known as sheep-kill or lamb-kill. Animals which survive plant poisoning often form an addiction for the plant concerned, and if farmers let such an animal back into the original field it will often lead them to the offending specimen.

Some animals undoubtedly get a 'kick' from specific plants, and nearly twenty species are recorded as regularly 'taking drugs'. Mongooses both eat certain roots and rub them over their bodies, which makes them fall into a stupor. Elephants eat the fruit of the umgana tree, which ferments in their stomachs to leave them 'staggering about, playing huge antics, screaming so as to be heard miles off and often having fights' (Ronald Siegel).

It is amazing how many plants do contain substances toxic to mammals. There are the 'obvious' ones such as deadly nightshade and hellebore, less expected ones such as laburnum, and surprises among many staple vegetables. Cassava, a tropical root, has to be specially treated to get rid of the cyanide it contains; the same chemical occurs in almonds, lima beans and apple pips, sometimes in alarming quantities; oxalic acid, again a killer in moderate doses, is found in spinach and rhubarb, the leaves of the latter being lethal even when cooked. Potatoes contain solanin at about 90 ppm, which is about a quarter of the danger level; this substance increases in quantity in potatoes allowed to become green in the light.

If we applied to vegetables and fruits the tests made on artificial foods and additives, such as those which resulted in bans on cyclamates, many might well be forbidden for human consumption. Coffee, for instance, contains trimethyl xanthine, methylfuran, chlorogenic acid, feruloylquinic acid, dimethylsulphide, tannins, and products of pyrolysis. That might make people think twice if they read it on the packet!

Bracken offers nothing to its fellow plants; instead its competition often eliminates them, both by physically smothering and also by producing phenolic compounds which accumulate in the soil and inhibit root growth. While it provides cover for some smaller mammals and supports a few insects, it does not contribute to any known food chain. Its stems contain sharp fibres which can cut deeply, and have further virulence within. Horses which eat a few mouthfuls have their vitamin B destroyed and rapidly die. Cattle suffer something closely resembling radiation sickness and apparently linked with cancer: their bone marrow ceases to function, white blood cells cease to be produced and the blood is unable to clot; they die from internal bleeding, even weeping tears of blood. Sheep have their eyes severely damaged by what is called 'bright blindness'. One might be tempted to think that bracken has developed into the perfect plant invader, inexorably annexing territory, choking any plant in its way and containing chemicals which destroy would-be grazers upon it.

Most of the poisonous plants seem to give no warning to potential eaters. In some cases, however, there are odours which probably act as warnings; for instance, most of the nightshade family have a typical pungent smell, as with henbane, thornapple and its relation the angel's trumpet, some species of which are so poisonous that South American Indians will not even doze in their shade. This reminds us of the fabled upas tree from Java, reputed to blight other vegetation around, to cause birds that flew over it to drop dead, and to kill any animal or man that ventured into its shade, so that the ground beneath was covered with bones. There is in fact a upas tree, *Antiaria toxicaria*, from which the natives used to obtain a very effective poison for arrows and blowpipe darts, as those in South America do from species of *Strychnos* which provide curare; but no tree has the effects suggested by the old travellers' tales.

A number of tropical trees give out more or less unpleasant odours; there is a Malaysian timber tree which smells of garlic and smaller shrubs in the tropical forests of the same area which stink like ordure. *Gustavia* is a South American tree with very handsome white flowers, the smell of whose trunk and leaves have earned it names like corpse tree and stinkwood. But there is no suggestion that such trees are poisonous.

One may ask why poisons exist in plants, but there seems to be no real answer. It might be claimed that, since they are harmful or even lethal to animals that eat the plants, they act as protection and, in evolutionary terms, they are a selective adaptation. But then there is the puzzle of very closely related plants with quite different capacities. Hemlock is deadly, as the Ancient Greeks well knew; but many very similar, related plants such as the cow-parsleys are not. 'Toadstools' are widely reputed to be very poisonous but in fact very few of these elementary plants are, and we find the death cap in the same genus as *Amanita caesarea*, a favourite dish of the Roman emperor Claudius. Indeed, so fond were the Romans of fungi for the table that they employed men especially to collect them, making Seneca exclaim in disgust 'Good God! how many men labour for a single belly!' Certainly the symptoms of death cap poisoning are sufficiently unpleasant before almost inevitable death for the fungus collector to exercise the greatest care.

One reason for considering plant poisons as merely fortuitous is that they seem generally to have no effect against small predators, slugs and snails, and the host of sap-sucking insects and internal parasites, such as eelworms, which because of their vast numbers are almost as harmful as the larger grazing animals.

Some insects actually make use of the poisons within plants. A Middle Eastern grasshopper which feeds upon the very poisonous latex of *Caloptropis procera* not only becomes nauseous to any predator rash enough to eat it, but stores up poisonous fluid entirely based on its food plant's toxins in gun-like 'repugnatorial glands' from which it can eject a 'death jet' to 60 cm! There are also many caterpillars which similarly feed on poisonous plants and actually store the poisons in their own bodies, making them deadly to the predators, mainly birds. These often develop protective coloration to give warning of their deadliness or bitterness, and there are some non-poisonous insects which mimic the genuine ones and escape attack in that way. But that is another story.

However, evidence is accumulating that some plants do repel attacks by predators of various kinds by chemical means, sometimes of a mysterious nature. A Pacific kelp, *Alaria marginata*, has been found to contain phenolic compounds like tannins which deter marine snails, and these compounds are more concentrated in the vital spore-bearing fronds than in the rest of the plant.

Silicon is a mineral which acts as a deterrent to predators in several plants, notably grasses. In rice, some strains contain much more silicon than others and it has been demonstrated that those with most prevent the development of the larvae of the striped rice borer and also of rice blast fungus. Slugs also are greatly deterred by silicon. This capacity seems to be due to the hard, abrasive physical properties of silica, the oxide of silicon, mentioned on pages 65–6 in connection with bamboos and horsetails, which are themselves largely impervious to pests and diseases (slugs certainly do not attack horsetails).

The possibility that sticky hairs on plants not only trap insects but ingest their juices has been mentioned in Chapter 24. The wild potato *Solarium berthaultii* has another way of dealing with aphids: they are covered with glandular hairs which release a liquid when a passing insect touches them. This liquid very shortly turns into a kind of glue, and the aphid becomes stuck to the leaf surface, unable to move. More extraordinarily, this potato gives off a warning odour or pheromone like that produced by aphids when alarmed, which keeps the insects away.

The sap of certain conifers and of *Melia* species contains a 'juvenile' hormone which makes caterpillars into Peter Pans, preventing them from ever maturing, so that no adult capable of reproduction develops.

Tomatoes and potatoes produce materials which upset the digestive capacities of insects eating or sucking their tissues. This inhibitor is normally present in low concentrations, but as soon as any leaf is damaged the concentration is greatly increased. In the same way cotton seedlings rapidly build up resistance to spider mites if attacked when their cotyledons develop.

The most amazing example of plant resistance to predators is as yet only part-proven. It relates to acacias and other scrubby African savannah trees grazed by giraffes and other large animals. As more and more of the leaves are chewed off, it is believed that the remainder become increasingly unattractive to the grazers due to increase in unpalatable chemicals, probably tannins. This is on a par with the previous examples; the extraordinary part is that this 'message' is apparently transmitted to neighbouring trees, possibly by a chemical transpired from one tree and taken in by another through the breathing pores.

It is also becoming apparent that many plants contain substances which help to repel fungus invasions, and as with the insect-inhibitors just described the production of these antibiotics is probably stimulated by the fungus attack. These substances, which are possibly based on proteins known as phytoagglutins, and are known as phytoalexins, remind one of antibodies in animals.

These 'warding-off' compounds have been shown to prevent further growth of invading fungus disease hyphae. Sap in beans immobilizes fungus spores within twenty-four hours of initial infection. Besides leguminous plants, solanaceous ones like tomato, potato and tobacco, and a few others, appear so far to produce phytoalexins. It seems that these are synthesized by the interaction of the metabolic system of host and parasite – it reminds one slightly of a binary nerve gas in which two harmless substances react in deadly fashion when mixed.

Not only fungi but bacterial and viral diseases cause the formation of phytoalexins in some plants – the latter of particular interest since no man-made chemicals affect viruses.

These anti-viral factors have been likened to human interferons, and the possibility of close similarity is strengthened by the fact that human blood interferon applied to tobacco leaves infected with mosaic virus stopped the spread of the disease, while the three proteins that appear to be anti-viral factors are triggered in both plants and animals by polyacrylic acid.

A new twist in this research has revealed that the injection of aspirin into plants also promotes production of anti-viral factors and has halted the development of tobacco mosaic virus. The basic ingredient of aspirin, salicylic acid, is a natural plant product, and has various peculiar attributes as described a little later and in Chapter 28.

There are certain chemical exudations from plants, which, as I have mentioned briefly in Chapter 19, can be toxic to other plants. Apart from the daisy family already cited, flax is inhibited from growing by the root antagonism of spurges and gold of pleasure, oats by certain thistles, rape by couch grass, spinach by radish and vice versa. During the establishment of rubber and coffee plantations in the Far East during the 1930s, it was customary to grow leguminous plants first to boost the nitrogen in the soil. Unfortunately some of these legumes proved also to exude substances very toxic to the young trees planted subsequently. Black walnut roots again exude a toxic substance which has been known to kill apple trees planted nearby, and to restrict growth of other plants. Grass roots are also toxic to apple trees, but in lesser degree, and so grass cover is used in orchards to inhibit root growth, as well as competing for nutrients, and thus provoke fruiting at the expense of shoot growth.

One compound released by roots of desert plants to inhibit the growth of others in the struggle for moisture is salicylic acid, referred to earlier as an interferon-like chemical – apparently this makes the membranes sheathing alien root hairs unable to retain water and nutrients.

Many seeds release inhibitors which prevent other species from germinating and curiously enough some of these are of present-day crop plants which inhibit weeds: thus beet prevents corn cockle from germinating, wheat and rye grass inhibit chamomiles and mayweeds. Exudations from roots may have the same effect upon germination, and possibly explain why cacti are so widely and apparently artificially spaced in semi-desert conditions. There is an American desert shrub, *Encylia farinosa*, whose fallen leaves effectively prevent any of the usual annuals from germinating under the sheltering branches, while in southern France, in country dominated by heaths, rosemary and similar shrubs, no annuals exist, because of root and leaf exudations.

Once again we may ask if plants are thus deliberately discouraging the competition of others for nutrients, light and water; and though this inhibitory power may often appear successful in doing so, we are then brought up by the fact that the dead roots of brome grass prevent its own seed from germinating, citrus seedlings are very difficult to establish in old citrus plantations, and the same applies to apples, peaches and roses. This suggests that toxic root and seed exudations may only be secreted by-products.

Most of this chapter has been about the repulsion of attackers or competitors by physical and chemical means. One last field, widely used in the animal kingdom, is that of camouflage and warning. Does this ever apply in the plant world? Warnings seldom exist. Those that seem to occur, in fruits and seeds, are invariably signals to appropriate distributors; the plant has no care about what happens to other animals.

But some plants do apply the principles of camouflage. One group comprises the so-called stone mimics of the South African mesembryanthemum family. These super-succulents grow in stony wastes, and some of them have undoubtedly developed both colours and patterns which very closely match the pebbles around them. The *Lithops* are the most remarkable group, their flat top surfaces almost flush with the ground; there are some fifty or more species of these 'pebble plants' or 'flowering stones' in shades of grey, brown, red and beige. Other mimic succulents of this family include the *Pleiospilos*, rough and speckled like pieces of greenish granite; whitish, warty *Titanopsis* imitating limestone; and several white-skinned kinds resembling quartz. Unrelated succulents include white crassulas and some extraordinary cubic asclepiads also resembling stones, and *Anacampseros* which remind me of bird droppings.

Certainly these plants are very difficult to detect with the human eye, and it is clear that they have grown to resemble the stones in their localities to an almost uncanny degree. But although I have heard of baboons being deluded, it seems that many desert birds including ostriches are not and they peck these succulent morsels out very accurately. During their short flowering their relatively huge blooms make them stand out, but during the dry season, when they become partly desiccated, they are certainly almost invisible. This seems to be one more only partly successful experiment of the plant kingdom, combined with or even provoked in this case by the adversity of the environment; if the experiment is no longer successful it may be because the birds have overtaken the plants in adapting to the circumstances.

A very subtle device against insects is that of some tropical passion-flowers which are as a group the food plants of caterpillars of *Heliconius* butterflies, which often defoliate them. When laying their eggs, the butterflies select the juicy shoot tips and space their eggs well apart so that the new-hatched caterpillars do not eat each other. To dissuade the butterflies, several passion-flowers produce 'dummy' eggs, so closely resembling the real thing and so similarly spaced that the insects never lay on these species.

This remarkable trick seems unique among plants, and in general terms it seems that spines, sticky hairs, thick cuticles, unpleasant taste and similar devices are better than mimicry against larger plant-eating animals. At the same time one must remember that plants are able to suffer much more damage and loss of parts than most animals because of their regenerative powers; and equally they make provision for loss by producing excessive quantities of flowers, fruit and seed, runners and tillers.

Part Two

26 · *The Exploited*

To exploit means, first of all, to make use of something, and in this sense plants have been exploited, almost since the beginning of time, by animals. Exploitation also means turning something deliberately or even unfairly to one's own account, and in this modern meaning the first real exploiter of the vegetable kingdom was man.

Such use occurred by imperceptible degrees. Early hominids became 'hunter-gatherers' who supplemented a meat diet with various vegetable material, or sometimes ate virtually anything edible. Some of the latter still exist, for instance the South African bushmen, some Australian aborigines, and a few central African tribes. The aborigines especially had to make every possible use of scanty natural resources in a very harsh environment; only rarely could they show a preference for a particular food. Any roots, seeds, fruits and edible shoots were gladly eaten, seeds sometimes being ground and baked.

The great period of the hunter-gatherers was from about 20,000 to 8000 BC, and as their experience increased so did the variety of fruits, vegetables and grains eaten and otherwise made use of.

However, long before that, probably 400,000 years ago in China, man's very early use of fire encouraged a different kind of plant exploitation. By involuntary conflagration this contributed steadily to the destruction of widespread forest, the creation of new habitats, and the localization of plant communities caused by the Ice Ages.

Fire-making remained essential for their very survival to those cave-dwelling people subjected to the glaciation of the late Pleistocene. Later on deliberate fire-making became man's first method of asserting his mastery over other forms of life both vegetable and, especially at first, animal. Fire could drive animals where man could slaughter them easily, open up the country for hunting and make prey visible, and it encouraged grassland at the expense of forest, thus stimulating the increase of meat-providing grazing animals.

At this point one might also emphasize that grass is the world's most important 'crop' as well as one of its most tenacious life forms – one which has been nicknamed 'the forgiveness of nature', and is capable of constant regeneration despite supporting vast numbers of animals. One square kilometre of grassland can support roughly 18,000 kg weight of animals of different kinds.

There are two main classes of domesticated plant: those which grow from seed, largely annuals, and those which increase mainly by other, vegetative means. The latter are almost all starchy root crops. Different climates may favour one or the other. Thus most of the roots are tropical – yams, manioc or cassava, sweet potatoes, arrowroots and taro; only the true potato and some other minor roots thrive in cold climates. Seed crops, which initially were largely grasses, including what we now call cereals, and many leguminous plants, occur most frequently in temperate zones. A vegetable diet based on seeds is more balanced than one based on roots; the latter is likely to tie its consumers to a specific habitat where there are adequate fish or animals to supply protein. Seed planters can move around more, and indeed may have to, because an annual crop means that the soil is bare much of the time and prone to erosion, loss of fertility and weed invasion. Root crops, being perennial, tend to stay where they are, and can indeed be grown on steep slopes.

Sometimes with roots active cultivation is hardly necessary; J. G. Hawkes quotes examples in Colombia and Venezuela where even today peasants grow potatoes close to the house where the plot is continually enriched by household rubbish and excreta from household animals, so that harvesting takes place continuously without any replanting.

Primitive people are notoriously untidy. Not only do they fling their rubbish on to middens among their habitations, but they are casual about dropping some of their gatherings as they bring them into the settlement. In this way there grew up around the homesteads of primitive hunter-gatherers all manner of plants of the kinds they liked to eat. Very often, notably in the case of grasses and leguminous pulses, these were plants of poor and disturbed soil where more permanent and slower-growing species could not get a foothold or could not grow well – what today we call weeds, making the most of difficult situations. They took readily to the trampled and disturbed areas around the encampments and along the approach paths to them. And as if by magic the settlers found the plants they wanted all around them.

Such situations are very good examples of the way plants half-way between weed and crop surround habitations. If there is a crop failure of the selected species, the weed species around will be harvested: we see how cultivation began with the gradual realization that some kinds of similar plants were better than others, and were therefore worth actively sowing and harvesting, while the weed species, as it were, presented themselves at the garden gate for possible acceptance, often showing unexpected potential in the rich soil likely to be found there.

Just as the change from gathering to harvesting and storing was extremely gradual, so was that from harvesting to active planting. This domestication of plants only occurred when the world had recovered from the effects of the last glaciation and when man had reached a social and cultural level of some sophistication and had settled in permanent villages or indeed townships. Plant domestication – agriculture, in short – is often considered to have begun around 7000 BC in Anatolia, Persia and Syria, with the cereals wheat and barley. Foxtail and other millets were the basis of an agricultural system in northern China from at least 5000 BC.

However, vegeculture of a different kind, based on the roots yam and taro, is now believed to have been practised in south-east Asia as early as 8000 BC, and that based on potatoes in the South American Andes about 6000 BC.

Let us consider first the cereal cultivation of the Near East where wild species of wheat and barley were involved, since the archaeological evidence is strongest for these. The seeds of these and other grasses seem originally to have been roasted, but they were later ground, when they had more potential in different dishes. In this connection it is fascinating to note that grinding was originally devised for milling ochre, so that man did not have technical problems to overcome when he realized that grain could be of more value as flour than roasted whole.

Wheat and barley, like other, later primary crops such as rice, soya bean, flax and cotton, were initially at a slight disadvantage if their cultivators moved to colder conditions, either north or into mountainous areas. In these circumstances certain weeds competed with them, enjoying the conditions provided by cultivation. These include rye and oats. Not only did these weeds compete, but they underwent selection processes which resulted in their mimicking the crop plants in various ways. Eventually some of the weeds were appreciated as crops in their own right, and these secondary crop plants sometimes superseded the primary ones in colder habitats, though barley is the most cold-tolerant cereal. In countries further east they included buckwheat, while various kinds of millet became an important secondary crop/weed in the warmer south. In rather different circumstances the tomato originated as a field weed of maize and beans in Mexico.

The most remarkable thing about cereals under early cultivation is that they rapidly underwent one very important change. In the wild it is clearly valuable for a grass to lose its seeds as soon as they are ripe. This creates a problem for the gatherer-harvester, for it restricts him to a one- or two-week period, which a warm dry spell can reduce to two or three days. In cultivation a cereal is needed which will retain its seeds in the head while the plants are harvested and until the heads are broken up by threshing.

The mutation to intact seed-heads from brittle ones probably occurred from time to time in the wild, but did not persist since it was not beneficial. But once man started gathering seeds he was clearly more likely to do so from plants with intact seed heads. Once cultivation arose from the deliberate use of stored seed, intact-headed plants were unconsciously selected. Such strains are of course largely dependent on man for their survival; present-day maize has absolutely no natural method of dispersal. Man presumably also tended to use the largest seeds for sowing, and so selection for bigger seeds took effect; another selective pattern would have been to plants without hard, stiff glumes – leaf-like organs which are found between seeds in a head, and make gathering painful.

It is also interesting to note that wheat and barley are basically self-pollinated, whereas most other grasses are cross-pollinated. Such sexual habits produce strains which breed fairly true, and pure lines can exist together without risk of being swamped or altered. In conditions where weed cereals abound it is only possible for self-pollinated kinds to be successfully maintained as strains, and this is one reason why wheat and barley were the first cereals to be domesticated.

However, these plants *can* cross-pollinate on occasion. This allows for some development and for the selection of the best resulting strains by man. There can also be occasional exchange of characters with wild relations, once again increasing cropping potential and adaptability to different conditions in new regions of cultivation. Evolutionary operations are greatly accelerated under man's domestication. Other examples include seedless oranges and grapes, and mangoes without the strong odour that made the wild fruits more attractive to their bat dispersers than to human consumers. Conversely, the original dispersers, like the bats spreading dates and birds so many fruits, found themselves in the role of thieving pests.

In the early millennia of agriculture, especially in the Near Eastern regions where a combination of unpredictable seasons and elementary methods made lean years likely, wild foods continued to be important. Early agricultural man also took to herding sheep and goats. These acted as an alternative method of 'storing' vegetable surpluses, apart from providing man with milk, wool and meat which needed no hunting.

It is interesting to examine diets of early Near Eastern farming villages. Detailed examination by F. Hole and K. V. Flannery of the remains of a village called Tepe Ali Kosh, dating from between 7500 and 5600 BC, on the steppe of south-western Iran, revealed seeds of over forty plant species. The most important of these were domesticated emmer wheat and two-row barley, together with wild caper (eaten for its fruits), pistachio, several wild legumes used for seeds, wild two-row barley, four wild grasses, vetchling (eaten for its pods) and shauk (*Prosopis*, a pod-bearing legume which can co-exist with cultivated cereals). Considering that remains of mammals, birds, fish and shellfish were also found, it is likely that the inhabitants of this early village ate better than the present-day descendants.

At the village of Tepe Sabz, also in south-west Iran, an early example of irrigated farmland dating from 5500 to 3700 BC, the proportion of domesticated plants has greatly increased. It includes three kinds of wheat, three of barley, linseed, lentils and grass peas. There are two kinds of wild grass, vetchling, shauk, caper, pistachio and almond. No seed of wild legumes was eaten – they are very small, though nutritious – but these plants were certainly collected as animal fodder, as indeed they are today.

Linked as it is to man's first settling down in communities, this seed- and herd-based agriculture of the Near East deserves special emphasis. Root farming is essentially a tropical occupation of primitive communities, often combined with culture of bananas and their relations the tropical plantains, which involves relatively little effort and forethought. The principles of seed selection seldom apply to root cultivation, although better kinds of root can be sought for cultivation. Basic methods of tropical root cultivation – as indeed of temperate potato growing – involve the planting of 'seed' roots or pieces of root, or of stem cuttings. These may go directly into the ground or, in a more sophisticated type of cultivation, into prepared, enriched earth mounds. Very often root growing is carried out in plots used for a year or two and then abandoned; this method is called *swidden* after the Old English word for a burned clearing. Though swidden may rely on only three or four vegetables, very many more can be involved: the peoples of Mindoro Island cultivate over 400 different plants, both from seeds and cuttings, by this method.

Not only does swidden give much lower yields than permanent tillage, but a community relying upon it needs something like thirty times as much land as one of plough farmers, and can probably only support an average of fifteen people per square mile. The real trouble with the method is that, if land is over-used by repeated burning, it is likely to end up completely useless – either totally eroded or covered with coarse, valueless growth like elephant grass.

The grain farmer has both to plan and to labour in order to carry through the cycle of tillage, manuring, sowing, weed control, harvesting, threshing and winnowing and finally storage. All these operations are capable of improvement, resulting in potential increase of the crop, which in any case has a very high yield per hectare compared with other crops or livestock.

Cultivation of roots is increasingly believed to be if anything older than that of cereals, as mentioned earlier. The late Neolithic Hoabinhians of south-east Asia may have been the earliest people anywhere to settle and cultivate, growing yams and taro among roots, with cereals like millet and, a little later, rice, which possibly originated as a weed in the swampy taro gardens. Carl Sauer suggests 'that this is the world's major centre of planting techniques and of amelioration of plants by vegetative reproduction'. Certainly the peoples of this area were socially, technically and artistically advanced; they may not have originated major civilizations but they were extremely competent and versatile at making the most of their environment. Plants provided the carbohydrates in a diet where fats and protein derived from fish and shellfish, and later from land animals.

It seems likely, in principle, that the Near East was one cradle of agriculture and south-east Asia another, and obviously each arose from very different circumstances. The Andean potato growers are a third group who found a locally native staple to cultivate.

It has been suggested that the peasant communities from which the important world civilizations have sprung – those of Europe, India, Egypt and northern China – were all mainly grain farming. Equally the Mexican and Peruvian civilizations were based on seed culture of maize, beans and squashes. In contrast, some writers, like I. H. Burkill in an article on the yam, go so far as to say that root farming 'advances that part of the population which contributes least to the common good'. This is surely an extreme view!

But in any case, my theme in this chapter is the development of plants under the impact of domestication; it is not materially affected by polemics nor differences in place or time.

It may well be that the Asiatic vegetative agriculture was a good deal more comfortable than the Near Eastern seed-based agriculture. It is in the latter that we can see how, even at this early stage, man and his few chosen plants and animals became virtually dependent on each other for survival. They were involved in what C. D. Darlington has called 'a matter of mutual selection, a kind of process in animal and plant adaptation where all parties to the transaction effectively modify one another'. Man became steadily transformed in his way of life; and as he moved into different habitats the conditions further modified the plants and animals he brought with him. In general, hunting and hunting-gathering became combined with cultivation, while at the other extreme a few groups became nomadic herders, or pastoralists. Cultivation was an irrevocable step away from earlier methods of finding food, and sometimes had unexpected effects on the society concerned. Thus, when the ancestors of the Mayas developed maize, their food problems were solved, and a man could provide enough food for his family by a couple of months' work a year. But an intelligent aristocracy arose from the more or less democratic hunter-gathering society, and enslaved the rest of the population to build palaces, temples and cities instead of being allowed to enjoy their leisure.

Quite early on the environment and the wild resources around centres of population became irretrievably altered by these two activities, so that man became set on his course of ever-increasing numbers fed from non-natural resources. The damage to habitats, as in the Oxus and Tarim basins, now desert, closed routes behind early migrants and resulted in the isolated communities which were the origin of nations. Larger wild animals were driven out of areas of settlement and often eventually made extinct. Weeds and pests of the crops moved in and developed. The whole evolutionary system of change, one of slow degree, under many checks and balances, was overset.

But at that time nature must have seemed inexhaustible, and man went on helping himself as his ancestors had done in their limited way, and indeed as all vegetarian animals had done over the millennia. It is noteworthy that few domesticated plants of major importance were not known and under man's control well before historic times: it can be said that man very early became a skilled economic botanist. From the wild he obtained all manner of other foods, bringing these into cultivation if demand required it. Such were fruits and nuts of many kinds, egg plants, beans, pumpkins and melons, asparagus, artichokes, onions, cabbage, lettuce, temperate roots including carrot and parsnip, sago, sugar cane and a few mushrooms.

Without understanding them primitive man was also making use of microscopic plants in food production, the yeasts and other fungi involved in production of bread, cheese, yoghurt, beer, wine and vinegar. In the tropics such micro-organisms attack food readily, and the resulting fermentation came to be very much to the taste of the inhabitants. Thus, in Indonesia, fermented soya bean and groundnut cakes make up a third of the protein requirement. Fermented soya bean paste (*miso*) is used daily by about three-quarters of the Japanese to make soup. In the Philippines bacteria produce a jelly-like substance on coconut milk and fruit juices (*nata*).

All kinds of plant, and every part of them, were examined by man for potential use, although large-scale cultivation of utilitarian plants on the whole came very much later. In the cradle of civilization the date palm was one of the most extensively used trees, providing nourishing food, constructional timber, thatching and walling material in its leaves and fibres, and so many other materials that there is an Arab saying that the palm has as many uses as there are days in the year.

In Ancient Egypt papyrus was another multi-purpose plant. Besides its immemorial use as paper (the pith was sliced into thin strips, overlapped, pressed together and dried) other parts of the plant were used for cordage, sandal-making, construction of boats and, in hard times, as food (from the submerged rhizome).

In the East the bamboo found quite as many uses as the date in food, structure, thatch, receptacles for food and oil, cooking utensils, weapons, farm implements, even musical instruments. The rattans or climbing palms are perhaps the most valuable tropical group, used in all manner of ways.

Apart from food and beverages made from plants, wood must have been the earliest vegetable product widely used – for building, weapons, tools or boats. With its vastly varied ranges of resilience, durability, and density from the very light balsa to ironwood and lignumvitae twelve times as heavy, wood is capable of a great diversity of employments. Besides all its obvious uses timber was later transformed into paper – one of its major employments today – and also, very much later, into rayon and plastics derived from cellulose. Some of its qualities were forgotten after the Industrial Revolution, but in the Second World War one of the most successful aircraft was the Mosquito with all-wood construction (including balsa), and it was followed by the Vampire and Venom jet fighters. Cars with chassis and body all of wood have also been successfully produced by an engineer who refers to wood as 'God's first plastic'.

Very early too came fibres, largely derived from the barks and stems of various plants, and separated from the tissues by 'retting' which often involves letting the plant material decay in water or in the open air. Retting in streams probably gave rise to the apparently quite early discovery that some plant juices, like derris, would stupefy fish. The most important fibre plant is undoubtedly cotton. Wild cotton has no lint on its seeds, and indeed lint-covered seeds are, like non-shattering seed-heads in cereals, unsuited for wild existence. It is possible that it was originally brought into cultivation for its fibrous stalks or perhaps for its oil-bearing seeds; the familiar cotton-filled bolls are entirely the result of domestication. Fibres provide textiles, cords, material for baskets, brushes, rugs, house divisions and filling materials.

Dyes were early extracted from plants, including lichens. So were essential oils, for medicines, flavourings, unguents and perfumes, and this developed later into great industries (200 million rose petals have to be distilled to make 1 kg of Rose Otto, or Attar of Roses). Today essential oils are used also in paint solvents, insecticides and as an ingredient in dozens of industrial compounds. Gums, pectins and resins find use in thousands of ways, including incense, varnishes, soap, all manner of cosmetics and creams, in food manufacture as in jams and ice cream, as thickeners and emulsifiers, and in many industrial materials as diverse as linoleum and printers' ink. Tannins are used for preparing leather. A familiar extract is latex, from which rubber is made, one of the staples of our modern civilization; many trees produce this as do numerous herbaceous plants; during the Second World War the Russians obtained rubber from a local species of dandelion. Plants yield many indirect food products such as sugars and starches, oils and fats; some of the latter, and also vegetable waxes, are used for a great range of industrial purposes.

One startling development of recent years has been the fermentation of plant material into alcohol on which to run motor vehicles. The alcohol is normally mixed with petrol and the resulting 'gasahol' only needs minor engine modification. New engines have been designed which will run on pure alcohol.

Sugar cane has been the most widely used plant for this purpose, which has been pioneered in Brazil where there are hopes of reducing petrol consumption by 40 per cent. Cassava is increasingly also being used for gasahol. Large areas of Amazon forest have been destroyed for gasahol farms, and further short-sightedness has resulted in pouring by-products of fermentation into rivers, resulting in widespread death of fish, instead of making use of it for fertilizer, cattle feed or biogas production.

Vegetable oils such as sunflower oils have also been used to power diesel engines – indeed Rudolf Diesel actually powered his 1900 prototype engine with nut oil. In the short term vegetable oil, either pure or in mixture with diesel, is successful, but in the long term it produces problems, notably excessive carbon desposits in engines. Chemical processes which could improve the situation are at present uneconomic.

The medicinal uses of plants are both very ancient and extremely numerous. Some of the earliest records, from China, the Near East and Egypt, are of herbal plants used in medicine – Sumerian tablets of 2200 BC list 1,000 plants with different uses. 'Primitive' people like South American Indians have an unparalleled knowledge of medicinal plants, and whole present-day cultures base their medical treatment largely on plants, including India, China and Vietnam. Even the scientifically minded West uses huge quantities of plant material in modern drugs, quite apart from specifically herbal remedies.

Though many plant specifics have been synthesized, or superseded by new synthetics, a considerable number have defied chemistry and are still extracted from the living plant, including some very important modern discoveries. A race is now on to examine tribal use of plants and to screen as many species as possible while both tribes and plants still exist.

Lastly among plant uses are insecticides such as derris and pyrethrum, antibiotics of fungus or lichen origin, some bactericides, and herbicides based on plant hormones.

Living plants may have special uses, as for instance in windbreaks, and especially in aiding the recovery of sterile areas. Various grasses in particular are used for binding sand-dunes and mud-flats, improving arid regions, colonizing rubbish tips and so on. Hidden ore deposits can be located by mapping the distribution of certain species and observing the altering growth of the plants concerned, as well as by analysing their tissues.

Once in a while a plant really scores over modern technology. The Fuller's teasel is my favourite example. The narrow oval flower-heads, armed with hooked bracts, are still used by the million in the cloth industry to raise the nap, or pile, on some top-quality dressed cloths. Mounted on boards, they make the pile rise evenly in one direction. Wire substitutes cannot produce the quality required.

A vegetable triumph over engineering, of less precision if more humour, was recorded by Henry Longhurst when describing the building of the 'Golden Stairs' road to the Gach Saran oilfield in one of the hottest parts of Iran. Local tribesmen were employed to drive the lorries, which were immobilized by over-heating for much of the day – all except one which 'was observed to make no concession to the heat of the day or the steepness of the road . . . One day the foreman in charge, and the field superintendent lay in wait and stopped the driver and demanded to be shown the engine. They found half a melon placed over the carburettor, the juicy flesh keeping it just cool enough to prevent the petrol from boiling.'

Vegetable technology is sometimes quite unexpected; Thomas Edison used bamboo fibres as filaments in his early electric lamps before tungsten was developed for the purpose.

Marine algae have a very wide range of uses. Many are important as food, particularly in Japan, and as fodder and manure. Agar and alginates are two products extracted from them which have a multitude of uses, the latter being sometimes used as a conditioner of heavy soils and a binder for light or sandy ones, and especially valuable in land reclamation.

And of course the remains of dead plants are of immense importance in the form of coal and lignite formed from the remains of the giant plants of the Coal Measures; while peat, formed by plants decaying in anaerobic conditions within a few thousand years, is also much used. Chalk, the basis of cement and agricultural lime, is largely made from the fossil remains of microscopic golden-brown algae containing calcareous plates, deposited at the bottom of Cretaceous seas; while antique diatoms form deposits of diatomaceous earths, used for filters, abrasives and refractory linings for furnaces.

As antique agricultural man migrated, he gradually spread the seeds, roots and bulbs of his crops. When enough people were spread about the globe to begin to understand that other communities existed and that certain plants and their products (as well as all manner of other natural and manufactured commodities) only occurred in one place or another, trade began. Slow and arduous though caravan traffic might be, there was an ever-increasing movement all over Asia, including of course China, and eventually spreading into Africa. Such caravans mainly carried such commodities as spices, medicinal plants, herbs and other improvers of daily life which were sought after as civilizations developed. They would not normally trade foods, but they might carry seeds, bulbs and roots. The first recorded plant-collecting expedition was indeed for frankincense trees, which Queen Hatsepshut of Egypt had brought by boat in 1482 BC from the Land of Punt (Somaliland) – thirty-two trees for her great temple at Deir el Bahri.

These early movements were certainly most important for the spread of plants, for exchanges occurred between Asia and Africa. The Romans spread Mediterranean crop plants to various parts of their far-flung empire. Later the Arabs imported into Spain rice, sugar cane and citrus fruits, the latter originating from China. The Polynesians would take breadfruit, taro, sweet potato and coconut with them on their long-distance colonizing voyages.

But it was the successful regular navigation of the Atlantic after 1492 which transformed the situation, making the interchange of old-world and new-world species possible. Among food plants Europe received potato, tomato, maize or sweet corn (and also tobacco) from South and

Central America; Jerusalem artichokes and so-called French beans arrived from Canada. Old-world tropical countries benefited in particular from the South American manioc or cassava. The Americas benefited with old-world plants including citrus, rice, sugar cane, bananas, yams, various peas and soya beans; and of course wheat, together with most of the European cereals, was brought to North America by the Pilgrim Fathers and their followers. Curiously enough it was Irish emigrants who brought the potato into North America. The coffee for which South America became famous reached it from Arabia and Ethiopia.

Portuguese ships brought coconuts and rice from Asia to west Africa. The notorious but horticulturally dedicated Captain Bligh moved bananas, breadfruit and the thick-stemmed sugar cane from Tahiti to Jamaica. Peruvian *Cinchona*, for quinine, was established in Ceylon and the Far East, and so was Brazilian para rubber – from seed smuggled out about a century ago. Serious botanical exploration in North America discovered timber trees such as the Douglas fir – now one of the most commercially important in Europe – while the Monterey pine, which in the wild is now confined to a small group of wind-contorted specimens in southern California, has gone forth to be one of the most valuable, fast-growing softwood trees for warm climates. About the same time, at the turn of this century, species of *Eucalyptus*, endemic to Australia, were brought to every country with a suitable climate to be used for timber and also for ornament.

The ornamental use of plants has absorbed man since quite early times, though at very different dates in various parts of the world: it can only develop when there is a combination of leisure and culture. Then, to quote Ruskin, 'flowers seem intended for the solace of ordinary humanity'. The Ancient Egyptians grew flowers and trees for ornament, and also used more incense than any other nation since. Advanced peoples such as the Chinese were cultivating flowers while the British were daubing themselves with woad. There is a Chinese proverb which says, 'When you have two pennies left in the world, buy a loaf of bread with one and a lily with the other.'

Gardening developed by degrees in most parts of the world. But whereas most civilizations had contented themselves with cultivating and improving what already grew around them, it was the British who, once they got started, sought plants for ornamental use most actively from far countries. As with food plants, it was the European command of ocean travel which opened up the world for ornamental plants, and among the earliest brought over in this way were the passion-flower, tuberose, marvel of Peru and 'French' and 'African' marigolds, by the Spaniards from Central America. Seeds and bulbs were at first the main means of introduction, but live plants were brought across the seas as better methods of keeping them alive were devised.

Some ornamentals probably owe their very existence today to their beauty. The spectacular flamboyant tree is only known wild from one area of Madagascar. Its equally striking relation *Amherstia nobilis* has only been seen in the wild two or three times, and plants in cultivation originate from a specimen in a Burmese monastery garden. Likewise the maidenhair tree, now so widely used as a street tree in North America, has apparently never been seen in the wild; it was grown in temples and gardens by the Chinese.

One of our houseplants, the so-called friendship or aluminium plant, has only once been found wild, in Indo-China; all the plants in cultivation derive from the original collection. Another ornamental only known in cultivation is the jade plant, the original trade specimen of which was bought in a Mexican market and which the Mexicans grow in their homes.

Thus in some cases exploitation of a plant will save it from extinction; in other cases it will be vastly improved and, though it becomes unable to fend for itself in the wild, will thrive in its artificial environments. This applies especially to individuals selected for abnormalities such as flower doubling and leaf variegation.

There are still many plant products that civilized man does not understand, but which are potentially of great value to him. Some of their potential has been worked out by 'primitive' tribes such as those of South America, of whose unique herbal knowledge a very little has been gleaned by anthropologists, yet which 'advanced' man is busy rubbing out of existence. Anthony Smith has described how the modern Brazilian says of his forests and wildernesses, '"Here there is nothing". But conversely the Indians will say, "The forest has everything. It supplies all our needs, the food we eat, timber and thatch, wood for bows and arrows, fruit and medicine." The forest is either *nada* or everything. It depends who you are.'

Modern forestry usually contemplates forests either in terms of timber or of land to be freed of trees for agricultural use (the effects of such destruction are described in Chapter 29). But a forest, properly managed, is an enormous resource in its own right. At the 1974 Commonwealth Forestry Conference it was emphasized how 'western-orientated foresters indiscriminately hew down the trees, while the native cultures which have the knowledge of how to gather a multifarious and sustained crop are rapidly becoming extinct. One Peruvian forest tribe, the Agurana, for example is known to be familiar with some fifty species of palm trees alone, each of which has its own name, characteristics and uses.' (I quote from Jon Tinker's report of the conference in *New Scientist*.)

The Indian forest conservator T. Krishna Murthy described how his forests were active producing areas of

'plants yielding medicines, oils, spices, gums, resins, tanning and dyeing materials, forage, beverages, insect and rodent poisons, green manures and rubber, as well as animal products like honey, hides, ivory, feathers and musk . . . There are oils from cedars, junipers, cinnamons, palmarosa, sandalwood and vetiver. There are resins from pine and other trees which are used in inks, lacquers, linoleums, paints, varnishes, plastics, chocolate, paper and pharmaceuticals . . . India's leather-tanning industry depends heavily on the forest for its 250,000 tonnes of vegetable tannings and dyes; on myrobalan nuts, wattle barks and mangroves, for example. Bamboos, the Indian substitute for scaffolding, are also used for brushes, fans, baskets, ladders and bridges, while 14,000 tonnes a year of canes and rattans are needed for ropes, cables, chairs, ski sticks, umbrella handles, and cricket bats . . . The indigenous peoples who in the Amazonia are threatened with extinction are in India the indispensable agents of rational use.'

If I have quoted this report at some length it is to demonstrate the inexhaustible, abounding and amazingly varied potential of the tropical forest which, as I elaborate on page 383, is disappearing at nearly 22 hectares *a minute*. Where shall we obtain these products when the forests have gone?

Our ever-increasing needs depend on improving our skills in manipulating cultivated plants; at the same time we must protect what remains of the wild more carefully, for it can provide more than we yet know, preserves the surface of the earth, supports wild animals we regret losing, and provides solace sought ever more avidly by the over-mechanized, boxed-up human race.

27 · *Man the Manipulator*

The moment man began deliberately to sow seeds, or to push cuttings of vegetative crops into the soil, he was in control; indeed, he *had* to be. Sowing seeds or growing roots, tubers, bulbs, suckers and so on involves clearing the ground and usually stirring it up in some way to provide better conditions for the plants. Thus there arose a work cycle, geared to the seasons, involving cleaning and turning the soil, manuring and weeding, sowing or planting, and ending in harvesting and post-harvesting operations. Over the millennia these operations have all been slowly but steadily improved and refined and have become a specific kind of evolution in their own right. This is the first line of control.

The second line is that of reproductive material. Seed needs cleaning to avoid weeds among the crop, and careful storage to give maximum viability and to ensure optimum subsequent growth. Nowadays seed is harvested with special care, may be stored in controlled temperature and humidity, and to increase its life is often packeted after extraction of much of its water content.

Reproduction by cuttings implies knowledge of the best time at which to take and 'strike' these, as the gardener calls it; what part of the plant to use, and how much of it. In the tropics very many plants root from cuttings pushed into the ground with little preparation, and often at almost any time. Temperate plants are usually more choosy about increasing in this way.

In various aspects of plant manipulation and cultivation increase by cuttings had widespread agricultural use limited to relatively few species for many centuries. Increased complexity in the practice, including greater use of root cuttings, was largely the result of wishing to multiply ornamental plants in the absence of seeds, which take too long to rear, do not appear in cultivation, or are useless because of their hybrid origin. Modern refinements include closed propagating cases, which may be warmed and irrigated from below, and the use of 'mist' by devices which fill the air with a very fine spray at intervals controlled by the drying out of the plant material, often using an 'electronic leaf' as the control mechanism. Powders based on plant hormones are widely used to speed up rooting, and there are special techniques, including deliberate wounding of the base of the cutting, which also accelerate rooting in some plants. The sheer multiplication possibilities with modern techniques are remarkable; I know one European grower who produced 20,000 individuals of the carpeting plant *Pachysandra* from one original plant in a year.

Another specialized modern method is the use of meristems, the microscopic growing tips of plants. This involves the excision of the meristems under a dissecting microscope (they are only a half to 1 mm across), and their growing on in sterile cultures until they are large enough to handle and treat more or less like orthodox cuttings, or seedlings. The need for this method stems from the vast increase in virus diseases, which exist in the very cells of the plant and hence are transmitted by sap-sucking insects. But the meristem normally escapes them because the virus can seldom keep up with the cells of the growing tip. Some of these viruses are lethal or visibly crippling;

others, much more insidious, simply reduce the growing and cropping potential of plants. Virus-free rhubarb, for instance, makes 70 per cent more growth than an infected plant. Apples are very often infected with hidden viruses which greatly reduce their cropping capacity; while it is well known that strawberries, which are very prone to viruses, as indeed to the aphids that transmit them, usually have to be discarded after three years. The use of meristems, laborious though it may be, allows growers to build up 'nuclear stocks' from which disease-free planting material can be distributed. Plants which have successfully been treated in this way include fruit trees, strawberries, potatoes, dahlias, carnations and chrysanthemums. Meristem culture may become an important method for speeding up the distribution of new varieties of normally slow-growing trees.

Meristems are started in growth in test-tubes on nutrient jelly, and if these test-tubes are slowly and continuously rotated, or in some cases periodically vibrated, the resulting growth forms numerous buds, which can later be cut apart and grown individually. This method has been seized upon by orchid growers to multiply their most attractive, highly bred varieties, which in normal conditions would take years. It has been estimated that, if the buds were regularly divided and restarted, four million infant plants could be obtained from one orchid meristem in a year. Naturally this greatly reduces the price of these exotics: indeed, production of numerous specimens of very rare orchid species will hopefully reduce the pressure from poachers if the market price can be brought down.

By contrast, grafting is an antique skill, although it also involves much care and precision. Its basic purpose is to create a new plant, usually a tree, whose upper parts bear specially good fruits and whose root system is vigorous and healthy. Thus wild root-stocks can often be used, while the scion, or top growth, comes from known, selected forms which, if hybrids, cannot be reproduced true from seed. Many plants likewise do not strike readily from cuttings, or take too long to do so. In many cases the root-stock is chosen to influence the final size and vigour, as with fruit trees, or sometimes to help the plant resist disease, as with tomatoes. Only closely related plants can be used, rather as organ grafting in human beings is controlled by blood groups.

It is possible that the idea of grafting was first derived from observation of natural grafting, in which young shoots or roots in close, well-anchored proximity will fuse together. This is 'grafting by approach', and can be emulated by the cultivator. It is certain that grafting with scions is a very ancient practice; there are many biblical references to it and it was well known in Ancient Greece. Despite the old wives' tales that then surrounded the practice, grafting was in commercial use, together with some understanding of the influence of rootstock upon scion, very early in the Christian era, and St Paul refers to the practice in his Epistle to the Romans. A scientific approach imbued grafting far earlier than it did most agricultural and horticultural practices, and a great number of different methods arose. Until the eighteenth century grafting was virtually confined to fruit trees; today it is frequently used for ornamental shrubs and trees, notably roses.

With annuals, seeds are the only effective means of propagation. Other plants may also increase better from seeds than from cuttings or grafting. Seed, however, can produce variable offspring. Normally, a species or a fixed horticultural strain breeds true from seed, but even the offspring of the 'pure lines' that result from continuous inbreeding, which are genetically identical, show considerable variation. Within pure lines also, variations can and do occur due to mutations and sometimes to some genetical instability. If seeds are of hybrid origin the offspring will vary greatly, and the next generation will be even more varied. Seed therefore presents constant opportunities for selection and improvement; vegetative means of propagation guarantee that the progeny are identical to the parent, for they are of the same tissue. A parent and its vegetative progeny are known as a clone, and clones can exist for hundreds if not thousands of years, as is possibly the case with olives, without material alteration. Convenience plays a considerable part in the method chosen.

Once man has mastered methods of cultivation and propagation, he seeks to improve his plant material. In some cases he selects the best forms from the wild; this is especially so in the case of plants grown from roots or cuttings. With seed, at the simplest level, he chooses the biggest and healthiest parent plants, or those with the biggest grains or fruits; this must have been how early man began, virtually unconsciously, to improve his cereals.

Familiar plants, each derived from a single species, whose cultivated forms have until at least very recently largely arisen as a result of simple selection of variations, include tomato, cucumber, maize, rice, beans, carrot, celery, lettuce, onion and beet.

The search for desirable characteristics is at its most obvious with ornamental plants. One of the simplest mutational 'jumps' is that of double flowers, always prized by the gardener. The arrival of the Spencer sweet pea, characterized by large flowers with undulating rather than flat upper petals, followed a mutation which occurred in four separate places between 1899 and 1902. The original Russell lupin, in which the florets radiate nearly horizontally from the stem instead of downwards, giving the spike a much denser appearance, was observed among his plants by an amateur lupin fancier, who spent years breeding from this one 'freak'. The excelsior foxglove is a comparable 'freak' in which horizontal florets transform the flower spike.

Once a seed strain has been established it has to be maintained. The simplest and oldest method is the removal ('rogueing') of inferior or nonconformist individuals in a crop. Simple as this sounds, crop yield improvements of between 20 and 50 per cent can be achieved within four generations by this mass selection. This procedure usually maintains several 'lines' within the general specification of the strain. Purer lines are maintained by more demanding methods.

From plant selection man turned to plant breeding. Often nature did this for him, and he found new variations within his crops, the result of chance matings or of mutations – like the Russell lupin – which would have been swamped in nature. Later he bred deliberately, crossing (hybridizing) varieties of the same species to allow for recombination of genetic characters, or something crossing distinct species if they were compatible. (Species with different chromosome numbers cannot normally be crossed.) Thus the present-day strawberry has arisen by the crossing in England of the North American species *Fragaria virginiana* and the Chilean *F. chiloense*. The first step towards our modern repeat-flowering roses was the introduction of a china rose, which held the repeat-flowering character, for hedging to the island of Bourbon or Réunion, where it mated naturally with a form of damask rose also used for hedging, giving rise to the ancestor of the Bourbon group. Such examples demonstrate how improved communications transformed the potential of plant breeding by making forced marriages between species normally living far apart.

Hybridization between species may result in sterile offspring. However, subsequent doubling of the chromosomes, which quite often occurs in nature, usually results in fertility, and the offspring becomes a new source of potential variation. Plants with more than twice the normal chromosome complement are known as polyploids; those with twice the number are tetraploids, and these often exhibit greater size, vigour and yield. Crossing of distinct strains may result in 'hybrid vigour', and the F.1 hybrids listed in seed catalogues (maize, brussels sprouts and tomatoes are examples) are the result of such crosses. They are also extremely uniform. With perennial plants such as lilies, seed-derived strains of this sort avoid the virus diseases which can be perpetuated by the normal vegetative propagation. F.1 hybrids do not themselves breed true and the basic cross must be repeated annually, a process normally carried out by hand.

One of the outstanding successes of F.1 breeding is pearl millet which occupies nearly 30 million acres in India. Within three years of beginning a breeding programme, a hybrid was obtained which increased yield by 88 per cent. Part of the success of this particular hybrid was the obtaining of male-sterile plants, so that crossing of the two parent strains was not upset by self-pollination. If the plants chosen for pollination (i.e. the 'mothers') are not male-sterile, the anthers have to be removed by hand, which is a very tedious operation. Obtaining male-sterile plants for this purpose is now an important aim of breeding in many plants.

One aspect of plant breeding is the deliberate search for a character. Lupins, lush, quick-growing plants, seem an admirable source of animal fodder. But wild lupins contain alkaloids which stop animals eating them. Breeders argued that once in a while there would be an alkaloid-free mutation. Sure enough, after examining 1½ million plants, six such individuals were found and bred from.

It is only in the last century that breeding has become scientific. Before that period even the reproductive processes of plants were not adequately understood, while it was not until the turn of this century that Bateson elaborated the science of genetics, and indeed coined the word, following rediscovery of the famous work of Gregor Mendel.

It is now possible to be purposeful in breeding rather than relying on chance. Crop yield, or flower size in ornamentals, are the usual primary objectives. New races of the 'big three' cereals – wheat, rice, maize – have been created in the last few years, including a promising wheat/rye hybrid called triticale. These have increased yields at least three or four times, perhaps as much as fifteen under experimental conditions. Breeding is also concerned with increasing protein content.

It is not only grains whose yield can be increased, but timber as well. By selecting the most promising trees in a plantation, increasing them by cuttings, and cross-breeding the results, timber trees can have their volume at a given age increased by at least 25 per cent. Hybrids of selected European and Japanese larch are bulking up to 180 per cent more than their parents. Even the constitution of the actual product can sometimes be improved by breeding, as with oilseed-rape, new varieties of which can produce oil free of the unpleasant and harmful erucic acid, present in dangerous proportions with older varieties.

Yield increase is also dependent on improving photosynthetic efficiency. Even the best fodder grass only uses 4 per cent of the light energy which reaches it. Recent breeding has pushed this figure up to 5 per cent and increased dry matter production by a third. A rice strain with nearly double the photosynthetic rate of older varieties has also been produced.

Breeding of produce for marketing has been directed towards uniformity and often simultaneous maturity, as with tomatoes and strawberries, to permit the best use of mechanical picking. With strawberries this is accompanied, in new varieties, with very long fruit stalks. The canning and freezing industries create a similar aim, as with brussels sprouts and peas. The latest pea hybrids have virtually no leaves; their stems are adequate for photosynthesis and the increased sun reaching the pods makes them ripen more evenly, while absence of leaves again makes for easier mechanical picking. Such requirements and others, such as that of toughness to withstand packaging, ought of course to be matched by good flavour and food value, which some present-day examples lack. The consumer should not be the last of the breeder's priorities.

Other recently achieved aims have included breeding short-stemmed cereals for ease of combine harvesting, which are also valuable because they do not 'lodge' or collapse easily; 'semi-dwarf' soya beans, for the same reasons; fruit trees with narrower branch angles to withstand heavier crops; fruits which detach easily when picked mechanically; mustard with tall, stiff stems and seeds that stay in the pods till harvested. In several crops new varieties have been bred to withstand colder or dryer conditions than normal. A peach, for instance, normally needs over 1,000 hours below 7°C if it is to initiate flowers and produce fruit. Breeding in warm Louisiana is aiming to reduce this period to 250 hours or less, so that the hotter southern United States can grow their own peaches. New plants to withstand mineralized or salty soils are also being bred.

One of the most important aims in breeding is resistance to the fungus diseases which are probably the biggest problem facing the crop cultivator today, because they are difficult to control chemically and are able to wipe out entire crops in the ideal conditions for fungus development presented by cultivation. Thus in 1947–8 an outbreak of wheat-stem rust in New South Wales destroyed a year's potential food for three million people – one example among a great many since the Irish potato famine.

Unfortunately, fungi are even more flexible than plants in their breeding adaptability and mutation potential. If a crop plant is developed with initial resistance to a disease, the fungus very frequently sidesteps this advance by presenting a new race to which the plant is not immune. Some of the most virulent strains of fungus disease, for instance wheat black rust, are almost certainly the result of breeding for resistance in the cereal. In the past breeding tended to be aimed at keeping one jump ahead of the fungi, but it is increasingly being discovered that plants may have an overall capacity to resist any strain of fungus; research is concentrating on developing these plants.

This kind of breeding can be achieved either by selection from resistant plants in the field or, after intricate evaluation of the plant's separate genes, by using a wild species or sometimes a particular variety unaffected by the disease. In the latter case breeders are able, by a process called back-crossing, to slot the disease-resistant character of the often otherwise worthless wild species into the make-up of the man-bred crop plant. As an example, resistance to the debilitating tomato mosaic virus, which reduces yields by up to 20 per cent, exists in three wild species, and is being bred into modern varieties.

In the same way, resistance to pests, including the intractable subterranean nematode worms, is an important breeding aim, despite the chemical and biological means of dealing with them described in the next chapter. Successful examples of inbred resistance have been the production of a hybrid rice resistant to the plant hoppers which are normally a serious pest of this crop, and another rice strain resistant to grassy stunt virus. The gene for the latter was found in only one wild strain out of thousands which had been preserved.

Man, in addition to manipulating specific chromosomes or single genes in such ways, has also sought to accelerate breeding programmes by using special growth chambers where controlled environments, aided if necessary by hormonal growth control, reduce the interval between plant generations. A remarkable example has been the production of cone-bearing pine trees in only two years from seed. Special pollination techniques have also been devised to speed up generation, including direct insertion of pollen into the ovary. Such controlled conditions make it possible to breed, say, a short-day and a long-day plant together which would otherwise not be in reproductive condition at the same time. The use of treatments to break seed dormancy and ensure even germination, described later, are again important in breeding work.

At one time it was hoped that artificially induced mutations might produce useful new plant varieties. The application of colchicine (an alkaloid derived from the autumn crocus) was the first artificial mutation-maker, and its use can increase the chromosome number. It may create triploids with three times the basic number which, though sterile, often develop greater size, or in which the lack of seeds, as in fruits, is a positive advantage. Or it may create polyploids which, with four or more times the basic chromosome count, usually show greater size and vigour. Bombardment with X-rays and various atomic radiations, and subjection to ultra-violet light and such chemicals as mustard gas, have produced a handful of useful new strains. These include barley Milns Golden Promise, created by gamma-rays, which has a very short, stiff stem; peanut NC4X (X-rays), which crops heavily and has a strong pod resistant to harvesting damage; and rice

Reimei (gamma-rays), which has many qualities, notably good resistance to 'lodging'. A few curious new ornamentals have also arisen through irradiation. X-ray irradiation of pollen has resulted in the breaking of self-sterility in fruits such as sweet cherries. However, as in nature, artificially induced mutants have proved more likely to be harmful than useful.

A complex example of the use of different techniques concerns the transference of mildew resistance from wild oats to cultivated forms. After an initial cross, an infertile hybrid was treated with colchicine to double the chromosomes and hence produce a fertile strain. Repeated back-crossing fixed the mildew-resistant character with minimum added characters from the wild oats. The resultant hybrid was irradiated and a stable, resistant oat was selected. This has now been crossed with several modern commercial oat strains, with the hopes of 30 per cent crop increases in temperate areas where mildew is a serious disease.

Radiation may, however, be valuable in breeding. Pollen can normally only fertilize its own species; incompatible pollen makes the stigma produce a 'contraceptive' which prevents the pollen grains from sprouting. But it is possible to make a female flower accept pollen from another species by mixing this with irradiated pollen from its own kind. The latter is sterile but its auxins remain active and carry its 'go-ahead' signal.

Another method of achieving this is to remove the outer coat of the pollen grains, which consists of exine proteins, by washing them in an appropriate material. These methods may eventually lead to the successful crossing of incompatible species, although at present genetical problems prevent the crossing of species from widely differing genera, let alone families. However, a whole new range of possibilities has recently opened up by methods one can lump under the heading 'genetic engineering'. Still very much in the experimental stage, these are described in Chapter 30.

Before leaving the field of breeding it must be said that in the hands of commerce, excess uniformity is often created. In what has been called the Green Revolution, man now tends to cultivate vast areas of a single pure-breeding plant variety, chosen for increased yields, various resistances, and so on. Most of the flax grown in Turkey is of one variety. All the wheat grown in Australia derives from only half a dozen parent strains. All the important herbage grass *Digitaria decumbens* grown in Central America and the West Indies is of a single clone. Most of the South American coffee trees started as seedlings derived from a single tree in Holland. The potentialities for total disaster from disease in such situations are enormous. Indeed, one can point to the Ceylonese coffee industry, built up from one stock just as in South America, which was totally wiped out by a rust disease in 1860.

The vital potential for variability is missing from these monocultures, while the wild-crop ancestors and their descendants, from which new varieties could be reconstituted if disease wipes out present strains, are rapidly being destroyed. The increasing use of weedkillers and advanced cultivation methods makes sure that no related plants approach the crops, while bigger and ever-extending fields are simply squashing out the old plants. Thus the antique spelt wheat, which twenty years ago existed in thousands of forms in Iran, has now vanished; in Greece the proportion of semi-wild wheats used has diminished from 80 per cent some forty years ago to 10 per cent today. The thousands of older flax varieties in south-west Turkey are fast disappearing, and so are the ancient maize varieties in Brazil. One authority (B. P. Pal) has said that monogenic breeding threatens 'all the plant breeding work carried out by nature over thousands of years'.

In some cases the breeding potential of species of wild forms so far untried, as well as of old, discarded varieties, has been seriously neglected. Thus it is only in the last decade that short-day forms of wild potato – not previously used because they will not form tubers in Europe – have been bred with good existing varieties, resulting in yields increased by 50 per cent and in some cases high resistance to potato blight. Again, breeders are trying to work the insect-trapping and pheromone-producing capacities of *Solanum berthaultii* (p. 311) into orthodox commercial potatoes.

Such potential will have to be tapped increasingly if plant productivity is to keep up with expanding world population, and if we are to continue using orthodox foods. This implies preserving genetical variability of food crops in every way possible.

There are two ways of doing this. *Ex situ* conservation involves the setting up of gene banks where seeds and roots are stored. There are now large numbers of these including some vast international collections, such as the International Maize and Wheat Improvement Centre in Mexico, holding 120,000 entries from forty-seven countries; the International Rice Research Institute in the Philippines, with around 60,000 strains; the Asian Vegetable Research and Development Centre on Taiwan, and the National Vegetable Research Centre in Britain, the world's first comprehensive vegetable gene bank.

Storing seeds and roots is fraught with problems and entails regular 'growing out' to ensure viability, though in some cases this seems to mean that selection for successful storage capability is unwittingly occurring.

Politics has also reared its ugly head; Third World countries accuse the West of collecting their wild crop races and breeding from them without any return, and some countries, like Mexico, have embargoed collection of any plant material for export elsewhere. There are, in fact, many problems in *ex situ* conservation, essential as it is.

The other method is *in situ* conservation, which implies setting aside reserves in which wild crop relatives grow, and having facilities for collecting and studying the resulting germplasm. But this too is a minefield of problems, starting with exactly where and how large the reserves should be, and especially how to prevent the destruction of apparent 'wild weeds' by peasant cultivators.

Once again current problems of technology and politics have taken us away from the main theme. Having, for better or worse, selected or bred desirable races of plant, man can control them in other ways. Seeds bring in a mass of related problems. Man must first learn when to harvest them; in principle premature harvesting must be avoided as it decreases viability and leads to poorer future crops. There are matters beyond the control of man, of course, notably the weather. Better viability follows a year of above-average sunshine and below-average rainfall, although exceptionally fierce sun and drying winds in the period just before harvesting can reduce it again. Some areas of the world which habitually have hot, dry weather around harvest time have become important for seed production.

Problems of mechanical damage have to be taken into account when harvesting; this has a direct effect on seed germination, and also an indirect one if it permits infection by bacteria and fungi. Logically enough, spherical seeds are the least likely to be injured in harvesting or handling.

Correct storage is the next stage which must be kept under control to ensure good seed; deterioration can even affect its genetical characters, viability and, in some cases, good growth following germination.

Although seeds vary in this requirement, as in so many others, it can be said that in principle dehydration to 6 per cent or even less is desirable. A few seeds must undergo desiccation before they will germinate at all. One British seedsman who dries and stores seed to around this level and seals it in foil packets guarantees its viability for three years. Certain seeds are further improved if sealed in various gases, notably nitrogen and carbon dioxide, rather than in air, but the use of these is experimental at present. Alternate moistening and drying of mature seeds can cause severe damage, especially with large seeds like beans and cowpeas where this causes internal fractures. Storage conditions which are too moist or too warm, or both, will reduce viability very quickly: one can say that the seed ages abnormally fast. For small dry seeds a combination of cold and dry is best, with a temperature between −10° and −20°C. Larger moist seeds like acorns must be kept just above freezing. The actual age of the seed is much less important than the storage environment.

In bulk storage conditions attack by bacteria and fungi and equally by insects must be avoided: even when the seed is not mechanically damaged these organisms can attack with serious results, not the least being the excessive heat which fungus activity can create.

Having harvested and stored seed in the best possible way, man now has to sow it, and he comes up against nature's own barrier: the inbuilt dormancy of many seeds, as outlined in Chapter 16. Man is in too much of a hurry to wait for the seed's own good time; he has, very often, to 'break' its dormancy for his own ends. This applies almost entirely to temperate crops and plants; in the tropics seeds are apt to germinate immediately if the warmth and moisture are adequate.

Breaking the dormancy of a temperate seed is called vernalization; the classical experiments on the subject (and also much of its present use) were on cereals. The winter varieties of cereal, if sown in autumn, flower the following summer after a cold winter period, but if sown in spring will not normally flower till the following year. Vernalization consists basically in giving the seed to be sown in spring a cold period (probably from two to six weeks) with a small amount of water; the latter causes the germination processes to start, even if infinitesimally. Gardeners often do this by mixing the seeds with a half and half mixture of moist peat and seed, placing this into small polythene bags and putting them into a refrigerator at just above freezing point.

A great many ornamental plants, notably shrubs but also alpines, are much more likely to germinate in the spring following autumn ripening if they are kept cold and moist during the winter. In the gardening practice of stratification the seed is spread in layers in pots filled with peat or sand. This has the added advantage of breaking down seed coats.

Various chemical substances can supplant climatic vernalization. These include gibberellin and seaweed extract which contains auxins and antibiotic substances. Germination of hard-coated seeds can be speeded up by attacking the coats with various substances including, rather surprisingly, dilute detergent solution. Really tough ones are sometimes treated with sulphuric or nitric acid. Pistachio seeds, notoriously slow to germinate, are treated with concentrated sulphuric acid till their shells are paper-thin, when they will sprout readily. Such treatments avoid the need for chipping seed coats, which can lead to damage. Certain protea relations can have their germination stepped up from about 20 per cent to nearly a 100 per cent by incubating the seeds in pure oxygen.

Vernalization is a technique widely used in agriculture and horticulture. Once the seeds have begun to germinate there are certain ways of giving them a 'tonic', but they have only been tried experimentally at present. Substances including vitamins of the B group, ascorbic acid and cytokinin have given beneficial stimulus to seed germination and initial seedling growth, and this may well become standard practice in time. The anaesthetics chloroform and ethanol have also been found to trigger germination, causing the same effects as plant hormones and light.

Most crops have to germinate in soil under field conditions, and here good tilling and adequate fertilizing are essential. The practice of surrounding individual seeds with a pellet of food material was originally devised for the sowing of trees and other plants in desert conditions; it is now quite widely used for both crop and flower seeds, when the pellet is useful mainly for separation when sowing and water absorption. The sowing of pre-germinated seeds can greatly speed up plant development, ensuring even growth and reducing failures, especially since the seeds will develop at lower temperatures than normal. One method of handling these is to suspend them in a kind of thin jelly which can be forced into the seed rows.

There is far more control of conditions when seeds are sown on a small scale in pots, boxes or greenhouse beds, and a good deal of work has gone to devising standard composts. Some glasshouse crops such as carnations are fed by irrigating beds of inert material with carefully balanced feeds. This hydroponic technique – originally done with the plants' roots suspended below netting in fertilizer solutions – has been particularly valuable in desert conditions, and was quite extensively used in the Second World War to provide troops with green vegetables where it was impossible to grow them normally.

A development of this is to mix desert sands with vermiculite (expanded mica): this substance, which has been used for hydroponic culture, can hold up to 200 times its own weight of water and, if properly mixed with the sand to about 50 cm, will create suitable conditions for plant growth as long as water is available. Other substances which can be mixed with sand, potting composts or, for that matter, free-draining soils to improve water-holding capacity greatly are alginates and acrylic co-polymers.

Feeding plants in normal agricultural conditions today is largely a matter of applying granular fertilizers. Though some of their ingredients may be natural in origin, these tend to be called 'artificial'. Fertilizers exist in many permutations of the essential nitrogen, phosphorus and potassium, and in many strengths per unit weight, according to crop and situation. Special mixtures are used for soils deficient in any trace element, though mixed fertilizers usually contain sufficient of these for normal use.

Whereas not long ago animal manures would regularly be added and every three or four years a grass or clover crop would be grown and ploughed in, very little organic matter is applied to the soil today. Nowadays when stubbles are habitually burned (swaled), ostensibly to destroy lurking fungus diseases, even that source of organic matter is lost, and all that remains for decay in the soil are the roots of the harvested crop. Swaling (which has been said to mark its proponents as primitive as Indians) also oxidizes certain soils – as in the Cambridgeshire fenlands – and contributes to their increasingly rapid disappearance.

It has been said that fertilizers feed only the plant and not the soil. However, too much manure *can* lead to 'soil sickness', and one must emphasize that the end results in terms of yield and food value are virtually identical, for the compounds provided by fertilizers for plant absorption are the same as those produced in the soil by biological processes. A few fertilizers may have a tainting effect on crops such as potatoes, especially if used in excess, but there is no sound proof that 'organically grown' crops are any better as food than those fed with 'artificials'.

But, as I shall describe in Chapter 29, it is all too easy for this tendency to apply nothing but fertilizers to lead to breakdown of soil structure, eventual loss of topsoil and erosion, especially as soils low in organic matter tend to become badly compacted by cultivating and harvesting machinery. Excessive or unbalanced use of fertilizers can make plants less able to withstand disease attack, and more liable to damage by rain. There is little doubt that the developed countries use far too much fertilizer in any case. Well-farmed soils have been shown, for example, to contain two to four times more phosphate and potassium than is needed by most crops, following past years of regular application. This surplus to plant requirements is clearly evidenced by the problems of excess nitrates and phosphates draining from farmland into rivers and lakes, which are also mentioned later. In contrast, Third World countries tend to be desperately short of fertilizers (partly, of course, due to their own practice of burning animal dung).

Many methods exist of stabilizing soil to prevent or reduce erosion, especially on steep slopes where desirable cultural methods such as terracing are impracticable. These involve spraying with various mixtures. One consists of latex, oil and water; another is based on colloids; another on epoxy-resins, and yet another on bitumen, which stops sand movement and retains moisture below. Once the 'binder' is in place trees or grasses can be planted, establish themselves and bind the soil before the artificial material disintegrates.

An improvement in the future may conceivably follow from synthesizing crop plants to make them capable of fixing nitrogen in the soil – one of the possible 'things to come' dealt with in the final chapter. The culture of nitrogen-fixing organisms used as a live fertilizer is feasible: trials with cultured blue-green algae in Japan, for instance, increased rice yields initially by 20 per cent, with a progressive rise in following years.

Feeding through the leaves has also been developed, although it is not being used on a large scale. It has been found that leaves can absorb suitably formulated food solutions (they are often based on urea) very rapidly, enhancing their own photosynthetic activity as well as benefiting the rest of the plant. Foliar feeding is specially valuable at times when there is strong competition for food supplies within the plant, as between root and shoot, leaf and bud or fruit development. If foliar feeds are applied at the same time as necessary insecticides and fungicides, the cost of application is virtually eliminated. Continental grape growers have greatly improved both the yield and health of their vines in this way, and it is this sort of specialized crop for which foliar feeds are probably best suited.

When man tries hard fantastic results can follow intensive feeding. Thus the world potato-growing record (a British one) is of 771 kg from six 'sets', one of which produced 150 kg. A pumpkin of 93 kg has been recorded, a cabbage of over 55 kg, and a radish of 7 kg (a 'normal' one as opposed to the hybrid giants sometimes grown in Japan, which can reach over 15 kg without special attention).

Encouraging growth does not depend only on feeding. Experiments have shown that spraying with that amazing substance gibberellin can make plants grow two to five times faster than normal, and also increase crops. Apart from its obvious value this quick growth may make it possible to grow plants in colder areas than normal because their growing season can be compressed into a shorter summer. Biennials can be made to flower in their first year, which can speed up breeding programmes.

Gibberellin makes many plants grow abnormally tall, and its absence within the plant may be the reason for naturally dwarf forms. In most cases, however, there is little practical use for such extended growth, although it might be valuable to extend the stems of fibre plants, speed growth of fodder grasses and increase the edible stalks of celery and rhubarb. Unfortunately the actual dry weight of the product may increase little, since the effect is basically to elongate existing cells.

The reverse effect, dwarfing, is now widely applied to ornamental plants, and is achieved by various synthetic materials. The dwarf pot chrysanthemum is the best-known example. In some cases dwarfing compounds may increase the flowering potential, as with azaleas and poinsettias, and possibly with fruit trees. These compounds are too expensive at the moment for agricultural use, but they have been used experimentally to reduce the height of inconveniently tall varieties of barley, tobacco and tomato. Treated plants are also remarkably more resistant to drought and subsequent wilting. In one experiment bean plants in pots without any water survived ten days longer than untreated plants.

Man has produced further scientific techniques to manipulate crops at various moments in their growth. Hormone sprays can for instance modify the sex of flowers, improve their 'setting', induce frost-damaged fruitlets to grow and make seedless fruit, thin out flowers and fruit if produced too freely, and conversely prevent excessive pre-harvest fruit drop as well as controlling dropping at maturity. Hormones and substances which release ethylene can accelerate leaf fall which may be desirable for easier harvesting. This is done with tomatoes in the United States; successful experiments have also been made to defoliate tabasco plants in order to exchange difficult hand picking for mechanical harvesting. The material used, in proportions of less than 1 ppm, also makes the fruits detach themselves more readily, which is especially desirable with tabasco.

Again, potatoes can be prevented from sprouting in store, or alternatively started into growth even when newly dug and naturally dormant. With fruit, temperature manipulation and the use of ethylene gas delay ripening and lengthen storage life. Other fruits need to have ripening accelerated, which can be done with hormones or, as with bananas brought to temperate countries in cold-storage ships, warming up in controlled ripening chambers.

Other specialized applications of hormones and allied substances promote quick flowering of pineapples and ornamental bromeliads, stop celery and lettuce bolting, control growth of coarse grass and hedges, and increase latex flow in rubber trees. Spraying with absicic acid – normally involved in leaf fall and dormancy – allows plants to withstand a period of drought or to grow better in arid conditions, for it keeps the stomata closed for up to nine days and hence curtails transpiration of moisture. Spraying plants with acetysalicylic acid – aspirin, in short – has been shown to make plants much more drought-tolerant. A silicone spray seals up the pores mechanically with similar effect, and apart from use on growing crops has been applied to prevent loss of water on plants in transit for replanting and to stop needle drop in Christmas trees.

Another widely used manipulation of a different kind is to induce flowering at a time when the plant does not normally bloom. The chrysanthemum is again the classic example. Not long ago we always considered these flowers the harbingers of autumn, for they will not normally bloom till the longer nights begin. I can recollect a case where an amateur chrysanthemum exhibitor sued the local council for erecting a street lamp where it cast light on his plants at night and prevented them from flowering. This was because the chrysanthemum, as outlined in Chapter 10, is a short-day plant. Now, however, the chrysanthemum is produced all the year round. It can be induced to flower in the long days of summer by blacking-out to give artificial short days, and its flowering can be delayed in winter by lighting at night giving artificial long days. Black-out is provided by completely opaque blinds of various kinds. Night illumination can be given with quite low-powered tungsten bulbs, and these do not have to be on all the time. Indeed, in experimental conditions some plants can be inhibited from flowering by a single short-duration flash in the middle of the night. With chrysanthemums, however, between ten and fifteen minutes' light are given every thirty minutes between ten p.m. and three a.m. The science of producing a succession of blooms of these AYR (all-year-round) chrysanthemums, which is sometimes combined with dwarfing them, is now very precise, and is the main commercial result of the work on photoperiodism. Other short-day ornamentals induced to flower at specific times include the poinsettia and kalanchoë, while the long-day fuchsias can be flowered in mid-winter.

All this control of timing implies the use of greenhouses or other structures. Modern greenhouses are a far cry from their ancestors, the seventeenth-century orangeries; they now have equipment for automatic control of temperature, ventilation, watering and insecticide dispersal. Temperature, incidentally, can be as important as light in controlling flowering. Extra light, too, can be provided, usually by fluorescent or mercury lamps. A recent American development is 'pulsed light' from gas discharge lamps, used for only 20 per cent of the time but giving results claimed to be equal to those from continuous light. Another suggestion is to use very short bursts of laser-beam light, scanned over the crop area.

Once lighting is being used it is possible to use opaque structures, which are cheaper to construct than greenhouses and more easily insulated. This has given rise to 'growing rooms' where, especially in their early stages, plants are given intensive treatment, tailored to their needs. Tomato plants in a growing room can in three weeks reach the size they would normally attain in a greenhouse in six to nine weeks; chrysanthemum cuttings production has been stepped up 90 per cent above normal.

A further refinement is to introduce carbon dioxide. Not only does this replenish the carbon dioxide used up in a closed structure, especially if conditions preclude ventilation, but increases its level to improve photosynthetic performance. Growth increases of at least 20 per cent are possible. The value of carbon dioxide was first observed by J. Priestley in 1772. He noted that mint, growing in air fouled by putrefaction, throve 'in a most surprising manner. In no other circumstances have I ever seen vegetation so vigorous as in this kind of air, which is immediately fatal to animal life.'

Simpler devices improve local climates. Light structures covered with plastic film are widely used to keep out wind and trap sun-heat in spring and autumn; these may be greenhouse size or low tunnels, the latter a development of the old French bell glass and the more recent glass cloche. 'Bubble houses' of strong plastic, inflated by a small continuous air pump, can cover a crop for an appropriate period and then be moved quickly to another. The use of black polythene on the ground, through which plants are inserted, helps to warm up the soil, conserve moisture, and also to keep down weeds.

While environment can be more or less closely controlled in the various transparent or closed structures, often using synthetic composts or inert aggregates with artificial fertilizers, large-scale crops outside must still be grown in soil and take their chance with drought, wind or storm. One climatic problem is frost, which can play havoc with a potential fruit crop at flowering time. The 'smoke-pot' – basically a controlled fire which gives out plenty of radiant heat, and is now relatively sophisticated – is still widely used. Another method is to use overhead irrigation lines to deliver a fine mist-like spray during periods of freezing. This coats the plants with ice, but as long as the spray is kept going the latent heat of freezing keeps the ice, and the plant within it, just below freezing point, which allows fruit buds to survive. Huge air-moving fans have also been used experimentally to bring warmer air down from above on nights of radiation frost.

Irrigation techniques are becoming more sophisticated, and intensive crops may be supplied with water via huge rotating sprays or fixed overhead lines if the weather pattern has created a water deficit in the soil. This implies plenty of annual rainfall and good underground water supplies. In countries like India where rain is seasonal, work is going into building storage dams to catch the monsoon rains and allow them to seep steadily into the subsoil to act as a water store and raise the whole local water table. Experimental farms in the most arid parts of Israel, where rain is very erratic and often slight, work on the principle of collecting 'run-off' from watersheds on slopes around; this is channelled down to the growing area and thus almost every drop of water is brought where it is most needed. This method has produced remarkable results. It is interesting that the modern experiments are based on the ancient Nabatean methods, in which crops were raised in areas where the rainfall averages 8 cm a year – over 50 cm is usually considered the minimum for any desert cultivation. Crops grown with profitable yields are not just cereals but all manner of fruit and nut trees.

Irrigation can, however, produce major disasters, like that in the 1950s which made saline desert of hundreds of square miles of the Indus plain in Pakistan; it has been estimated that it would take twenty years and cost two billion dollars (more than was spent on the original irrigation works) to return the land to productivity. Such problems are due to cumulative build-up of mineral salts in the soil. Where soil or water are saline, the irrigated area has to be flushed with water to remove the excess salts and allow them to drain away. It is difficult to do this where water is already in short supply. On some soils this heavy 'leaching' should be accompanied by applications of gypsum and deep subsoil cultivation, both of which are again hard to carry out in primitive communities. Sometimes salinity can be overcome. In Israel no problems have followed using actively saline water. This is in dune sand which drains very readily. The Israelis have also, however, tried out various crops to see which will stand salinity, as described later. They have also made remarkable advances in cutting the cost of desalination.

Desert irrigation has also been accused of providing excellent breeding conditions for locusts and of encouraging animals carrying diseases such as bilharzia and malaria, but there are precautions to be taken against these.

America seems to have succeeded in modifying the weather within certain limits, during research connected with the Vietnam war. Although there is at present a strong lobby to prevent weather and climate modifications being used as a weapon in war, there are obvious opportunities for the use of some of the techniques to assist agriculture, as long as weather control in one place does not affect neighbouring areas or countries, or indeed the globe's whole weather pattern. 'Cloud-seeding' from aircraft or rockets could increase world rainfall by an estimated 15 per cent. It can also reduce the likelihood of excessive hail: the Russians claimed in 1976 that 4 million hectares of crops were currently being protected against hail. Radar is used to detect the hail-forming zones in clouds at 40 km distance, and guns or missiles then deliver suitable substances into them to stimulate the crystallization of super-cooled drops, which prevents the formation of large hailstones. The substance most used has been silver iodide, but Israeli scientists have discovered a frost-provoking bacterium which may be cheaper and more efficient. This technique is claimed to be reducing hail damage by a factor of four or five. But tempting as it may be to call down rain or disperse clouds at will, there are considerable dangers in tampering with systems that affect everyone on earth.

An early legal case in the San Luis Valley in Colorado showed how cloud-seeding designed to prevent damage to barley by hail was believed by cattle ranchers to have created severe drought, dried up the grass and harmed their animals. The barley growers estimated that natural weather conditions gave them a good quality crop once in twenty years; but the ranchers seemed to get by most of the time. It is certainly never going to be possible to please everyone all the time.

To conclude, man – having assured vigorous, trouble-free growth of specially bred plants, and applied whatever technical manipulations are thought necessary – is still up against the great bugbear of any kind of cultivation in that its artificial conditions are perfect for the invasion of weeds, insect and other animal pests, and all manner of diseases. This problem is the theme of the next chapter.

28 · *War of the Worlds*

Fust time I see de boll weebil he sitting on a square,
Nex' time I see de boll weebil he got his family dere.

The old American plantation song is observant. And indeed, what could be better for pests and diseases than the well-fed host plants, often so far removed from nature that they lack basic resistance, massed together and giving endless opportunities for spread? What better for weeds than the nicely stirred soil supplied with fertilizers, and the well-spaced crop plants?

Continuous cultivation of the same crop inevitably leads to massive build-ups of pests and fungi. These are very much rival worlds to our own, and have been with us since the earliest days of agriculture, as remains of beetles in grain and bread from Egyptian tombs up to 4,000 years old have shown. Their inimical attentions to our crops, whether growing or stored, have been estimated to cost us at least a third of what we labour to produce – certainly enough to make good the present world food deficit. In certain parts of India over half the food production is lost; it was calculated in 1977 that in Uttar Pradesh – the biggest state – alone there were 600 million rats, compared with 90 million people, and that the rats ate enough of the food produced to support 100 million people.

Insects are more destructive than fungi; of the estimated 900,000 species over one in ten is a pest and about 3 per cent are serious ones. As this means about 27,000 different species, the size of the problem can be appreciated. Mice, rats, birds and other larger creatures can also become very serious pests.

Man's main reply to these unwanted guests at his feast is the use of chemicals. Weed control starts with cultivation, whether from hoe or mattock or sophisticated machine; but in the well-to-do, technologically developed countries at least chemical control is highly developed, with materials capable of selecting weeds and leaving crops unscathed. Most of these very specific weedkillers are based on growth-control substances or 'hormones': the selective weedkillers we use on our lawns are familiar examples, leaving the narrow-bladed grasses but causing the growth processes of broad-leaved weeds to go berserk and more or less literally make the plants tear themselves apart. It is now also possible to destroy grass among broad-leaved plants selectively, and indeed there is steadily increasing precision in selecting the target weed, as for instance in the remarkable Australian achievement of destroying mistletoes on gum trees by the injection of a selective weedkiller.

One major problem is the eradication of wild forms among cultivated varieties. Thus the wild oat is very difficult to dislodge from cereal fields; it infests nearly 70 per cent of arable farms in Britain. Not only does it compete with the crop but to remove the wild seed from the cultivated after harvesting is a serious problem. This situation is largely due to lack of crop rotation. However, chemicals have been devised to kill seedling wild oats before the crop is sown, or to destroy them selectively in the growing crop. Yet others protect the crop against weedkillers which are used against its wild relations. Quite a number of cereal weeds have, like wild oats, successfully 'self-selected' for seeds indistinguishable in size and weight from those of the crop, so that they are harvested and resown with it.

There are, of course, many broad-spectrum weedkillers, some of which kill every plant in the ground and may be more or less persistent for up to six months; others, like paraquat, nicknamed 'the chemical hoe', which scorch off surface growth; and yet others which, forming a film on the soil surface, kill any weed seedlings which may emerge. Most of these, one is happy to state, have little lasting effect on soil texture, and are claimed to be harmless to soil organisms, although there are some fears that they may adversely affect valuable soil bacteria.

Selective weedkillers have, however, come under a cloud recently. Though some appear to be virtually harmless to humans handling them or to animals eating sprayed plants, there are unpleasant reports of malformations in human babies following the use of the selective 2,4,5-T as a defoliant in Vietnam. It must be said that this chemical was used there in strengths vastly exceeding those in normal horticultural practice, and had not previously been suspected of ill effects. It also appears that the dangers lie not in the chemical itself but in the lethal impurity doxin associated with manufacture which is very difficult to prevent and is absorbed by fish and shellfish and thus insidiously passed on to human beings after the use of defoliants has ended.

There are other dangers accompanying weedkiller use. The enormous areas which need dealing with have resulted in mass spraying techniques, including the use of light aircraft and helicopters, and a very small breeze can cause the spray to drift on to surrounding vegetation, other crops and gardens, with sometimes disastrous results. Some plants, such as tomatoes, are exceedingly sensitive to selective weedkillers, and the merest suggestion of these is enough to cause distorted growth. Careless or excessive spraying, which may occur if spray swathes overlap, can damage the crop plant, and in cereals encourage a weakening disease known as 'take-all'.

Other approaches to the weed problem may be desirable. In the first place it may not be best to kill the weeds outright, because in their early stages they can prevent soil from drying out and blowing away; if a material is used which merely retards weed growth, the growing crop will eventually overgrow the weeds and destroy them owing to lack of light. An alternative method is to pre-treat crop seed so that it grows away more quickly than the weeds once the soil has been disturbed at sowing time.

Another refinement is to attack the weeds before the crop is sown at all. This means breaking the winter dormancy, or dark-induced dormancy, of the weed seeds by light ploughing followed by exposure to a compound which releases ethylene. The resulting seedlings then perish in the winter weather. With couch grass, which is very difficult to kill and springs up again even when the roots are cut up into fragments, research has been aimed at cultivating infested ground at the right out-of-season moment to make the root fragments grow, and then spraying with a specific herbicide.

Pest and disease control present more serious difficulties. Some of the worst problems follow accidental introduction of an organism which finds a new host to its liking: it was an Asian fungus that destroyed the North American sweet chestnut; the Bermuda cedar was virtually wiped out by two accidentally introduced scale insects; while the notorious Colorado beetle reversed the process when the potato was introduced to it.

I have already pointed out briefly that fungi adapt themselves to overcome genetical resistance in plants; and they do exactly the same to overcome toxic chemicals. Insects are equally adaptable in this way, and we know that some species which were originally wiped out by DDT now appear to thrive on it. Some of the smaller pests, such as red spider mite, are exceptionally able to develop resistance in this way. At least 200 major insect crop pests are known to be resistant to all existing pesticides, and in 1984, 447 were recorded as resistant to at least one class of pesticide. Indeed, by using such materials the scientist, to quote Alec Nisbett, 'succeeds only in creating an almost artificial animal, a strange creature whose major function is to break down unnatural chemicals in its attempt to re-infest the man-made world that first shaped it as a pest'. Man's main concern now should be to ring the changes on the materials he uses so that pest and disease populations do not have time to develop a resistance to any of them.

In nature, of course, pests of this sort are likely to be kept in reasonable check by the action of predators upon them. One great disadvantage of chemical pest control is that it all too readily destroys the natural predators as well as the pests, especially since 'broad-spectrum' chemicals capable of killing a wide range of pests are generally used. A few specific materials are being developed, like pirimicarb which kills only aphids. Another complication is that the use of a chemical against one pest allows another one to build up, as occurred when DDT was first used to combat codling moth on apples; as a result the woolly aphis and the hitherto rare red spider mite reached epidemic proportions.

It is not only insect predators that can be killed. Bats eating insecticide-contaminated insects have suffered much population reduction; loss of their habitual nesting sites has also cut their number. But bats are great insect-destroyers. The British great bat daily eats nearly half its own weight in insects. Bats in Texas have been estimated to consume 6,600 tonnes of insects a year; these are guano bats which also produce quantities of very rich fertilizer. The Russians are actively encouraging bats, setting up suitable forest breeding places for them. If there is eventually a reduction in pesticide use this will encourage the bats.

Our basic reaction to excessive build-up of a pest for whatever reason is to step up the amount and frequency of insecticide applied. All this does is to speed up the pest's ability to resist, increase the likelihood of side effects on consumers, destroy useful organisms, affect associated organisms, and frequently encourage other pests, just as the excessive use of sprays against the cotton 'boll weebil' of the song has encouraged a local malaria-carrying mosquito which has become resistant to pesticides.

Then we are all familiar with the much-publicized biological chain reactions, and their long-term effects, which occur after the use of very persistent pesticides. The villains most often pilloried are DDT, the first organochlorine compound, and its successors such as aldrin and dieldrin, extraordinarily successful as killers of often intransigent pests like soil caterpillars, but which persist in the soil. The chemicals are then transmitted along food-chains: worms ingest them, birds eat worms, affected pests and treated seeds, and these birds, as well as bird-eating birds of prey, may die. This may be a direct result of accumulations of chemicals in their tissues or affect their capacity to produce fertile eggs. Fish can take up chemicals which run off in water and this can create other food-chain problems: many sea-birds are affected. Not the least problem is the possible effects on ourselves, not very clearly understood as yet, but likely to be deleterious. When nursing mothers in the United States have so much DDT in their milk that, to quote a scientific humorist, 'in strict terms of federal law, it is illegal for them to carry their busts with them across one state line to another', there may be cause to worry. (Recent research has suggested that, apart from toxic effects on the babies, such DDT accumulation can also cause abnormal behaviour in the mothers.) But we must never forget the 'risk-benefit' ratio: these compounds have done and are still doing vast good, and as in every aspect of life the good and harm need to be balanced with great care and thought.

We should obviously try to ensure that any chemicals we use for agricultural purposes have a relatively short persistence in the environment, where surpluses should rapidly break up into harmless materials. One line of development is that of systemic chemicals, which actually enter the plant's sap, destroying sap-sucking insects but no others, and persisting there for a period. Apart from the obvious advantages, there are some situations where a systemic is the only possibility – as in the case of swollen shoot virus of cacao, which is transmitted by sap-sucking mealy bugs. However, it is impossible to attack these by direct sprays since ants, 'pasturing' the bugs, build impermeable tents over their colonies!

Systemic fungicides which kill the invading organism as soon as it enters the cells are also beginning to appear, and are in principle highly desirable, for surface films of fungicide very readily wash off. Systemic fungicides can be sprayed on to leaves, injected into soil in the root area, or laboriously injected direct into trunks: it is possible to control Dutch elm disease by annual injections. The ultimate chemical treatment is, perhaps, a combined systemic containing insecticide, fungicide, bactericide, nematocide and conceivably viricide against viruses. Systemics are, however, as liable to genetical evasion as contact sprays.

Materials of natural origin are always believed to be much less likely to harm other organisms than synthetic chemicals. Of the main ones, nicotine is poisonous to mammals and derris to fish; pyrethrum appears to be harmless to mammals; and none is persistent. An extract of garlic with oil and water has been successfully used again insect pests in Spanish experiments. Caffeine prepared from tea or coffee killed insect pests in experiments by affecting their metabolic processes. An extract of cedarwood has again been shown to kill many insects.

More attention will certainly be focused on naturally occurring insect repellents. One of these which has recently been developed is called azadirachtin and is obtained from the neem or margosa tree, as well as the Persian lilac, chinaberry or Indian bead tree, species of *Melia*. First, this extract is distasteful to insects, as with old-fashioned extract of quassia; but if insects do overcome their dislike and attack the plant it upsets their cycle of growth and, by preventing regular moults, eventually kills them. The great advantage of such materials, which are obtained from trees widespread in sub-tropical countries, is that they can be made very simply on the spot by peasant farmers, so that supply and expense are no problem. This is equally true of garlic extract, and the familiar garden geranium (pelargonium) contains 'anti-feedants' which deter slugs. It seems possible that algae may contain useful insect killers – two species have already been discovered which kill mosquito larvae – and the quest for such substances continues.

Plants also contain fungus repellents and one, known as wyerone, which has been isolated from beans, is effective in very low concentrations against fungus organisms on many kinds of plant. Garlic, already mentioned as a pest-killer, has been shown to repel some fungus infections.

Mycorrhizal fungi, those that associate with plant roots as described in Chapter 22, are often lethal to their disease-producing relations, due to their production of natural antibiotics. Such mycorrhizal fungi associated with two species of pine were tested against 200 root-attacking fungi, and reduced the onslaughts of about ninety of them. One mycorrhizal fungus was effective against 92 per cent of possible fungus enemies. Its effect on the fungus causing littleleaf disease of pine was to prevent within twenty minutes the latter's free-swimming spores from moving – they normally swim for six to eight hours – and stop their germination altogether. Some of the antibiotics are specific against a single disease-causing fungus, and their extraction and use may well be of considerable future value. Certain yeasts have been shown to reduce attack by fungus diseases very considerably.

A remarkable development has been the discovery that inoculation, at seedling stage, of a mild, artificially produced strain of virus disease can prevent full-scale attacks. The British experiments in this field were on the tomato mosaic virus, a disease which greatly reduces crops though it is not lethal. This notable parallel with human inoculation against, say, smallpox, has given encouraging results in field trials.

Though pests and diseases can presumably develop resistance to 'natural' materials as quickly as they do to synthetics, such materials can cut out expensive manufacturing processes and are much less likely to taint food crops, as happens with a fairly wide range of synthetic materials.

Pesticides and fungicides of any kind are normally sprayed on to the plants. In most cases the finer the spray the more leaf area it will cover, depending of course on the kind of crop, and modern machinery has been steadily refined for this purpose. But even with the most sophisticated equipment and materials designed to make the sprays 'stick' there are losses from drifting and general over-spraying, and there is also the difficulty of reaching every part of the leaves, especially their undersides. One recent and successful technique is to use the pesticide or fungicide in very fine particles – spray or powder – and propel them on to foliage electrostatically, carrying a charge. They then cling to the naturally earthed plant surfaces. It has been found that drifting is much reduced and the electrostatically deposited material is much less easily washed off by heavy rain.

Quite a different line of control is to employ natural predators against pests. Indeed, before insecticides were used, it was found that the average pest had a two-year cycle. In the first year it might reach serious proportions, but this also encouraged its natural enemies, which would wipe most of them out in the second year. Since most plants over-produce flowers, or tillers on cereal crops, this did not necessarily reduce yields in the bad year. The devastating scale insect of citrus trees was one of the first pests to be deliberately attacked by a natural predator, in this case an American ladybird. On a smaller scale gardeners are aware that ladybirds and their larvae are well worth preserving to keep down aphids on which they feed with gusto. The Canadian winter moth, first noted as a serious pest in 1949, was by 1969 wiped out with the aid of two European insect parasites. Under glass the very resistant white fly and the red spider mite have both been kept down by predators. Many other examples could be cited of insects used successfully against other insects. In some cases it is desirable to use predators which are genetically resistant to specific pesticides, so that both can be used together as needed, as has been done with a mite against the red spider mites on almonds in California.

Other organisms can also be used. An eelworm has been employed to attack the maize root worm, a beetle larva. It is possible to produce the eelworms – cultured on a mixture of agar and dog food! – at the cost of only a dollar per million, and this million will control root worm and many other soil insects over several acres; moreover the eelworm can be cold-stored for five years.

It has also been found possible to infect pest insects with fungus, bacterial or virus diseases to which they are prone. Such diseases have the advantage that they will normally only attack the specific insect concerned, although their use at unnatural rates can induce genetical resistance to arise in the target. A recent example was the experimental spraying of 400 hectares of spruce forest infested with budworm, a caterpillar which rapidly defoliates the trees, with a mixture of live bacterium and the enzyme chitinase. The latter damages the skin of the caterpillar, making wounds through which the bacteria can enter. Ten days after spraying, over 80 per cent of the budworm caterpillars were dead. Another fungus has been found which will decimate sap-sucking aphids.

It may also be possible to use fungi to combat fungus diseases. Thus potato scab is attacked by a penicillium species. Gardeners sometimes work lawn mowings into their potato patches, which encourages growth of the penicillium and reduces scab attack. Heavy 'green-manuring' – the turning in of a growing crop – has been tried with potatoes and even cereals.

Some fungus and virus diseases are spread in pollen: fire blight of apples and pears is one. Scientists have cashed in on the habits of bees by placing streptomycin in hives in such a way that the bees will distribute the fungicide.

Another modern method which has been used on several important pests is to release very large quantities of sterile male insects. This involves breeding the insects and irradiating them with gamma rays, after which they are released (at various stages of their life cycle). Though sterile, they are sexually active; their numbers swamp the wild male population so that most of the wild females are mated by sterile males. The result is either no eggs or sterile ones, and if this procedure is followed over several generations a very great reduction in pest population can follow.

Breeding of the kind described to insert a specific character into a plant strain is also possible with insects, in this case the aim being to insert a dominant 'lethal gene', usually one inducing very low fertility, into a race and then releasing quantities of these insects into the wild population, where the effect can be fairly rapid extinction.

As with plants the development of an insect is controlled by hormones. If the transition from larva to pupa can be disrupted by applying the wrong hormone, or too much of the right one, the insect will perish. One such hormone, the 'juvenile', prevents larvae from pupating. Although it has proved very difficult to extract these hormones from insects or to synthesize them, certain plants contain the same substances, whose presence is presumably due to long-term evolutionary adaptation against such enemies. It is notable that the plants concerned, including the yew and other conifers, and the 'fiddle-head' ferns, are of very ancient lineage. Thirty grams of dried yew leaves contain as much insect hormone as does a tonne of silk worms. One of the hormones is an ovicide so powerful that when mating a male transmits it to the female and makes the resulting eggs sterile. If hormone-treated males were bred and released like the sterilized ones described earlier, the same effects would result. But it must be stated that at least one pest, a flour beetle, has already developed some resistance to treatment with juvenile hormone and some important natural pest parasites may be destroyed.

Other substances manufactured by insects themselves are being experimented with. These are the pheromones, in effect transmitters of messages between insects, notably those informing males that the female is sexually mature. Such materials are effective in minute quantities, and, synthesized if necessary, can be used to lure insects to centres where they are destroyed. Cotton bollworm adults have been experimentally controlled in this way in Ghana.

Cultural methods can also reduce the threat of pests and diseases, the oldest being crop rotation, which prevents a massive build-up of enemies by not providing the same host more than once every three or four years; it is also effective against weeds. Considerable checks on pests can be achieved by dividing up the crop. In Cuba recent experiments involve interplanting maize and sunflowers in strips 8 m wide. Because this placed foodless barriers in the path of the maize armyworm, the pest's invasion and effects were reduced by 50 per cent at no expense and without unduly complicating mechanical harvesting.

In the United States Dutch elm disease has been checked by felling trees around centres of infection to create swathes across which the fungus-carrying bark beetle cannot fly.

Again, it is possible to avoid certain pests by planting the crop at a time when the pest is not active. Thus even in eelworm-infected land, very early planting of potatoes will allow plants to make adequate growth before the eelworms are active. Another cultural practice illustrated by this pair is that of leaving some potatoes in the ground after harvesting the early crop. These tubers start to sprout in the late summer, and the eelworms are hoodwinked into thinking that another season has begun, for they respond to exudates from the potato plant. But the late-sprouting potatoes are killed by winter frost and the eelworms involved die out, since they have not returned to the safe stage of the resting cyst.

A strain of wild barley was recently discovered to be resistant to eelworms. Breeding transferred this character to cultivated barley. Apparently the plant chemical involved prevents the female eelworms from reaching maturity and being able to breed, so that the gains are cumulative. The new barley, called Sabarlis, has yielded up to 20 per cent more than eelworm-susceptible strains.

It may be possible to put off eelworms by using other plants with powerful exudates. Thus it appears that the tall annual marigold *Tagetes minuta* blots out the exudation of a host plant such as the potato, so that the eelworms do not emerge from their cysts and a proportion perish. Recent Scottish experiments showed a natural 'wastage' of potato eelworm, in the absence of a crop, of 17 per cent a year; with a crop of *Tagetes* there was a wastage of 55 per cent, which in fact suggests an altogether more active effect from the *Tagetes*.

Growing a crop in company with another plant may reduce disease infection; an old example is of cotton whose mortality from root-rot fell from over 50 per cent to only 2 per cent when grown with the bean relation moth. This was attributed to a lowering of soil temperature but was probably due to antibiotic root exudations.

When a fungus disease such as wheat rust has two hosts – the second is the barberry – control can be achieved by destroying all the plants of the second host. This can give rise to clashes of interest, as with the attempts to get rid of fire blight disease in Britain. This aptly named disease, almost unknown on this side of the Channel until a few decades ago, attacks and rapidly kills apples and pears. (It is believed to have reached Britain in pear boxes from North America which were acquired by Kentish growers.) It also attacks other members of the rose family such as hawthorns, roses, etc. But Ministry of Agriculture attempts to destroy ornamental rosaceous plants in the vicinity of orchards met with much opposition from gardeners. Long-term hopes of combating fire blight are based on the possibility of breeding the remarkable disease resistance of a Chinese wild pear into cultivated varieties.

Occasionally grafting can deal successfully with a serious insect pest. The classic example of this is the phylloxera aphid which decimated European grape-vines late last century. Eventually a cure was found by grafting the desired grape varieties on to American vine stocks, which are resistant to, or at least fully tolerant of, the insect. Although this saved the French and German vineyards it was also responsible for allowing downy mildew – before then a disease endemic to America – to take a hold on European vines, where it has remained a serious complaint ever since.

Biological control has also been applied to weeds, where introductions, as with pests and disease, can rapidly get completely out of hand. Here, the classic example is the destruction, in 1925, of the introduced prickly pear in Australia. Brought in as garden curiosities, various species of this cactus were introduced at the end of the last century and found Australian conditions so much to their liking that by 1900 they had spread over 4 million hectares and by 1925 over 24 million. After trying biological control with cochineal insects, which were attacked by a native ladybird, the Argentinian moth *Cactoblastis cactorum* was tried. A mere 3,000 eggs were brought in, but they proved sufficient and by 1938 the moth had decimated 10 million hectares of prickly pear and the land was restored to productivity.

One of the worst weeds in the world is the beautiful water hyacinth, accidentally introduced into the U S A in 1884 following its collection in Venezuela, and since then to other tropical areas. It can double its numbers in eight to ten days. In Louisiana alone it occupied over 400,000 hectares in 1975, having doubled its extent in two years. Thirteen hundred kilometres of the Nile in the Sudan are totally blocked by it, and large areas of Zaire are unapproachable and have been effectively abandoned in the last fifteen years because of the weed on the Congo river. At least 3,000 km of irrigation canals in India are useless because of water hyacinth, and in some places, like Bangladesh, the weed cannot be stopped floating into and increasing in rice fields, where it crushes the crop. Hydro-electric machinery has been rendered useless also.

Biological control with insects and with micro-organisms is currently being evaluated, and sea-cows or manatees are now being seriously considered as control agents. Small-scale experience in Guyana over many years has suggested their efficiency in other places. Manatees can also be used against submerged water weeds; another promising method of dealing with these is the introduction of herbivorous fish. In a recent trial in southern California, where three species of submerged weed were clogging irrigation canals and drains, the African tilapia fish almost entirely solved the problem within four months. Both fish and manatee could conceivably be 'cropped' as an extra benefit to man.

The black sage of Trinidad has been deliberately attacked, with good results, by leaf-eating and seed-eating insects; the parasitic witch weed of central Africa is being partially controlled by a weevil; and the Mediterranean milk thistle, now a widespread weed in California, and other species of thistle which have become a nuisance in the United States, have been successfully controlled by a seed-eating weevil after careful testing to check that it will not attack cultivated thistle relations such as globe artichoke. However, efforts to destroy ragwort in English pastures, using the cinnabar moth habitually found on it, have not been successful, and as far as I know there is no habitual crop weed which has been dealt with in this way.

The skeleton weed is another accidental introduction (to Australia from the United States) which has become a nightmare to cereal growers, and has been described as the worst weed ever in that continent. Its deep roots with their regenerative buds cannot be destroyed chemically. Although some promising insect parasites were found, the most valuable natural control has been achieved by a specific fungus rust disease. While this is a radical new development, there are clearly dangers in using plant diseases in this way – once the basic weed has been eliminated they might possibly turn to other plants.

The 'Tagetes effect', mentioned earlier, has been used against weeds also. There seems little doubt that planting *T. minuta* 15 cm apart in a bed of ground elder previously sheared to the ground will kill the weed in a season; it has also been used in this way, with varying results, against bindweed and couch grass. At present it is a small-scale remedy, but possibly a pointer to the future.

Virus diseases remain very intractable. Meristem culture and the building up of virus-free stocks, as described in Chapter 27, are the basic methods of outwitting them, although in a few cases – strawberries are one – controlled heat treatment destroys the virus without harming the plant. Since viruses are almost all spread by sap-sucking insects, the successful control of these will restrict these diseases.

In considering pest and disease control one must remember that in most natural habitats the number of species present, of plants and all other organisms, usually results in a complex system of checks and balances, making for stability: indeed, the most stable natural ecosystems are almost always very complex. Agriculture simplifies the species structure and ecological system so much that balance becomes impossible, and a disastrous jolt is likely if anything goes wrong. Biological control is the logical outcome of applying ecological experience to such problems. At present, as in some mythological tale, the more we cut down the enemy the greater his numbers seem to become. To quote Horace (translated by Conington):

> Drive Nature forth by force, she'll turn and rout
> The false refinements that would keep her out.

Both insects and fungi have, indeed, an incredibly long history of making the most of changing circumstances, and our sometimes makeshift, hit-or-miss methods seem perhaps rather pathetic in face of this enormous evolutionary know-how and the fantastic fecundity of these organisms. We are, however, beginning to understand the repercussions that may follow unthinking use of, say, a pesticide. Quite apart from pests developing resistance there may be unexpected developments, as with the build-up of fruit tree red spider mite after the successful control of codling moth with DDT. Conversely, the introduction of a biological control organism must always be circumspect, or the control may conceivably be worse than the original problem.

Destroying one class of pests may result in another organism changing its habits and becoming a pest in turn; this has happened with birds who have turned to attacking fruit and tree buds since there is now so much less insect food available. Destroying too many weeds and hedges around cultivated areas may force pests from such natural harbours on to crops, and equally reduce the population of natural predators. *Some* pests must be left alive if their predators are to survive also, not to mention the essential insects such as wild bees which in many places, especially in the United States, have been so decimated that certain crop yields are falling due to lack of pollination.

We do not worry about destroying caterpillars, aphids, rust fungi, or rats; but in Britain, at any rate, our emotive attitude to birds means that any attempt to control them, even the pigeons and the sparrows which have been dubbed 'avian rats' for their destructive habits, is likely to meet with a vast outcry; though this is not the case in America and elsewhere.

Chemical control materials themselves have become a source of emotive dispute, and there is certainly no doubt that they have often been used excessively and carelessly, and without thought as to where the surpluses of often very persistent substances may end up. I shall be discussing this further in the next chapter. Here it is sufficient to suggest that *balanced control* – what one authority has called 'pest population management' – must increasingly be invoked, perhaps under the authority of the International Organization for Biological Control of Noxious Animals and Plants set up in 1956.

This integration of chemical and biological methods is already occurring in many places, including Russia, California, Peru and Canada. Integration implies the control or at least advice of ecologically trained biologists, in contact with growers and familiar with the problems in the field. It may reduce the freedom of the individual grower but is highly desirable for both natural and economic reasons. The cost of chemical methods alone is increasing all the time – and consider that temperate fruit growers may apply fifteen different sprays in a season.

Here vested interests in the shape of the chemical industries will inevitably rear their heads. In some countries their salesmen have an enormous hold, as in Holland, where quite excessive amounts of chemicals are used by brainwashed growers. Politics also play their part, controlling the quantities of crops grown, fixing prices, giving subsidies to growers of essential but unrewarding crops and others to prevent the growing of crops in surplus. A grower can sometimes even make money by making his crop unsuitable for human consumption but usable for animals.

Even more subtle and alarming is the example of a major chemical firm which markets grain sorghum in the form of coated seeds. Two of the coatings are chemicals which protect against grass invasion; the third protects the crop against the company's herbicide which is lethal to sorghum!

I am certainly not suggesting that we can do without chemicals, nor am I underestimating their enormous potential for doing good: in India malaria was reduced from 75 million to 5 million cases in a decade, and in Britain alone the use of pesticides is estimated to save crops worth over £200 million every year. But it is important that agriculturists should change their emphasis from total chemical control to balanced control so that fewer chemicals may be expended per hectare. This has already been done successfully on a number of occasions, as witness the Russian use of a fungus disease plus DDT against Colorado beetle: the DDT was used at only one-third the usual rate because insects weakened by the insecticide were more susceptible to the disease. In Peru the cotton crop is now protected from its several pests by a combination of predators and mineral insecticides.

The chemical manufacturers may even end up quite relieved at the prospect, for it now takes an average of five years and over £5 million to develop a new material in order to test it for efficacy, lack of persistence and every kind of side effect.

Balanced control should not just be concerned with pests and diseases. It should give advice on new agricultural projects and, one hopes, help to avoid failures such as the original East African groundnut scheme or the Queensland sorghum scheme, in both of which environment and land potential were inadequately assessed. It should ideally extend to every aspect of cultivation, such as improving soil structure by use of more organic materials.

Aircraft and satellites will probably be used to monitor the health of large-scale crops, locating infections and infestations by studying colour deviations from the normal. Such 'earth resources technology satellites' will equally give help in finding minerals and water resources.

There can certainly be no let-up by man in his fight against his insidious, fecund adversaries; it is a fight for our own survival. But man's manipulations in this field must be increasingly careful and precise, for the problem must be considered in relation to conservation of wild life and habitats, natural resources, and the one earth we all inhabit.

29 · *An Incestuous Relationship*

'Mankind has an incestuous relationship with Mother Earth' – so ran a graffito recorded at the University of Michigan in 1970. The despoliation of the earth's resources by mankind has been so much discussed and written about that one hesitates to do so yet again. But as we have seen the plant kingdom is, in one form or another, so vital a part of the world, so important in terms of resources for man, that man's effects upon it must at least be summarized.

Not that comment upon this 'incestuous relationship' is anything new. In Isaiah 5 we read 'Woe unto them that join house to house, that lay field to field, till there be no place, that they may be placed alone in the midst of the world.' Plato, in the fourth century B C, was lamenting about Attica: 'Compared with what it was, our land is like the skeleton of a body wasted by disease. The plump soft parts have vanished, and all that remains is the bare carcass.' In China, erosion began to follow deforestation due to the expansion of cereal cultivation during the Ch'in dynasty of the third century B C.

I will content myself with one more apt quotation – from Thomas Love Peacock's *Gryll Grange* of 1860:

> They have poisoned the Thames and killed the fish in the river. A little further development of the same wisdom and science will complete the poisoning of the air, and kill the dwellers of the banks . . . I almost think it is the destiny of science to exterminate the human race.

In past aeons the globe certainly suffered its own natural disasters. The whole arrangement of seas and land masses has entirely altered. Mountains have been pushed up and eroded away, rivers have cut their gorges. Hot and cold periods have altered the climates, ice masses have moved to and fro across the land. But most of these enormous changes have been infinitely gradual and slow, and the natural world adjusted itself gradually and slowly around them – although there is an increasing belief that various gigantic cataclysms occurred in the earth's history, with consequent sudden alterations of climate and life forms. Today, at any rate, we have a situation where changes occur with appalling rapidity, are made with very little long-term thought, and are often irrevocable.

The first acceleration of natural change by man was caused by his use of fire nearly half a million years ago, and once this began there must have been very many accidental fires as well as deliberate ones. In combustible forest a few hours' conflagration can reduce hundreds of years of growth to ashes, and entirely change the vegetation of the landscape as well as its animal life.

Cultivation meant removing trees and scrub, while tilling broke down the soil into small light particles, easily blown or washed away. Parallel with this, as we have seen, wild grazing animals were encouraged, and domesticated ones kept in ever-increasing numbers, destroying the more succulent plants and browsing on trees and shrubs. An uncountable number of trees was cut for building, firewood and agricultural clearance. It is hardly possible to visualize some areas as they must have once been: Crete entirely covered with coniferous forests, of which a remnant exists in the White mountains; the Sahara a prosperous, forested area, full of animals, as we can tell from the rock paintings in the now arid Hoggar and Tassili mountains; Mount Lebanon clothed in cedars, of which today some 400 remain – the

Emperor Hadrian was an early conservationist, for he had warning plaques placed around the cedars.

It was only in 1620 that the Pilgrim Fathers landed in Virginia. Early travellers recorded that, in the American prairies, 'the strawberries grew so thick that the horses' fetlocks seemed covered with blood' (Lewis Mumford). What more piercing image could one have of a virtually virgin land of which it has been said (one hopes with exaggeration) that only 1 per cent now remains as those first settlers found it?

Any kind of cultivation impinges upon wild plant life, and the more intensive and extensive this cultivation the less chance the wild plant has – even the carpet-bagger weeds. The cultivated plant is a fifth columnist to the wild one, its cultivator a hired assassin. In this chapter I am concerned with cultivation solely as one of the many uses of land which blot out natural vegetation, and this includes single-species planted forests as well as food and utilitarian crops.

Man has always found it difficult to appreciate the delicacy of quality and texture of soil, and to realize that it is not inexhaustible, and that some processes are virtually irreversible. Savannah takes over from forest destroyed for one reason or another, steppe follows savannah after casual cereal cultivation and heavy grazing, desert creeps into steppe.

The causes of change are sometimes more subtle and unexpected. Thus in the seventeenth century there was an unexpectedly large number of Indians – about five per square kilometre – in the American chapparal or sagebrush zone. Although they were mainly hunters they also ate a lot of acorns, and this made them preserve the forests. At that time grizzly bears kept grazing animals at a reasonable population. But when the white man came into the area he steadily destroyed the grizzlies; the herbivores therefore became too numerous and the trees and much other vegetation were eventually destroyed.

If climatic change adds impetus to this vicious circle the results are serious indeed. We are witnessing this today in that zone of Africa north of the equator and south of the Sahara, from the Atlantic to Ethiopia, known generically as the Sahel. In the early 1960s new pumped wells and mass cattle vaccination extended the lives of the cattle. Since cattle mean wealth to tribes like the Tuareg, and are killed only for ceremonial feasts, their numbers steadily increased, eating up the available pasturage more quickly than it could regenerate. The people grew crops like sorghum and millet without rotation. The rains have failed in successive years, the Sahara desert has moved steadily south, and the people have had to do the same, leaving their dead cattle in the deserts behind them. In recent years, as will be all too familiar, the disaster has resulted in massive famine in many of the Sahel countries, with original homelands becoming treeless, eroded deserts and being abandoned. A downward trend in rainfall seems at present to be continuing.

Erosion occurs when soil structure is destroyed and the resulting dust is blown away in the wind and washed away by rain. It is likely to be most severe on even the gentlest slopes of shallow soil. Run-off from rain then cuts into the surface, forming ever-deepening gullies exposing useless subsoils and shales.

There is probably no country where the results of erosion cannot be observed. The dust-bowls and badlands of the United States were the first terrible warnings to advanced nations, but they have not been heeded. Very large fields encourage wind erosion, as one can see in places such as Texas and in Britain's once-rich fenlands. The removal of hedges to make these vast fields (19,000 km a year in Britain alone in the 1970s) destroys many kinds of plant and reduces essential animal habitats.

Recent data from the UK Soil Survey suggest that 44 per cent of Britain's arable land is currently at risk of further erosion, and coincidentally 44 per cent of United States farmland is now losing soil faster than it is being replaced, at a rate of around 1·7 billion tonnes a year. The US Soil Conservation Service has stated that 'the US crop surpluses of the early eighties, which are sometimes cited as the sign of a healthy agriculture, are partly the product of mining soils'. The main cause of this erosion in the US is intensive monoculture.

Average topsoil losses in good farming with yearly crop rotation in American prairie country are over 1 tonne per hectare. Reduce the crops to one, the loss trebles; leave out the traditional fallow year and losses rise to 7 tonnes per hectare. In the tropics the figures are even worse. In Nigeria, 3 tonnes per hectare for a cassava crop on almost flat land goes up to 87 tonnes per hectare on a 5 per cent slope, and to 221 tonnes per hectare on a 15 per cent slope.

Capitalists and communists mismanage alike. Humus content of the richest soils in the USSR, over-used by wheat monoculture, has dropped to 5 per cent and on marginal lands the soil, virtually sterile, holds 0·2 per cent humus.

Third World erosion problems stem in part at least from population increase, which means ploughing on less and less suitable land, including steeper and steeper slopes with no anti-erosion precautions. In conjunction with overgrazing and the cutting of trees and bushes for fodder and firewood, these are ingredients for disaster, and for ever-spreading desertification.

Cultivation has virtually always started with the removal of forest or scrub. Whether this is cut or burnt is initially immaterial. A forest can withstand swidden, or slash-and-burn, techniques, and indeed maintain complete stability, if the human population density is very low (four or five per square kilometre). This is true even of the delicate ecological balance of a tropical rain forest, where mature trees will return to small cleared patches in 150 years, although it takes 400 years to reassemble all its vegetable and animal components. But if large areas are destroyed and cultivation or grazing imposed on them, stability is completely destroyed.

Of the world's many types of forest, the tropical rain forest is the most at risk; once gone, it is impossible to regenerate this amazing web of life, this great natural ecosystem whose total vegetation weight is estimated at roughly 1,000 tonnes per hectare, including nearly 50 tonnes of lianas, epiphytes and parasites. Broken down, this averages 2 per cent leaves, 71 per cent trunks, branches and twigs, and 27 per cent roots. It has a virtually closed cycle of growth and decay, both on the surface, where an average of some 59 tonnes per hectare of vegetable litter is annually deposited – only 6 per cent of the total biomass – and below, where movement of water steadily weathers fresh layers of rock which are gradually penetrated by the roots. This rock is typically remarkable in being largely devoid of nutrients; all the food for the forest plants comes from that small amount of litter.

If the trees go, the soil lies unprotected from the ferocious attack of tropical rainfall. It is washed away by the tonne and in a very short time. Where, as frequently happens, the residues are rich in iron and alumina, exposure to the sun alternated with further rain eventually hardens these into sharp, useless rock formations called laterites. In temperate countries similar activity may reduce once-forested areas into a waste of naked peat.

Removal of forest today is mostly done by clear-felling or burning. In some places, like Java, foreign timber companies use special machinery which removes every scrap of woody flora, mainly to produce wood chips for the production of chipboard and paper pulp. In such situations some secondary growth *may* occur, though it will be of very second-rate quality, because the original rain forest trees need the special environmental conditions to germinate and grow. Even if replanting was envisaged, these native trees have proved almost impossible to nurture artificially. Other trees can be grown, though haste is essential if the soil is not to be irretrievably damaged. In temperate climates it is usually possible to replant; the Chinese are notable in doing so almost as soon as they have felled conifers.

Apart from ruining the ecosystem and its soil, removal of forest enormously increases the run-off of rain. This occurs on the flattest terrain but is made even more serious if trees are cut on slopes and especially watersheds. Topsoil disappears, gullies form; the loss of the natural 'sponge' of the tree roots, holding both water and soil, results in silt-laden floods descending precipitately on lower areas whenever it rains. The consequent flooding is horrific, covering huge areas of land, sweeping away houses, livestock and people, and clogging dams and irrigation works. An Indian government advisor has said, 'Once it took a month of the monsoon to produce floods; now vast areas are flooded after the first rains. Wherever the forests have been left intact there is water [available locally]; wherever they have been cut down there is a crisis.' The devastating floods of recent years in India, Bangladesh, South Korea and China are due to deforestation on the upper reaches of rivers, but such results can be seen world-wide. In the USSR uncontrolled felling over huge areas has led not only to over-full rivers but, to quote a newspaper account, 'widespread wind erosion reaching catastrophic proportions in some places: drifting sands shrouding fields and meadows, many rivers drying up . . .' In the 1960s, floods affected 5·2 million people a year; in the 1970s the figure was 15·4 million, and it continues to rise. Topsoil loss is horrific: 10 per cent of the world's annual soil loss is carried out to sea by the Ganges alone.

As an example of forest destruction, Brazil is at present the scene for a process which has been described as having all the makings of an ecological disaster on a gigantic scale. This process was started by the opening up of the interior made feasible by the Trans-Amazonica Highway which allowed penetration of one of earth's last big unexploited natural habitats, the Amazon rain forest. To quote Anthony Smith in *Mato Grosso*,

> that streak of a road, fired like an arrow through the heart of the jungle, is being lethal to the old kingdom. One day they will wonder, as they already do in so many other parts of the world, where the forest used to stand. And whether it is true that there used to be trees as far as the eye could see.

The process of destruction began in the 1960s, and eye-witnesses like Anthony Smith reported a continuous band of smoke and flames eating ever deeper into the green canopy of trees. Burning must take place in three consecutive years to stamp out finally all the regrowing potential of the forest. Then the jungle is dead, with not a coppice left. It contains some valuable timber trees, among them ebonies, teaks and mahoganies, but it is not thought worthwhile to save such individuals from the general incineration; it would impede the speed of clearance. It is not difficult to fell the average 30-m Amazonian tree – five or six axe blows are usually enough; a bulldozer will knock it over in one pass.

Burning forest typically releases nutrients but these are likely to be available for about two years only, owing to the mineralization of the soil and its inability to retain the nutrients, so it is quickly left sterile. Burning destroys all the leaf litter and humus, the forest's only reservoir of food beside the much larger proportion actually contained in its living plants.

In the third year the owner, if he is well informed, will plant a special type of tall wiry grass; otherwise natural grasses, not always suitable for cattle, will appear. Once the grass has established itself the cattle are brought in.

As early as 1969 Anthony Smith observed the beginnings of 'hideous erosion with gullying', and reported no plans by the new landowners to deal with this in any way.

The whole adjustment of the area is too enormous for anyone to predict precisely what will happen, save that . . . following the destruction of the forest and its razing to the ground, the forest will not reappear even if all the developers were to pack their bags and leave. The land would then be scrubland as far as the eye could see, much like certain wastelands of the United States. The Americans found the rich prairie grasslands and often made them poor. In Brazil, the story could so easily be repeated with the forests.

Since then new roads have been built. Ranching continues, many smallholders have moved in to struggle as best they can, and prospectors are finding oil and many mineral deposits. The jungle is going steadily, and with it the South American Indian tribes who once found a total way of life within it. Africa across the Atlantic, Brazil has insisted that Amazonia is her own national concern and no one else's.

Amazonia is not, of course, the only area of tropical rain forest. It is just part of a global disaster. Ecologists and conservationists are totting up the destruction of rain forest and realizing how much has already vanished of this fragile, totally irreplaceable ecosystem. The figures are truly appalling. Rain forest is being destroyed at the rate of 11·4 million hectares per year – no less than 21½ hectares disappear *every minute*. A hectare can easily be cleared within a couple of hours with power machinery. The total area of rain forest was originally nearly 16 million square kilometres. In 1985 nearly half of this had gone. The figures for different continents at that time were roughly: Africa, 52 per cent destroyed; Latin America, 38 per cent; south-east Asia, 40 per cent; Indian sub-continent, 64 per cent. If the current destruction continues at the same rate, there will be virtually no tropical rain forest left in thirty years at the outside.

The rain forests have been described by Robert Allen of the International Union for the Conservation of Nature (IUCN) as

> the most exuberantly variegated assertions of life on this planet. No other plant community can match them for diversity and complexity . . . There are estimated to be more than 25,000 species of flowering plant in the tropical rain forests of south-east Asia, and forty-nine per cent of the genera are found nowhere else in the world . . . Tropical rain forests are essential for watershed management, as reservoirs of genetic diversity, as laboratories and schools, and as living emporia of beauty, interest and fun.

Replanting with crops, even of trees like rubber trees and oil palms, is no substitute for rain forest in provoking rainfall. It cannot be emphasized too strongly that tropical forests are natural life-support systems.

Forest cutting also provokes the 'albedo effect' in which sunlight reflects straight back from unprotected ground, instead of being absorbed by vegetation; this creates hot, dry weather patterns in which rain clouds do not form.

Removal of forests of other kinds, or the open woodland characteristic of many parts of Africa, typically results in savannah grassland which can support grazing animals. It is, however, extremely easy to over-graze. Where cattle are over-grazed in tropical regions, the first results tend to be ever-coarser and less palatable grasses as the most edible ones are finished off. Grass round waterholes tends to be completely grazed out, especially in drought years when some waterholes dry up, and all the cattle – which need to drink daily in striking contrast to many antelopes – collect around the few that remain. Once the rains do start severe soil erosion follows. In temperate climates the grasses tend to be eradicated to be followed by unpalatable or spiny weeds; the ground may become virtually disintegrated before erosion sets in. Africa is a sad example of long periods of uncontrolled maltreatment, where forests have been extensively replaced by deserts.

An indirect effect of this deterioration is the reduction of suitable terrain for large herbivores such as elephants, hippopotami and ungulates. The activities of too many animals of this kind, especially if encouraged by a reduction of natural predators or if enclosed in too-small national parks, wreck the remainder of their habitat far faster than natural recovery can repair the damage. One hippopotamus could eat a swathe of grass 8 km long in a night: a concentration of the beasts will consume all the available grass and reduce the area to mud. Hippopotami also toboggan down sloping banks creating gullies which form starting points for local erosion.

Destruction of forests and woodland is carried out for many reasons: political expediency in moving people out of towns; a false belief that better returns can be had from ranching, say, than from proper management of forest resources; the simple macho aspect of destroying what seems inimical, which operates in Brazil; the get-rich-quick basis of logging for timber and wood pulp. Often this is compounded by duplicity based on greed, as in the Philippines where the logging companies manipulate their records and fudge estimates of damage, as well as intimidating the local inhabitants who often have title to the land. Sometimes the results appal in a different way, as with Nigeria, once a major exporter of timber, which now has to import 100 times more wood than it exports.

At a rather different level, trees and scrub are torn up all over poor countries for fuel. A Worldwatch Institute paper stated that 'For more than a third of the world's people, the real energy crisis is a daily scramble to find the wood they need to cook dinner [and also to keep warm] . . . all too often, the growth in human population is outpacing the growth of trees.' In some places people have to travel 40 or 50 km to obtain firewood.

The use of animal manure as fuel, so common all over Asia, is another cause of severe imbalance in productivity: in India alone 300 to 400 million tonnes of cow dung – which is reduced to 60 to 80 million tonnes when dry – is burned annually instead of being used to replenish nutrients and, especially, organic matter in the impoverished cultivated fields.

The all-consuming goat is the prime enlarger of deserts in Asia and northern Africa and had caused the most intensive plant destruction of all before power machinery was launched into forests. It has been said that the Roman Empire owed its decline and fall to the goat. Such destruction is botanically irreparable especially on isolated islands which tend to have very high proportions of endemics; the Hawaiian Archipelago, for instance, has 96 per cent endemic flowering plants. In addition to their accidental destruction, Hawaii offers one particular example of the almost complete extermination of a species through demand and total lack of replanting. This is the sandalwood *Santalum freycinetium*, which once made huge stately groves. During the late eighteenth century the Hawaiian monarchy established itself and based the island's economy largely on selling this wood, mainly to China and Polynesia. By 1925 the groves were decimated and export virtually stopped. Today this tree has partly re-established itself on the island of Oahu. Another species of sandalwood was exterminated on the Juan Fernandez Islands for the same reason. Other nations were more far-sighted, and *S. album* from Indonesia and India, the most frequently used of these aromatic sandalwoods, has been so much planted throughout its millennia of demand that no one can be certain where it originally grew wild.

Goats were put on islands by pirates and explorers to build up sources of food against their return. Captain Cook was responsible for Hawaii's goats among several other places during his South Seas voyages in the 1770s. Goats increase at phenomenal speed: the most lurid example is of three goats released on to Isla Pinta in the Galapagos in 1959. By 1971, the population had risen to an estimated 20,000, and in the eradication campaign which followed more than 40,000 goats were shot in the next six years.

Other islands whose floras have been decimated include Phillip and Norfolk Islands east of Australia, Robinson Crusoe's Juan Fernandez Islands, and St Helena.

Goats can, of course, be harnessed as biological controls: they have been used in several areas to reduce the amount of unwanted scrub on rangelands, where the net result is less scrub, more grass. But mostly they increase far too fast if left on their own.

Forest burning and cutting, over-grazing, and feral goats, sheep, cattle and pigs have combined to destroy most of the forests on the lower levels of many Polynesian islands. Losses of forest range from 30 per cent on Tahiti and Moorea in the Society Isles to between 60 and 90 per cent in seven of the Marquesas Islands. Denuded areas show the familiar pattern of rapid soil erosion.

Very often man introduces foreign plants, for crop or ornament, which spread rapidly on goat-ruined islands. Some of the settled Galapagos Islands, for example, are being swamped with yucca and New Zealand flax, to the detriment of the unique original flora.

One important outcome of cultivation and altered land-use is indeed the spread of introduced weeds, which prevent local plants recolonizing. In Java, for instance, there were few plants naturally adapted to become weeds because these seldom occur in heavy forest, so that once arable land was opened up there was an astounding invasion of foreign plants. In 1968 there were at least 300, over three-quarters of these being American or European natives. Much the same has occurred in

Malaya. Canada has a high proportion of European weeds; California suffers from eastern Mediterranean intruders; New Zealand has well over 1,000 aliens, which have indeed completely altered the vegetation balance; and in parts of central Portugal introduced Australian species like acacias have crowded out the native trees and shrubs.

Even the sea can suffer from uncontrolled exploitation. In Japan, where a great deal of seaweed is used for food and for the iodine industry, collection is normally by grappling hooks which wrench off the algae with their holdfasts. Serious depletion of colonies then occurs. Algae are collected for food around Hawaii, in Indonesia and elsewhere in the Pacific; in the Bay of Biscay a calcareous type is dredged up for preparation as a soil conditioner. Increasing uses for algae continue to be found and, as with forests, these undersea resources ought to be managed rather than ravaged. There is some hope that collection methods based on cutting, to leave a holdfast which will regenerate, will be increasingly used.

Besides deterioration due to misuse of natural resources, man's increasing numbers cause total sterilization of land by ever-spreading towns, roads, airfields, subterranean mines with spoil-heaps, open-cast mines, and industrial developments of all kinds. Industry, towns and vehicles emit vast quantities of abnormal chemicals into the air both as gas and as minute particles, while our waste products from every source impose increasing pressures on the natural world.

Pollution in general has been thoroughly dealt with in many contemporary accounts, and we are all familiar with its harmful aspects and its accompanying problems of wastes and surpluses. Here again I am only concerned with its effects on wild or natural plant life, not on man, his creations nor his crops. While pollution which affects larger plants and animals is serious enough, the worst is anything which harms micro-plants and other micro-organisms. With their vast capacity for rendering pollutants harmless, they are the most important part of the world life cycle to mankind but equally they are very vulnerable.

Most pollution is created by household sewage and garbage, animal wastes in intensive farming, industrial effluents of every kind and the disposal of highly toxic surpluses and waste by-products, both chemical and those derived from nuclear reactors. Some of them are unnamable because many industries do not know exactly what they are disposing of. This matter disperses into the air or into soil and watercourses. Even where solid wastes are buried underground, toxic materials are likely to leach into the soil and find their way into watercourses. Radioactive fallout certainly affects plants; it has apparently killed lichens in Siberia and increased their radioactive content alarmingly in Canada and Alaska.

Sewage, which has been called social effluent, and industrial wastes tend to be poured directly into rivers, lakes, estuaries or the sea. Sewage is sometimes treated, industrial effluent less often. Sewage can be recycled into fresh water, as it is in South Africa; treatments include the use of algae, of photosynthetic bacteria which can operate without oxygen, or the 'reverse osmosis' process pioneered in California, in which semi-permeable membranes are used.

In addition to chemical compounds, industry frequently has to dispose of water heated up during the industrial processes. If it is not cooled down in cooling towers or serpentine canals, but emptied straight into a river excess heat will destroy plant and animal life. Some of this waste heat could conceivably be transferred to domestic and greenhouse hot-water systems, and even be piped underground to encourage field-grown crops.

When lakes are used as sewage dumps the lower layers accumulate organic debris and become incapable of supporting life, while the upper layers lose oxygen which is consumed by the bacteria breaking down the sewage. This has occurred in the major Swiss lakes, which are incidentally so full of disease organisms that swimming and consumption of fish from them are forbidden. In the Great Lakes of America the sewage problem has been accentuated by the vast quantities of nitrates draining into them from surrounding farmland over-fed with nitrogenous fertilizers. The amount of these used in the United States has risen ten-fold in twenty-five years. This is partly due to increased cropping, partly because current cultural practices make soil structure deteriorate and the soil less able to retain nutrients.

Not only are excess nitrates a health hazard in water used for drinking and irrigation, but they can encourage 'blooms' of blue-green algae. These use up even more oxygen than the sewage-degrading bacteria. Excessive phosphates, such as those released by fertilizers, livestock manure and detergents, can increase the algal blooms. If the algae use up all the lake oxygen all organisms including themselves perish and add to the rotting sediments. Larger plants such as Canadian pondweed can also be stimulated into excessive growth by such circumstances.

A few years ago there were fears that the North American Great Lakes were 'dead' but today, thanks partly to improved precautions including treatment of sewage, the lakes certainly 'survive', although the balance of fish populations has altered and it is sometimes dangerous to eat the fish.

Other hazards to lakes include amenity use. In Lake Constance, which is only 70 km long, many hundred tonnes of petroleum residues from motorboats have been deposited on the bottom.

Excessive logging can also lead to trouble. Lake Baikal, often cited as one of the purest and most lovely inland seas, holding an estimated fifth of the world's freshwater reserves, is now suffering from wood processing residues, especially from a cellulose plant and also from some 1½ million cu m of wood which have sunk in the lake over the last ten years, as well as from other industrial residues.

In both these cases the micro-plants in the lakes – the start of the food-cycle for larger animals – have been seriously affected. The fishing in Lake Baikal and in nearly 4,000 km of some fifty rivers deriving from it has been ruined.

Rivers have always attracted industry for the convenience of transport and of waste disposal. Add to these industrial wastes hot cooling water, municipal sewage, and fertilizer run-off and rivers can very rapidly become entirely sterile, with every aquatic plant and other organism killed; pollution can be as extensive as the river is long. In 1972 a Netherlands newspaper reported that the combined waters of the Rhine and Meuse at Rotterdam were so full of chemicals, and so warm, that it was possible to use the water to develop a photographic film into a recognizable picture! The 4,800 km long Mississippi is, not surprisingly, equally liable to pollution; a 1975 report announced that the drinking water of New Orleans contained sixty-six organic pollutants alone, many derived from pesticides. Metals get into rivers very readily, and a recent report on the British Tees showed cadmium, copper, lead and zinc at concentrations around 1,000 parts per million. Water plants suffer damage from copper alone at 11 parts per billion.

Man has relied on the traditional self-cleansing nature of moving water for too long.

Holland is probably the worst-off country involved in the Rhine – a river recently described as 'Europe's majestic sewer' – on which it is very largely dependent for clean water. Too much salt in the Rhine water percolates into the high water table and into the Dutch farmlands, just as higher up the river it affects boreholes at a considerable distance.

Great river 'sewers' such as this continue through estuaries (where their load of muck is added to by sewage outfalls) sometimes polluting beaches and harming algae and associated animals, and into the sea. Further sewage and garbage are towed out in barges to be emptied a few miles off shore.

Although sewage and nitrates can enhance the growth of seaweeds and marine plankton, both plant and animal, most wastes that go into the sea donot. Much natural nutrient sinks rapidly to the sea bottom; it may be made available again if currents bring it up. Toxic wastes of many kinds are dumped in deeper waters, supposedly at 'safe' distances from land, sometimes in containers which will not, supposedly, corrode for an 'indefinite' time. The dispersal and persistence of both solid and liquid wastes in the oceans is still far from known, but it is certain that large concentrations of persistent materials dumped at one point will in time become identifiable thousands of kilometres away. Among the most worrying of such materials are radioactive wastes and polychlorinated biphenyls (PCBs) (by-products of paint and plastics manufacture). Already these PCBs have been identified in Antarctica. The quantities of material pushed into the sea in one way or another are not far short of astronomical.

Sea pollution may be both in solution and on the surface. When Thor Heyerdahl crossed the Atlantic in his primitive reed boats in 1968 and 1969 he recorded visible pollution from coast to coast, largely in the form of oil, tar and plastics. A recent survey of the Sargasso Sea revealed an average of 3,500 plastics particles per square kilometre.

Oil is a major pollutant of the sea, as we are especially aware after the sinkings of supertankers and the leaking, over a period of many months, of an ill-cased underwater oil well off Santa Barbara, California, in 1969, which ruined many kilometres of the adjoining coast, and was one of the first major events in the United States to rouse public opinion about oil. Similar events have occurred in the Gulf of Mexico and, due to war, in the Persian Gulf. Apart from such exceptional cases oil is discharged all over the oceans all the time by ships cleaning out tanks, and near the coasts by leaks and spillage at oil terminals. Oil on coastlines is capable of destroying every kind of seaweed and the organisms that live among them; the detergents used to clean oil off shores are almost as lethal to them.

Enclosed seas such as the Caspian and the Mediterranean are especially vulnerable to the build-up of oil and other toxic wastes. In the south of the Caspian oil has literally turned the sea bed into tar, while industrial and sewage pollution in the tideless Mediterranean have caused damage and reduced tourist appeal in many areas.

Oil pollution is not confined to seas and lakes. Oil has been discovered in Peruvian and Ecuadorean Amazonia and there is every likelihood of scores, if not hundreds, of oil rigs springing up in the jungle, their wastes spilling into the rivers to destroy vegetation and animal life.

Mercury provides a striking example of an insidious poison whether in lake, river or sea. In Japan many people died in a very unpleasant manner (from 'Minimata disease') from eating fish caught downstream from plastics factories using mercuric catalysts, and malformed children have subsequently been born. In Sweden it has been shown that mercury from seed dressings continues to accumulate in fish-eating birds, and probably in humans, despite restrictions on the dressings. A major hazard of mercury is that even in concentrations as low as 0·1 parts per billion it can greatly interfere with photosynthesis and multiplication of plankton, both in fresh waters and sea. Similar effects have been established with PCBs, which persist unchanged in seawater for at least a month; 13 parts per million cause a 50 per cent reduction of planktonic photosynthesis, while young fish can also be killed.

During the last twelve years at least twenty species of plankton have shown alarming population reductions due to these substances and to oil tainting. With plankton we are back at plant life, the first, vital stage in the marine food cycle.

Returning to land, the spread of the 'concrete jungle' severely affects wild plants. The problem is more acute in small heavily populated countries; a recent projection for England and Wales – based on current growth rates without counteraction – estimated that by AD 2800 every square metre would be in urban use; the current figure is more than 11 per cent of the country's area built over, and the expected level a century hence 25 per cent. Holland and West Germany will reach total urbanization even sooner if population checks do not occur. Let us remember, however, that extrapolation is not always to be trusted: in the 1880s prophets of doom were forecasting a traffic standstill in London early in this century, with the Euston Road three feet deep in horse manure!

Property development can affect a wide radius of surrounding countryside, even where it does not involve actual building. Thus in Florida drainage of swamps resulted in a constantly falling water table and vast changes in the natural balance. The peatlands which stored summer rain dried out; many areas caught fire and burned literally for years, sometimes exposing bedrock. The character of the whole area and its plant and animal populations, especially of the Everglades which, with its unique ecological qualities, is one of the great tourist attractions, rapidly altered for the worse until recent protective measures began to slow changes and control water supplies. But even though certain parts of the Florida Keys have been protected, a vast amount of tourist development – both fixed commercial sites, and mobile homes with their individual sewage problems – are fast creating new problems for this unique landscape.

Lowered water tables may be caused by massive drainage schemes, as in Florida, but more frequently by tapping of underground water supplies. Deeper and deeper the boreholes have to go, while the upper levels steadily dry out. Not only does this create mechanical problems, as in London, where the shrinkage of the drying clay causes buildings to tilt and crack, but it affects all wild plant life; trees deprived of their habitual water supply perish and the balance of vegetation changes.

The Aral Sea sank 176 cm in 1968 alone. Lake Sevan, in Armenia, has lost 179 sq km of its original 1,416 sq km of surface area since 1960. It is clear that one of the major problems in the future is going to be water shortage.

Modern technology in developed countries has tried hard to overcome these problems. In California water is moved on a massive scale for crop irrigation and human use, with rivers turned backwards and water pumped in unnatural directions. In 1985 one part of this water manipulation, the State Water Project, was consuming 4 billion kilowatt hours of electricity per year more than was generated. California is losing 25 billion cu m of groundwater per year, and in places like the San Joaquin valley the land has sunk up to 10 m because of this assault on the reserves in the aquifers below. Here too thousands of hectares have been lost to cultivation due to salt build-up after irrigation.

Desalination is undoubtedly a possibility, if an expensive one; large-scale pilot schemes exist in Mexico and Abu Dhabi. This need not apply just to water; soil can be biologically desalinated by growing salt-absorbing plants like tamarisk and sea-blite and harvesting them. Despite disasters due to ill-conceived irrigation, it is even possible to use saline irrigation water, as experiments in Israel have shown; not just with basically salt-tolerant crops like esparto grass but with plants such as date palms, tomato, lettuce, rice, barley, maize, sugar beet, sunflower and cotton which have been found, if specially managed, to tolerate chloride levels up to twenty times higher than previously thought feasible. Breeding is also being aimed at increasing plants' salt resistance. Plants irrigated with saline water seem also to have an improved capacity for withstanding periods of drought.

Air pollution is yet another facet of the waste problem. The pollution is composed of burnt residues, including those from incineration of solid wastes, and by-products from industry, both gaseous and of solid particles. Aerial pollutions can travel thousands of kilometres in the air before being 'grounded' by rain.

Short-range pollution can be very severe, especially when it involves heavy metals, fluoride, and various organic horrors. These residues render soils unfit for many plants, apart from making them poisonous to consumers, often via grazing animals. Local denudation of vegetation around industrial plants may result in soil erosion. A U S Congressional subcommittee found, in 1985, that almost all chemical plants in the country were releasing 'staggeringly high amounts of hazardous chemicals'; many do not appear to know just what they are releasing. In general, these chemicals are posing health problems to humans, not plants.

As far as plants go, the current disaster in the 1980s is due to the long-range emissions grouped together as 'acid rain'. These are sulphur and nitrogen oxides, waste products of burning both live forests as in Brazil and fossil fuels as in factories, power stations and vehicles the world over. These oxides, in company with ozone, change in sunlight into dilute sulphuric and nitric acids. First effects to be noted were in lakes, where these emissions, deposited as rain, began to acidify the water. A healthy lake has a pH of about 7 – neutral, in fact. Below pH 5 the variety of invertebrates declines, and some fish are affected. Below pH 4, most fish die; as acidity increases there is a decline in the number of higher plants in favour of algae and mosses. The record for acidity is rain at pH 2·4 which fell in Scotland in 1983; this is about six times more acid than vinegar.

Acid rain affects foliage directly, especially that of conifers, but the major and possibly irreversible effect is upon soils, especially thin and rocky ones. The effect has been described as transforming forest soils 'from a pool of nutrients into a cocktail of toxic metals'. According to the German biochemist Professor Bernhard Ulrich there are three phases in the process. In the first, the nitrogen in the emissions is dominant and, acting as an extra nutrient, causes the trees to grow more vigorously. This is why the emissions of the Industrial Revolution had no apparent effect.

Phase two occurs as acid accumulates further and further down in the soil which becomes progressively less capable of neutralizing it. The sulphate in the acid rain combines with calcium, magnesium and other nutrients which then leach from the soil; the yellowing of pine needles at this stage is an indication of magnesium deficiency. Growth of trees is slowed and the wood becomes damaged.

After this the sulphate starts to combine with other metals in the soil, notably magnesium. This is often present in large amounts, but bound up with organic materials which keep it harmless. Once combined with sulphate, its toxicity is unleashed and trees begin to die positively.

Phase three starts at this stage. As the ratio of aluminium to calcium increases, aluminium poisoning steps up. It inhibits cell division in the roots and reduces the trees' ability to resist disease infection. The inner root cells are next damaged, and this allows the poison to invade upwards, dooming the tree to death by poisoning, disease and starvation.

Aluminium is also harmful to the soil bacteria which decay dead matter and return it to the soil cycle, after which the soil may actually begin to generate acid on its own. At this point the process appears to be irreversible.

Acid rain first focused on Scandinavia – recipient of emissions from Britain and Europe – and is now a world wide problem, with North America severely affected, and beginning to arise in Latin America and from the Middle East and India to the Far East as new factories spring up. Utilitarian conifers are the worst affected – in the Black Forest probably half the spruces and firs are showing the signs, while around Los Angeles the ponderosa pines are bearing the brunt. Here the problem is accentuated because of topography which creates an inversion layer so that the city is often shut under a 'saucepan lid' of photochemical smog which includes ozone and peroxyacetyl nitrate, both very harmful to plants let alone humans. Such smog is now to be observed over Tokyo, Melbourne, Mexico City and Ankara, while Calcutta, Lagos and Jakarta are examples of cities under smog from charcoal burning.

The world production of air pollutants is measured in hundreds of millions of tonnes a year and increases all the time. Their continuous monitoring is now carried out in many areas of serious concentration. In Britain and the Ruhr, for instance, wild lichens, which are very susceptible to sulphur dioxide, are being used as an early warning system; in the United States tobacco plants are used to check ozone levels; small nets of sphagnum moss have been used to check airborne pollution from metals like cadmium.

Acid rain has become a political problem, with governments accusing each others' countries of creating the pollution and demanding control measures. Some governments are still stalling, asserting that the case is not proven while their chimneys go on blasting these evil vapours into the air; many are reluctant to fit sulphur-removing systems, new burners to cut emissions, or to use low-sulphur fuels, while there is equal reluctance in some quarters to impose apparatus for lowering the nitrous oxide emissions of cars. Great Britain, one regrets to say, is among the worst offenders in dragging feet on these issues, while the country's own conifers and those of Scandinavia and other countries steadily die.

No survey of man's effects on the plant world could omit mentioning his deliberate attacks upon it in war. A thousand years ago and more, invaders from the east laid waste many countries from Afghanistan westwards, destroying dams and irrigation channels as they did towns and palaces. The present barrenness of Iraq in particular is still largely due to this ancient succession of invasions. Large areas of France and Belgium were churned out of recognition by shellfire in the First World War, and even now there are fields round Verdun where nothing will grow, but this was a by-product of direct fighting. In the Vietnam war the Americans aimed directly and deliberately at vegetation to reduce both cover and food crops available to their enemy. The primary weapon was air-sprayed herbicide, the selective weedkillers 2,4-D and 2,4,5-T used in very heavy concentrations and sometimes in mixture; cacodylic acid, an arsenical compound; and the very powerful and persistent picloram which is a hundred times stronger than 2,4-D and which is banned in the United States. Several million litres were sprayed over roughly 1·7 million hectares of the Vietnamese countryside with the expected devastating results.

Fire was also tried, to burn away the dead trees left after herbicide spraying, but not extensively, apart from one five-week blaze which destroyed 4,000 square kilometres of very dense forest.

Destruction of vegetation has been furthered by over twenty-six million bomb craters averaging 9 m across and 4½ m deep; the explosive equivalent of one Hiroshima bomb was released over Indo-China every five days between 1965 and 1973. Over 15 million tonnes of munitions were expended. In addition to normal bombs large numbers of 'daisy-cutters' were dropped. These 9,000 kg bombs, designed to create helicopter landing areas in jungle, explode just before touching the ground, clearing every tree in an area the size of a football pitch and destroying all plant and animal life over a kilometre radius. Finally 'Rome ploughs' – 33 tonne super bulldozers – were used to finish off enormous areas.

The result of this activity is total destruction of vegetation cover. Massive erosion follows, the ground becoming slowly colonized with worthless weeds. The most prominent of these are elephant grass and various species of bamboo which make impenetrable thickets 2 m high. The coastal mangrove forests attacked – nearly half of their original total – have shown no kind of regeneration at all, even ten years after spraying. These mangrove areas were very important to the economy, the trees providing most of the country's fuel in the form of charcoal, and harbouring vast quantities of crustaceans and molluscs. Their natural regeneration could take a century.

Even where trees have not been killed they are often much weakened by metal fragments; more than half the rubber trees left in Vietnam are liable to fall in storms, while timber trees are so full of splinters that, even if they are not invaded by fungus rots, they have become useless to man. Bomb craters fill with water where insect larvae breed, resulting in diseases such as malaria, dengue and haemorrhagic fever. Use of the selective weedkillers in extreme concentrations has caused a dramatic increase in human stillbirths, in babies with congenital malformations, and apparently in cancers.

Unexploded bombs and, far worse, innumerable vicious anti-personnel devices prevent the peasants making serious attempts to get the land back into any kind of fit state. Ironically, one of these devices, the Dragon Tooth Mine, is a minute object with a plastic wing based on the design of a sycamore seed.

Although defoliation around roads may well have protected South Vietnamese and American troops from short-range attack, the direct effects on the North Vietnamese seem to have been minimal. At least 10 per cent of South Vietnam's agricultural land has been put out of action for perhaps a century, and probably 2 million hectares of forest – the country's timber needs for about forty years – have been wiped out with little hope of regeneration. This did not prevent Brigadier General William Stone of the United States Army stating that 'we do not think that our use of herbicides has created permanent ecological damage in Vietnam'. Nor has it prevented the United States Administration from insisting that the use of herbicides is not within the meaning of the 1925 Geneva Protocol's ban on chemical weapons, and including their possible use in what are called 'scenarios' for potential future conflicts, including one against Cuba where herbicides 'would improve visibility in the sugar cane fields'.

The United States are not alone in using herbicides. Among others, the Portuguese and South African air forces used them against the fruit trees and cassava fields of the Angola guerrillas in 1971, and the Israelis against 'illegally planted' Arab cereal crops in 1972. But these were very small-scale attacks. In Latin America and the 'Golden Triangle' spraying has been carried out, not for war, but against crops of cannabis, coca and opium poppy being illegally grown.

War in the future may involve not just herbicides but deliberate infection of crops with virus and fungus diseases. The description of anti-vegetation warfare in Vietnam, which has been described as 'ecocide', has been elaborated not so much for its intrinsic horror and apparent senselessness, but because it presents a high-speed model of potential ecological disasters which could be caused by the ordinary activities of our civilization today. It also emphasizes that such disasters can arise through political reasons, and equally through expediency, industrial aims or simple inertia.

Such reasons prevent Brazil from working out the probable future results of the destruction of Amazonia. Close ties between the BASF industrial complex at Ludwigshafen and the local administration prevent control of the half-million cubic metres of untreated effluent poured daily into the Rhine. Property development overrides conservation in Florida and a thousand other localities. Big mining firms such as RTZ can without compunction gouge away great hunks of New Guinea, and propose open-cast mining in national parks. The Philippines government has permitted foreign logging companies to destroy the country's forests, turning a blind eye alike to overcutting and to abuses to the local population. It is politicians, big business and technical experts who decide on sites for industry and cutting forest, not ecologists or conservationists.

Even so, the more prosperous nations are being forced into a measure of control over business interests. More and more monitoring or 'muck-watch' agencies are being set up, frequently spurred on by pressure from other industries, and also by citizen action groups. More controls gradually pass into legislation, and more sewage and other wastes are being treated. The states bordering the Rhine have come grudgingly to accept that some control of wastes is necessary, although much of the agreement reached is confined to restricting effluents to present levels. Also, the law is often one thing and its exercise another:

the United States Congress may have taken note of waste disposal problems, but recent examination of rubbish tips all over the United States showed frequent unchecked burning, and equally rubbish and banned toxic wastes in direct contact with town water supplies.

Legal interests in the United Kingdom are apparently not concerned with environmental issues, whereas in the United States many public-interest law firms have arisen to cope with the public demand. In Russia a recent decree requires the Union Republics to plan more actively in support of conservation and the more rational use of natural resources, and against pollution; a campaign is also under way to educate the public on the importance of nature conservation. Marxism has not been an environmentalist's ideology until these recent beginnings. One of its creeds is that man is master of his fate and, by the same token, of his apparently inexhaustible lands. State ownership does not guarantee conservation any more than in capitalist countries. Apart from actually felling timber in incredible quantities, using pesticides too heavily, over-cultivating sandy soils without regard to possible erosion, irrigating deserts and building huge hydro-electric schemes with little forethought, Russia also has plans for damming the Bering Straits, melting the Arctic icecap and reversing rivers, which are likely to have extensive climatic repercussions. Natural landscape and its wild plants have suffered heavy toll in the face of such activities, which are as much in the grip of vested industrial interests as in any capitalist nation.

However, environmental interest *is* being shown in Russia, as in the United Kingdom and the United States, and the signs in general are that increasing amounts of money will be spent, if reluctantly, by industry and government on cleansing and treating, recycling and re-use of wastes, and on cooling of water. Sooner or later sea dumping of really toxic materials must stop: their containers will eventually corrode and create problems for future generations. These wastes will increasingly include radioactive materials, for more use of nuclear fuels is inevitable as we use up the fossil fuels of coal, oil and natural gas. Without adequate precaution it is quite conceivable that within a few decades such wastes may partly sterilize the sea, destroying its plankton, algae and fish, together with all their possible uses for human food. We have become too used to treating the sea as an infinite sink which will absorb everything and nullify its effects, but neither it nor the air has an infinite capacity for self-purification.

It has indeed been suggested that the atmosphere may suffer from longer-term effects due to pollution, which may only be a few decades away from a probably irreversible situation. These are summarized in the hypothesis of the 'greenhouse effect'. The concentration of carbon dioxide, and also of methane, in the atmosphere has been rising, the first largely due to burning fossil fuels, the latter being increasingly produced from paddy fields as well as unexpected places like termite mounds. A doubling of CO_2 concentration is estimated to cause a rise in overall world temperature of 2°C, and three or four times that in high latitudes. This warming could mean much drier conditions across North America, Europe and Russia, with the threat of new dust bowls and certainly of reduced cereal yields. Melting of the polar icecaps could raise sea levels considerably, flooding many major coastal cities.

Similar effects may follow damage to the very small amounts of ozone in the atmosphere. The ozone layer in the stratosphere acts as a reflector for heat re-radiated from the earth's surface and as an ultra-violet filter. Damage to it, which may be happening from release of fluorocarbons and other chemicals, would mean more sunlight reaching the earth's surface, more warmth low down and less high up. Apart from the heating effect, this would destabilize weather patterns.

It seems to many that climatic destabilization from these factors and, more immediately, the cutting of forests (with their vast 'buffering' capacity in holding CO_2 and water), has already begun, though there are very many imponderables in any examination of climatic change.

One global trouble is, that while the more prosperous nations are adjusting themselves to these factors, the less prosperous are prepared to put up with pollution in order to develop their resources. Spokesmen for the Third World – mainly the more vocal South Americans – are at present busy arguing that they must overcome underdevelopment at any cost. They regard their prime environmental problems as malnutrition, poverty and disease. Conservation is not a viable aim, considering the condition of their inhabitants; the ecological vigilance and concern being shown in the United States and Europe are not practised, and local inhabitants are unlikely to protest.

Unfortunately the tropics are much more prone to erosion and climatic change caused by new forms of land use. But the underdeveloped countries are unwilling to listen to the rich ones on such matters, and who can blame them? The cost of production in rich countries is likely to rise as pollution control is enforced, and poorer nations will take advantage of this to remain competitive – and polluted. Our planet will almost inevitably become vastly different in the next few decades before any stability is achieved in these multifarious problems; apart from anything else the momentum of habits and practices will be very difficult to re-direct.

I have spent this chapter looking at the gloomiest side of the problem. It must be remembered that a number of pollutants, such as sulphur dioxide and methane, exist naturally, and that nature is helpful if given the opportunity. Micro-organisms capable of degrading organic materials, such as those in sewage, are plentiful, able to work on a wide range of compounds and to adapt readily to new nutrients. Most of them need oxygen for this process so it is doubly important not to fill watercourses and lakes with substances which eliminate the available dissolved oxygen. Many molluscs also extract pollutants, including heavy metals, from water; shellfish beds could be used as natural purifiers for many effluents. Higher plants can act as air purifiers and bind contaminants composed of solid particles, and planting of tolerant tree species should increase – thus the ponderosa pines of Los Angeles could be replaced with sugar pines and redwoods. Green corridors into cities and green belts around them should be increasingly maintained, both against air pollution and for the solace of their inhabitants.

But these are rearguard actions. In terms of the plant's world as a whole there is little cause for optimism. This chapter will have shown how wild plants and their habitats are increasingly affected by man, both directly and indirectly, with the natural world under unremitting and very diverse pressures. It is a situation where man can be said to know the price of everything and the value of nothing.

30 · *Things to Come*

Accuse not nature, she hath done her part;
Do thou but shine, and be not diffident
Of wisdom, she deserts thee not, if thou
Dismiss not her.

These lines from Milton's *Paradise Lost* are very much a text for the future of the world, and have never been more apt. The kingdom of the plants has a genealogy of 4 billion years, during which barely conceivable period of time great dynasties, as it were, have risen and perished, innumerable experiments have been tried and discarded. The latest dynasty – that of the angiosperms or flowering plants – has itself seen a considerable range of trials and errors before arriving at the roughly 250,000-plus species that we know today. It is a sophisticated group in terms of general efficiency and ingenuity. Today the angiosperms, and likewise all the other groups of plants left at various evolutionary horizons, face an unprecedented challenge.

What indeed *is* the future of the plant kingdom? Has it actually *got* a future? There are two sides to this question – plants that are truly wild, in natural habitats, and plants which mankind uses, and has developed from wild species, often changing them radically in the process, and which could, mostly, never survive back in the wild. It hardly needs saying that this latter sort of plant is quite essential for the survival of the human race, primarily for food and almost equally for animal nutrition; for warmth and cooking; for constructional timber; for health, and a myriad other purposes which have been spelt out earlier; and up to a point mankind needs plant life for mental replenishment both in and away from our concrete and tarmac jungles.

Taking the world's necessities first, at least half the world's population is undernourished, and probably one in ten – nearly 500 million – are starving or on the verge of it. As I write in 1985, hundreds of people are dying daily in Africa, a situation brought on by unprecedentedly low rainfall, itself caused partly at least by man's own actions. In these circumstances people are unable to grow crops: figures tell a horrifying story – in 1970 the twenty-four African countries worst hit by the 1984–5 drought produced 150 kg of grain per head; in 1985 the level was 100 kg. That this is not just due to population growth is indicated by the fact that Africa is the only continent in the world to display this downward trend.

Some of this problem is political, some of it logistical; there are grain mountains in the world that could feed some of the starving if the will and the means to get it to them were there. There is still a horrific gulf at government levels between the haves and have-nots, as exemplified in a 1984 headline – 'EEC fears record grain harvest'. Fears, because large surpluses mean lower prices.

But it is not a problem that will go away. There are over 5,000 million people in the world, and their numbers increase all the time; no longer on the exponential curve which predicted 8 billion by the year 2000 – current UN estimates are 6·1 billion – but still in terrifying numbers. Life expectation in many countries is steadily increasing. Even to improve the lot of those alive the first necessity is to step up food production: it is the quantity of food people can put in their stomachs that is vital, although it has been widely suggested that protein is the key to correcting malnutrition and high-protein foods will certainly cut down the production needed in actual bulk terms.

Yield of our basic crops is continuously increasing, as previous chapters have described, partly by better cultivation methods, partly by improvement of varieties. Rice is an excellent example. It is the staple food of around 2,500 million people in Asia, where the population increases by three-quarters of a million a week. Over 280 million tonnes of rice are produced annually, and it is a sobering thought that much of it is still planted, attended to and harvested by people bending over water-filled paddy fields. But in Japan the yield is around four times that in Laos, and the discrepancy between more specific areas is perhaps ten-fold. Part of this is due to varieties differing in photosynthetic efficiency by up to a 100 per cent, but reluctance to accept new cultural methods, lack of money and consequently of fertilizers, as well as the use of land whose present fertility is inadequate, are all influencing factors.

On the breeding side rice varieties now in production can give three crops a year instead of the previous two, can crop more heavily, are more resistant to diseases and to physical weakness causing lodging, and are more adaptable than older kinds to being grown in different localities.

The other great cereal crops of the world, wheat and maize, provide similar potential. Such cereals contain 7 to 12 per cent protein; oats, with up to 20 per cent, will probably be increasingly grown in future.

There is no visible end to the improvement of our basic crops, strengthened by using new species for breeding – as long as we do not destroy these in the name of improved cultivation. Apart from increasing yields, aims will include pest and disease resistance, the capacity to grow in saline conditions, higher photosynthetic efficiency, and more efficient conversion of nitrates into proteins.

At present half the world's calorie intake is supplied by just three highly bred grasses – wheat, maize and rice, and another seventeen species of plants supply 90 per cent of the world's food needs. Thousands of traditional local food crops, many highly nutritious, capable of growing in difficult conditions, and – since they are not highly bred – often resistant to pests and diseases, are ignored. Some 10,000 species have been recorded as used by humans as food, drink and flavourings. There is every reason to bring more of these into cultivation, especially as they do not usually require the high energy inputs of the major crops in machinery, fertilizers and pest control. Some local crops are grown at least in part for their leaves, the eating of which would produce a more balanced diet and, for instance, prevent so many children becoming blind in tropical countries. Two teaspoons of cooked fresh leaves a day can prevent this.

The Peruvian quinoa produces seeds with one of the highest protein contents known. Another central American possibility is chaya, a shrub producing nutritious spinach-like leaves. For arid conditions where little else can grow the buffalo gourd, yeheb nut and morama bean are all suitable. Every part of the winged bean, which has been called 'a supermarket on a stalk', is edible, and because its roots hold nitrogen-fixing bacteria it actually improves the soil in which it grows. The Australian desert grass *Echinochloa turnerana*, already valued as fodder, offers possibilities as a grain crop for arid areas. A very different possibility is offered by eel grass, a flowering plant which grows in shallow seawater from the sub-Arctic to the subtropics and, as the Mexican Seri Indians long ago showed, has seeds which if properly harvested can be used as a good-quality flour grain. *Jessenia polycarpa* is an Amazonian palm with huge bunches of fruit containing oil very like olive oil. As an example of a plant that could be managed in its own habitat there is *Terminalia ferdinandiana*, a small North Australian tree whose berries contain vitamin C at around 3,000 mg per 100 g of pulp, compared with 50 mg for oranges.

These are just a few examples of a positive treasury of plants known often from relatively small areas which offer great potential as basic crops and, after breeding, enhanced yields. Many are at risk in their present habitats, jessenia and the yeheb nut, for example.

Besides potential food crops many other plants are worthy of cultivation and development for all manner of uses. The tamarugo, for instance, grows in the Chilean Atacama desert in soil covered with a salt crust several feet deep. Its pods and leaves can support sheep in an area where ordinary forage is impossible to find or grow. Sagebrushes and saltbushes also offer forage possibilities in arid areas usually considered useless. There are unfamiliar plants which produce wax, like *Calathea lutea* and the candelilla. The desert-growing Mexican guayule produces latex of a quality approaching that of the best rubber trees (*Hevea*). Again, examples could be multiplied of plants which could be grown in areas useless for conventional crops and seldom otherwise attractive to man.

The last few years has seen the production of hydrocarbons from plants – petrol and oil replacements, in short. Fuel alcohol can be made from trees, sugar cane, maize, molasses, castor oil seeds and other more or less waste products; almost any vegetable matter can be used, including many residues. Potential exists too in apparently 'useless' plants like the tree-like succulent spurge *Euphorbia tirucalli* and its relation the gopher plant. It has been estimated that twenty-five to 125 barrels of oil could be produced annually per hectare from *E. tirucalli* on a 'petroleum farm'; such spurges thrive in arid regions and resprout when cut.

Growing some of these utilitarian plants brings up the question of cash crops, which usually entail farmers growing plants they cannot eat for sale in distant markets. Such growers are at the mercy of price fluctuations in world wide trade; the EEC release of sugar surpluses from sugar beet in 1985 undercut the growers of sugar cane enough to threaten several tropical economies based on that crop with ruin.

Increased crop concentration must be one aim if all cultivatable land is not eventually to be given over to crops. Existing crops can be made more productive, and for some new techniques can be applied. With fruit trees, for instance, vertically trained trees can be packed closely together, and experiments have been made with mini-trees, sprayed with growth-retarding hormones to stop growth and encourage fruit bud production: trials suggest that a crop of 50 tonnes per hectare could be harvested in the second year, as against the 20 tonnes yielded by an orthodox orchard after seven years' growth. Intensive planting means also that it is much more worthwhile installing automated systems for irrigation, feeding and pest control.

There should be more use of soil potential by growing several crops in one season (relay cropping) or even at one time (intercropping), thus making use of different environmental conditions exactly as the gardener does in catch cropping. Another advantage of this idea is that it counters erosion. One approach has been to leave land unploughed, sowing crop seeds, with the aid of special machinery, into 'cover crops' of grass or cereal, which may be killed later by spraying with a contact weed-killer. This creates permanent protection against soil erosion, for the soil is never left untenanted, while the dead plants provide a mulch which keeps down weeds, raises soil temperatures and retains moisture. Most important, perhaps, it maintains the soil structure which ploughing destroys, as long as care is taken when using heavy harvesting machinery.

A typical cycle at present in use in Kentucky is to direct-sow maize into a winter crop of rye or barley. On harvesting the maize, the roots of which remain in the soil, soya beans are direct-sown. When the latter near maturity the winter cereal is sown from aircraft, to germinate while the beans are harvested. Such methods have transformed the productivity of many previously barely viable holdings.

Other recent examples of intercropping include the sowing of mung bean alongside maize; the maize yield went up appreciably since the beans suppressed weeds without providing excessive competition. A special problem arose in Mauritius, where 80 per cent of arable land is occupied by sugar cane. The only way to increase food production without curtailing the economy's prime crop was to intercrop. Quick-growing maize, soya bean, groundnuts and sweet potatoes occupy the space between the rows of cane during the three to four months which the cane takes to fill the space.

As already mentioned, the condition of cultivated land and the loss of topsoil is giving increasing cause for alarm. Apart from commonsense cultivation methods the large increases in fertilizer prices may well mean some return to the use of animal manures, at present regarded by intensive farmers simply as a tiresome disposal problem. Destructive practices such as swaling may be replaced by light ploughing in of stubble, which not only improves soil texture but will reduce the quantity of fertilizer needed. Another decade may well see such traditional practices firmly re-established where agriculture has become over-developed; unfortunately they are unlikely to bring back into cultivation the millions of hectares lost in the last few decades to erosion and salination. As a 1984 Worldwatch Institute report states, urgent action is needed; this quiet disaster, caused by treating soil as a non-renewable resource, is as serious as any other facing humanity.

The extraction of vegetable protein has been experimented on for many decades. This involves maceration of seeds or foliage and processing under careful control to avoid changes in the constituents. Various legume seeds have proved the best sources so far, but leafy vegetables may be used in future: they can yield up to 30 per cent protein and are probably the most productive orthodox plants in this respect.

Up to 90 per cent of their nitrogen can be extracted from some crops, and yields of at least 3,000 kg per hectare have been achieved in India from lucerne. The protein can be kept for some months in moist cake form or for several years as a dry powder. The fibrous parts of the plants left after protein extraction can still be used as direct, low-protein fodder. Such forage crops as lucerne have the added advantage that they remain in the ground, being periodically part-harvested, and thus form a valuable insurance against erosion, which a totally harvested crop does not. Vegetable protein is at present mainly used for feeding animals, but it is hoped to develop material suitable for, and hopefully palatable to, human beings.

There are also hopes that excesses of unwanted plants, such as the river-choking water hyacinth, could be used. The plant, processed in various ways, has in fact many possibilities as well as human and animal food materials – agricultural compost, bio-gas production, the making of paper and moulded products, the treatment of sewage and other wastes and the decontamination of water polluted with toxic heavy metals and other substances. Technologically, there is no end to the possibilities of water hyacinth!

If sea pollution can be minimized there may be much extended deliberate cultivation of seaweeds, especially in areas such as Japan where they form an important part of the diet. Experiments to cultivate giant kelps in the Bay of Biscay were halted owing to fears about their effect on the environment. Like other seaweeds, the giant kelp is important to the alginate industry.

Plans have also been made to harvest turtle grass, a marine flowering plant from the Caribbean, as cattle fodder. Experiments suggest that it could be cut twice a year, since it regenerates within two months. Turtle grass has a protein content of about 13 per cent.

Single-celled algae are already being extensively cultivated. The inhabitants of the Lake Chad area have since ancient times used the blue-green alga *Spirulina platensis*, which grows in shallow water of a certain chemical balance. Cultures in the south of France, using troughs in the open air, yielded at the rate of 25 tonnes dry weight of protein per hectare per year; in California, a region of high temperature and long days in which yield is continuous, 50 tonnes has been produced. This compares with 0·4 tonnes equivalent for wheat and less than 0·1 for beef. Dried *Spirulina* contains up to 68 per cent protein, and is thus a very rich fortifier of animal feeds and a rich potential base for human synthetic foods. There are several pilot installations now producing *Spirulina* protein.

Another promising single-celled alga, *Scenedesmus acutus*, has a high yield and can be processed into a bland, easily digestible powder, nearly as rich in protein as *Spirulina*, but without the mushroomy taste which is not quite so easily manipulated by the 'aroma artists' in producing algal-protein foods. There are already a number of small-scale production plants as far apart as Thailand and Peru. In the Sudan and Peru *Scenedesmus* powder has been given in milk to children suffering from the protein deficiency disease kwashiorkor, with good results.

It should be noted that it is the very smallness of these single-celled algae which makes them so efficient, because the smaller they are the higher ratio they have of surface to volume. The unicellular *Chlorella*, much used in early experiments, has a ratio of 10,000:1. If it were the size of a tennis ball the ratio would be 1:10 and its growth rate likely to be 1/10,000th that of the real alga.

Chlorella has been used by the Russians in research on closed life-cycle support systems for spacecraft. One man in 1974 spent a month in a sealed cabin with *Chlorella* to recycle his atmosphere and water; he also took 50 gm of the alga daily as part of his diet. The algae were kept in a 30-litre 'reactor' illuminated by ultra-violet lamps.

Such algal cultures, and their combination with certain bacteria, can be used to treat sewage and reclaim water and still produce a useful end-product. Other cultures could extract heavy metals such as mercury from industrial wastes. The Russians are already using marine bacteria to oxidize oil from polluted seawater; the Americans are using similar organisms to decompose petrochemical wastes in a process which should eventually result in full recycling of materials. Bacteria are also being used, so far experimentally, to digest the offensive part of concentrated manure, 'farm effluent', which has become a problem of intensive animal farming. A sludge is left which can be dried and spread on land. Single-celled algae can also be grown on wastes: in China farm and human wastes are fed into ponds to nourish algae, on which fish are fed. Similar but more individual experiments have been made in the United States; and there is a pilot scheme in Britain. Herbivorous edible fish are also being used to eat weeds in canals and irrigation ditches.

Sea plankton, both plant and animal, is another possible future crop. Although Alain Bombard and others have shown that man can subsist on plankton alone, its deliberate cultivation is likely to be to feed fish rather than humans directly. Success again depends on keeping the seas reasonably clean. Ingenious suggestions have been made for creating artificial upward water movements in the sea by releasing thermal wastes at lower levels, or even by building nuclear reactors there. This would bring nutrient-rich water to the surface layer which, it has been shown, could make plant plankton increase sixty times faster than normal. This would enhance animal plankton and fish and mollusc populations for our own food. It might equally be possible to pump the nutrient-rich water into 'aquaculture lagoons' where algae could be grown.

The cultivation of fungi is steadily increasing. Mushrooms contain good proteins and several vitamins, are palatable as direct foods without the need for processing, or can be readily dried. It is possible to grow 80 tonnes per hectare per year. They use cellulose directly, a method of transformation shared with the cow and other ruminants; the oyster mushroom can be grown directly on pieces of log or felled stumps, as it is commercially in some parts of eastern Europe. Mushrooms are especially valuable to vary monotonous diets, as in south-east Asia, where the paddy-straw mushroom is widely grown: increased productivity and improved strains are being worked on, as they have been with our more familiar European edible mushroom.

Fungi at the mycelium stage are also being cultured to provide protein-rich bases which can be 'spun' into various food substitutes or analogues. Known as myco-proteins, these can double their weight in the fermenting plants every five hours; they are fed on glucose syrup originating from wheat or maize.

Similar analogues made by spinning protein from field and soya beans have been less acceptable because of their inherent flavour.

In principle, these processes involve dispersing the fungus or vegetable matter in alkali, forcing it through a fine mesh into an acid solution, and collecting the precipitated filaments which result. These are stretched, a 'binder' is added and finally a coating of fat. The size and density of the filaments, the amount of fat, and of course the flavours and colours added simulate types of meat, chicken or shellfish very convincingly; also chocolate biscuits. Several kinds are already on sale, in the form of chunks or mince. In another process now in active use in the United States, the high-protein material is used as an 'extender' to be mixed with ground beef or hamburger, in the proportion of one to three. The resulting product, in which the extender takes on the flavour of the meat, is cheaper than pure meat and largely indistinguishable from it.

These extenders are being widely used in various prepared dishes, and in hospitals, schools and canteens.

Weight for weight such analogues are as rich in protein, carbohydrate, calcium and iron as actual meat or fowl, provide appreciably more calories, and have a fibre content like that of wholemeal bread. Some doubts have been expressed about their total nutritional capacity, especially in the formation of brain cells; but considering the millions who already subsist almost entirely on vegetables these doubts cannot be considered proven. In any case the addition of yeasts can overcome any deficiencies of leaf proteins. These analogues represent a great potential increase in food productivity, considering that it takes nearly 10 Calories/1 Calorie of grain to add to the weight of a steer grown for beef, and about 6 Calories for a pig.

As mentioned in the previous chapter, pastes of soya beans or soya with groundnuts are very important protein-providers. The Japanese fermented paste is now being made on a large scale in factories where the required conditions can be strictly controlled. Not surprisingly it is being found that varieties and strains of the organisms concerned are as important here as in cereals, potatoes or apples, and there is little doubt that there will be deliberate selection and breeding of new micro-organisms to develop such food substances. The controlled production conditions will steadily be improved to increase yield, and at the same time protect the useful micro-organisms from infection by harmful bacteria and fungi which on occasion upset the peasants' cultures. Apart from this such micro-organisms have a short 'cropping' time, use up little land, and can be grown with salt water, avoiding the use of precious fresh water.

It seems possible that similar pastes can be made from wheat, which might provide a useful means of absorbing wheat surpluses if the pastes were locally acceptable.

Another already productive development, which would have sounded like science fiction a decade ago, is the feeding of photosynthetic bacteria upon methanol derived from natural gas, and of yeasts upon petroleum oils. The microbial or single-cell protein (SCP) thus produced is at present largely destined for animal food, where it may replace fishmeal, skimmed milk powder and soya bean flour, which are becoming much more expensive. The latter, with its great potentialities for human food, will doubtless be grown in many more places in future to avoid the supply problems that arose in 1973. Yeasts have been added to human foods in the past: both the Germans and the Russians did this quite a lot during the Second World War.

Special strains of hydrocarbon-fed yeasts are being cultivated in many countries. It is cheapest to use gas-oil because the yeast production is a bonus in a process which improves the oil quality. Many other substances can be used, including molasses, sawdust, wood pulp, residues from the processing of sesame and carob seeds and the like, and the solution left after protein has been precipitated from leaf extraction. The use of waste products which would otherwise be burnt or damaged, on a local or village-level scale, to make animal foods, is already being carried out in many places.

Yeast is incidentally a very efficient converter of energy, capable of a production of 65 kg from 100 kg of sugar, compared with 20 kg useful production by pigs, 4 kg by beef cattle, 15 kg from milk, and 5 kg by fowls. It has been estimated that an animal weighing 500 kg produces the equivalent of ½ kg of protein in a day; the same weight of *Torulopsis* yeast in ideal growing conditions produces 2,000 kg in the same time. The average yeast culture doubles in volume every four hours. Some yeasts produce fats rather than proteins.

Another potentially valuable development uses a fungus, *Trichoderma viride*, to create glucose from insoluble cellulose-rich materials such as waste paper.

Such methods are undoubtedly a very fast, relatively cheap way of filling the 'protein gap'. Later they may conceivably be replaced by complete chemical synthesis. One might of course ask, why not just eat more beans (not necessarily soya, which are not to everyone's taste)? There is an answer, important in this time of dwindling land resources – the area needed to produce a million tonnes of soya beans is 2 million hectares; to produce the same amount of myco-protein, 100 hectares are sufficient.

Food technology of this kind (which has been called 'food engineering') tends to dismay those of us who are accustomed to eating well, with clearly recognizable meat and two veg., not to mention all the trimmings. However, the products appear to be absolutely safe to eat, they can be made almost indistinguishable in taste and texture from foods to which we are accustomed, and they are extremely nutritious. Doubtless the techniques of producing them will be refined as time goes on. Many people are somewhat conservative about food – the have-nots, curiously enough, often more than the haves – so no doubt each culture will need its own synthetic products; while fermented soya bean paste is already a staple in Asia, the British and Americans may demand imitation beef-steak, though they should remember that cheese, beer and bread are all produced by microbiology.

Food crops from single-celled organisms and fungi provide a slightly ironic reflection upon the millennia of plant evolution, in that the organisms of the living world's infancy are being found so important to its later age. Some of these organisms are also playing a vital role in new developments for creating brand new crop plants. In theory these can result in plants which could not possibly be produced by orthodox sexual breeding, as outlined in Chapter 27, which is restricted to close relations.

At present this still means using genes that exist in plants currently alive, and is no answer to allowing wild species and primitive land races of crop plants to be wiped out in whatever way. There are possibilities of engineering new genes, but they are far in the future. As it is, our knowledge of the actual genes that reside in our crop plants is at best very incomplete, and a lot of hard work in mapping these is undoubtedly necessary if the new methods are to produce worthwhile results and, as is predicted, vastly accelerate the 'numbers game' of conventional breeding and transform the possibilities for novel plants 'made to measure'.

What are these methods, that smack of some highly sophisticated Frankenstein at work? In the 1970s scientists were pinning their hopes on 'protoplast technology' in which protoplasts – plant cells denuded of their cells walls to expose their boundary membranes – from different plants were forced together. Re-forming a cell wall comes next, then the new-created cell divides and forms a mass of undifferentiated tissue. Then, if all goes well, this tissue, or callus, will form a new plant. A hybrid tobacco was created in this way in the 1950s. Unfortunately, although tobacco and its relations like petunias, potatoes and tomatoes proved amenable to this process, other important plants refused to co-operate with the techniques used.

This is one method of genetic engineering; the more hopeful and less chancy is 'recombinant DNA', in which strands of genetic material are transferred from one plant to another. Instead of fusing whole cells and hence whole chromosomes with characters both good and bad, manipulating DNA opens up the possibility of transferring only desired genes.

Some of the practice involved is very hard for the layman to understand. In many experiments, viruses and bacteria are used to pick up the genetic material it is desired to transfer, which can in due course by deposited in the chromosomes of the host cell. The bacterial plant pathogen *Agrobacterium tumefaciens*, which creates crown galls, and the gut bacterium *Escherichia coli*, are particularly used as transfer or vector agents.

One recent successful experiment has resulted in a tobacco plant which contains a bacterial toxin which kills attacking insects. This toxin exists in *Bacillus thuringiensis*, for long the major component of biological sprays which, used like any pesticide, kill insects by infecting them with a deadly disease – which is quite specific. In over-simplified terms, the researchers isolated the toxin-creating gene from the bacillus and inserted it into the Ti plasmid from *Agrobacterium tumefaciens*, which has been isolated as an ideal transfer agent since it can breach the cell walls of a plant.

During these experiments plant/animal fusion has been carried out. In 1977 one such chimera combined red blood cells from hens with yeast cells, and in 1984 a piece of human DNA was transferred to a petunia. These experiments will not produce 'green men', clucking plants nor leafy birds, but the various interactions of nuclei and cytoplasms show potential, for instance making plants produce proteins in excess of what they can manufacture on their own.

The aims of genetic engineering are in fact much the same as those of orthodox breeding. Higher protein levels, resistance to salinity or harmful minerals, better employment of fertilizers, more efficient photosynthesis, increased cold-hardiness are among them. An increase in the cold-hardiness of wheat by a mere 2°C could increase worldwide production by at least 25 per cent.

The most discussed pipe-dream, probably, has been the creation of cereals, and indeed other plants, capable of fixing their own nitrogen, either by fusing leguminous cells infected with the nitrogen-fixing *Rhizobium* bacterium with those of the cereal, or transferring the bacterium's nitrogen-fixing gene into the plant's cells. Nitrogen-fixing crops bring enormous advantages to the soil, enhancing fertility, reducing the need for fertilizer application and avoiding excesses of nitrates running into water supplies.

Unfortunately cereals are not prone to crown gall, and hence *Agrobacterium tumefaciens* cannot be used with them for gene transfer in this or any other respect. There are other problems with cereals as described later.

Nitrogen fixation is so important that it is worth diverging from the genetic engineering theme to mention other experiments which have been in progress on the tropical grasses which fix atmospheric nitrogen with the aid of the bacterium *Spirillum*, which does not form nodules but lives in close association with the roots. Varieties of maize, bred from standard varieties showing some promise as nitrogen fixers, have already shown a fixing capacity a hundred times better than their parents, and at certain stages of growth fix nitrogen at the same rate as soya beans. *Spirillum* is not active in colder climates, so temperate cereals are unlikely to benefit but there seems to be much potential with warm-country cereals.

Returning to genetic engineering, the step after introduction of desirable genes, or of protoplast fusion, is to create actual plants from the cells concerned. Sometimes the mass of proliferating cells does produce shoots and then actual plants, but others produce only roots, and some just continue to multiply without differentiation. There is no rational basis for the differing results, and unfortunately the monocotyledons, which of course include the major world crops wheat, maize and rice, are particularly unresponsive to tissue culture.

There are as we have already seen various other ways of regenerating plants. One is the use of meristems or growing points, while some plants, like citrus, naturally produce numerous embryos, the sexual embryo often being suppressed, so that their seeds will produce plants identical to the parent. Such embryoids can be produced artificially, and it has been possible to culture-grow one or a few cells, from a root or elsewhere, by a kind of false embryo development known as totipotency. The nourishment of these cells is very simple, the easiest method being in an appropriate aqueous solution which becomes, in effect, an artificial ovule. That aspect of *Brave New World* is a fact in the plant kingdom if not yet among animals.

Plants can also be culture-grown from pollen grains and equally anthers; these have only half the normal chromosome count, and if this can be artificially doubled the true-breeding result could be valuable in certain aspects of breeding.

Another aspect of genetic engineering with promise is known as soma-cloning. Tiny pieces of stems or leaves are placed on agar gel bolstered with nutrients and growth-inducing chemicals. Unlike the meristems from shoot tips, which are identical to the parent, these small agglomerations of cells then develop into plantlets which are likely to be different from each other in certain characters. British research in the 1980s has, for instance, resulted in thousands of genetically distinct clones of potato among which were isolated some with better yield and resistance to diseases and virus infection. The main advantage of soma-cloning over seed production is that plantlets testable for certain characters can be achieved more quickly; the problem of non-selectivity is the same.

So far the theme has been the creation of actual new plants which can be multiplied by orthodox methods. Something akin to the production of myco-proteins described earlier is the direct extraction of end-products from the undifferentiated cell culture. The value of the process is greatest where the product sought occurs naturally in very small quantities. Thus a Japanese company has built a plant to produce saponins, valuable as a source of the hormone oestrogen, from tissue cultures of ginseng. Another makes shikonin, a red dye with anti-bacterial qualities, from the roots of *Lithospermum erythrorhizon*; an idea of the cost of the process is given by its value of £2,000 per kg.

Drug companies are among those experimenting most assiduously in such extraction. Already the cancer-controlling alkaloids in the rosy periwinkle have been so produced on an experimental scale.

Work is also going on in extracting flavours, such as those of chilli peppers and saffron, both expensive – capsaicin costs £250 per kg and saffron £5,000. Possibilities exist for extracting the flavours of cocoa, coffee and tea, but sugar production by this means has been discarded as excessively expensive.

Many problems remain to be solved with the various aspects of genetic engineering, but equally it undoubtedly has the capacity for improving human life and, by creating more efficient crop plants, perhaps relieve pressure on natural habitats.

But solution of the world food problem cannot be restricted to increasing crop yields. Energy inputs and outputs must also be considered. One may, for instance, contemplate the fact that the food products now needed to feed 210 million Americans (and equally, better-off Europeans) would feed 1,500 million Chinese. This is because Western nations eat a lot of meat, and to feed meat-producing animals calls for over 900 kg of grain per person per year, each person eating only 63 kg grain in direct form; whereas each Chinese eats 160 kg directly as grain products and only 18 kg of meat. Nearly 80 per cent of the U S grain harvest goes to livestock for meat production.

At the same time we must consider how modern rich agriculture is dependent on high energy consumption, both for direct cultivation and harvesting and, via the petrochemical industry, for fertilizers. In Britain alone a fuel energy input of about 200 million therms in 1938 had by 1970 increased to about 800 million therms. Nearly 1,100 million additional herms are expended to produce fertilizers, insecticides, weedkillers and the actual machinery involved. In the United States the proportion of energy for fertilizers, etc., compared with that used up for tractors and machinery appears to be very much higher, in the region of eight or nine to one.

If this level of energy-intensive farming, as it has been called, were practised all over the world, beyond the 'developed' countries, it would mean that at least 40 per cent of total world energy potential would be absorbed. We hardly need reminding of recent energy crises nor of the rates at which energy reserves are being depleted, and a less highly energy-based agriculture may well be forced upon the world in due course. This may mean a lower level of productivity, which makes the population problem even more worrying, especially in countries like India where at least a quarter of the grain produced rots in store or is eaten by pests and vermin.

To talk about the areas of land still 'under-used' and thus potentially available for crops of whatever kind is sheer lunacy. This applies both in terms of a world in which any natural beauty is to remain – the pastoral scene carried to its ultimate is nothing but a food factory – and, as is increasingly being perceived, in terms of maintaining the very necessities we need: adequate rainfall, retention of topsoil, freedom from drought; and the possible 'greenhouse effect' and climatic destabilization summarized in the previous chapter.

More international co-operation on the distribution and prices of food crops could see the United States, for example, utilizing more of the land it pays farmers to keep out of production, and exporting more of the grain it grows so well. Starvation and undernourishment could quite probably be avoided at this moment if the nations of the world were able to organize their wealth and make better use of food surpluses which are sometimes destroyed and sometimes stored, with inevitable degeneration.

Around 43 per cent of the earth's surface is now desert or semi-desert – both natural and man-made – an area of over 57 million sq km. It is estimated that the area of desert created by human activity and carelessness is almost as great as that devoted to agriculture. Each year about 12 million hectares of farmland is being degraded to desert levels. Dr Baumer, a special adviser to the UN, has said that 'opening the desert to cultivation is a dream. There is a lot of evidence that the soils in arid regions are not very good and case studies show that the carrying capacity of desert regions, even when well managed, is low.' Alas, this kind of evidence cannot be faced by some people. A politician from Burkina, for example, attended a meeting which described such a case study of an arid region of the Sudan, comparable to much of his own country, which concluded that intensive agricultural development would simply result in the region becoming a desert. The official said 'We cannot accept such conclusions . . . and if the [1977] UN conference reaches such conclusions we will not accept them there either.' He went on to state that his country intended to expand from 6½ million population to 30 million, and to provide every one of its people with a standard of living like that of California.

As things stand today we have to overcome such pressures of necessity and ignorance, together with greed, politics, religion and national pride. Necessity is a matter of economics, and could be overcome rapidly, given perhaps less expenditure on weapons and space, and a measure of international goodwill. Greed is a matter of conscience and law. The last three are different matters. Overriding all, as I see it, is the need to stabilize a world population where life expectancy is continuously increasing, and this again calls for international measures and understanding.

When Christ was born there was an average of just over one person per square kilometre. Today there are about forty; by the year 2000 there will be at least eighty – and these are averages; in some parts of the world the figures are far higher.

I cannot express my own beliefs in this matter more succinctly than did Professor John Postgate in a 1973 *New Scientist* article:

> The rate of increase of the world's population now exceeds any conceivable rate of increase in mankind's ability to mobilise food and other resources . . . These problems [starvation and social instability] are not matters for the future, they are consequences of the population explosion which have overtaken the world today.

He adds that 'In both poor and rich communities, the consequences of the population explosion are distracting people from the problem itself.'

Survival itself is not enough, and in any case, as Paul Ehrlich has said, the danger of simply stepping up the 'green revolution' now is that we will probably end up with a much larger population facing starvation a little later. To quote from a still-valid 1971 interview with him published in *New Scientist*:

> My professional opinion of what could be done, if a very broad view, is that by putting perhaps five times as much money and energy into agriculture and disease control for the next 20 years, we could transition to sensible systems that would allow us to get today's results with half the energy and ecological threat. It's going to be an extremely sticky and expensive transition; you can't just stop what you're doing now and start doing something else . . . And during the transition to the new agriculture, yields are certain to decrease, which is why ecologists are so wild for population control. We are already stretched to the breaking point, and every person we add makes this hideous choice, between present and future needs, more difficult. We know that a lot of the things we are doing – like overfishing the ocean . . . – point towards long-term disaster, and yet the excuse for doing it is that we have a huge population to feed.

Is that doom-watching, or is it logical thinking?

There is, of course, the diametrically opposed view. Dr Norman Borlaug, who won a 1970 Noble Peace Prize for plant breeding work, has castigated 'hysterical environmentalists', stating that

> the green revolution is not a breakthrough – it is a temporary success in man's war against hunger and deprivation. Its continued success will depend on whether agriculture will be allowed to use fertilisers and pesticides. If denied their use . . . then the world will be doomed, not by chemical poisoning but from starvation.

No mention of population is made.

Most of this chapter so far has been about food and utilitarian plants which mankind must have for survival. We have seen how the wild plant in its natural habitat is very much at the mercy of short-term expediency, whether it is capable of direct exploitation or, increasingly, because it is in the way of 'development' and human resettlement.

One major cause of specific exploitation is human greed, for some of the grievously endangered species are ornamental plants which have been over-collected for garden use. Such plants can be big business. Thus there are no longer wild daffodils in large areas of Portugal; this occurred many decades ago. More recently wild bulbs of many kinds are an important export from Turkey, where around 350,000 kg are exported annually. In North America some twenty species of bulb are only to be found in national parks.

World-wide, orchids are at high risk. They have always been valuable commodities, even though their cultivation is no longer just the rich man's pastime. In Victorian days collectors in India cut down whole forests to pluck off the orchids growing on the trees. Today poachers operate even in national parks. Besides these, cacti are most at risk, and many rare species from Mexico and southern USA are close to extinction, while commoner kinds are being much diminished. In Mexico too the cycad genus *Ceratozamia* has been eliminated in many of its limited stands. Some plants are endangered because wild blooms are cut for the cut-flower trade, including banksias and 'kangaroo paws' in Australia.

There is now a Convention on International Trade in Endangered Species of Wild Fauna and Flora, known for short as CITES, which is beginning to show its teeth; but the rewards of poaching are so high that high risks are taken. A single plant of the cactus *Strombocactus disciformis* was in 1985 worth up to US$583.

Several ornamentals are primarily endangered by habitat destruction, for instance African violet species in East Africa, and proteas and their relations in South Africa.

There are estimated to be between 250,000 and 270,000 species of flowering plant in the world. No one knows how many of these 'modern' plants – in evolutionary terms – have been extinguished since man arrived on earth, but it is an ever-accelerating process, 1985 estimates being that we are losing roughly one species a day and that by 1990 the rate may increase to one an hour. One in ten of existing 'higher' plants is presently reckoned to be threatened with extinction – that is at least 25,000, compared with 357 mammals, 437 birds and 187 amphibian and reptiles considered to be in similar danger and recorded in the Red Data Books of the International Union for the Conservation of Nature. Moreover, it has been estimated that for every plant made extinct, between ten and thirty other organisms are likely to go with it.

Note too that whereas virtually all animals and birds (with the exception of insects) are known and listed, a great number of plants are not know to science. The species of probably four-fifths of the land surface are not yet catalogued.

Unfortunately much botanical work now being carried out is on temperate plants, whereas we are most ignorant about the possible values of little-known plants of the tropics, where at least 70 per cent of the earth's plant species grow (and, as I have earlier shown, it is in the tropical rain forest that most destruction is being done).

The plants in worst danger are the most local, for instance the South African blushing bride, the St Helena redwood and the jade vine from the Philippines. These are examples of extremely beautiful plants which deserve preservation for ornament alone, and many such are of course being cultivated and preserved in this way. I have earlier mentioned the maidenhair tree, the dawn redwood, the flamboyant and the *Amherstia*, now widely cultivated but either extinct or excessively scarce in the wild; the lovely Phillip Island glory pea has also been extinguished in the wild. Other cases provide the scientist with the incalculable value of a specialized flora, as with the almost entirely endemic plant population of the Hawaiian Archipelago – a positive 'laboratory of evolution' – where all the dangers that I have mentioned in this chapter exist to threaten the plants, except that widespread industrial pollution is replaced with unbridled tourism expansion, and introduction of alien plants. Hawaii already has a long roll-call of totally extinct species, and many hundreds more are considered to be endangered.

There are people who scoff at saving plants whose only value is beauty, or rarity, but such plants are surely as much to be valued as an osprey or an oryx, and in fact are likely to be seen by more people. But the 25,000 threatened species also represent untapped capital for, as already remarked they certainly include many with utilitarian value as potential food crops, or for supplying oils and fibres, or, most notably perhaps, for medicinal drugs. Consider in this context that one single plant, the rosy periwinkle, widely grown for ornament, contains fifty alkaloids, at least two of which are tumour-suppressive, and one used for killing leukaemia cancer cells.

A less spectacular but noteworthy plant is the Ethiopian endod bush, widely grown for hedging. This has been discovered to kill the snails which carry bilharzia, at a low dilution at which it appears to be harmless to animals and people. This is a cure which could be used at grass-roots level, which has its own importance.

Many other 'medicinals' could be cited, some known and used by 'primitive' tribespeople, others appropriated for use by Western chemists. Some of these – digitalin from foxgloves is a long-known example – have resisted attempts at synthesis, so the plant is extensively cultivated.

It is not only flowering plants that can be valuable: the humble lichens have provided us with a large number of powerful antibiotics. Gandhi once said,

> The unlimited capacity of the plant world to sustain man at his highest is a region as yet unexplored by modern science. I submit that the scientists have not yet explored the hidden possibilities of the innumerable seeds, leaves and fruits for giving the fullest possible nutrition to mankind.

Economically minded botanists and biochemists realize the truth of these words, all the more in recent years when the enormity of vegetation destruction has become apparent. Politicians, however, seem not to be taking it in. Now some of these plants can be cultivated, though the trees of the tropical forests are mainly very difficult to cultivate, especially as their seeds often have very short viability. But the odd plant preserved in a botanic garden begins to resemble the last dodo in a zoo. Not quite, of course, for plants have seeds which can quite often be preserved and germinated for dissemination, either to other gardens or re-introduced to the wild; and of course plants in 'captivity' are readily available for study. But some scientists regard few *individual* plants as worthy of such preservation. As Dr Peter Thompson has written, 'The great majority of threatened plants demonstrate no clear-cut or absolute independence on *ex situ* conservation for survival, nor any identified quality, which may be of economic or even aesthetic value.' It is more as a 'part of the whole edifice of botanical complexity' that any individual plant deserves conservation.

When he wrote those words Dr Thompson was in charge of the seed bank established by the Royal Botanic Gardens, Kew. In that position he had faced up to the problems. He asked, for instance, 'How much would it cost to make collections of a tree which occurs at a density of one individual in 10 square kilometres of dense tropical forest and is believed to produce seeds irregularly at unstated seasons?' In any case, 'the collection of seeds is only the start of the process. Experience from innumerable expeditions suggests that a large proportion of the species collected will prove unamenable to cultivation, and be lost within a few years.'

To save what is left of the earth's natural habitats there are three main avenues of approach. One is to make better use of the areas already under crops, hopefully with new-created, heavier-yielding plants which will do a better job than existing kinds in the same area. Alternatives to wild plants used commercially, notably trees, should be found, as is being done up to a point by the use of very quick-growing trees for paper and fuelwood, or trees which resprout when cut.

Second is the halting of present degradation of land, and the rehabilitation of damaged land, so that those who live in such areas can continue to do so. The better use of fertile ground everywhere calls into question the use of grazing animals. In principle there is a move towards allowing wild animals to remain in natural country which should be culled for food to balance the capacities of the vegetation, as reindeer herders do now in sub-Arctic country. This is preferable to changing entirely the nature of the country for domesticated animals, often with harmful effects on its quality, apart from the problems posed by diseases and insects to which the cattle may be prone. Antelopes in Africa, for instance, have a greater potential in growth, reproduction and meat production than cattle; and they do not rely on waterholes since they do not need a daily drink, partly because they browse at night when leaves are moist and in any case feed on leaves like those of acacias with a high water content. The great appeal of this, which is still not much more than a theory, is that it leaves land and its vegetation in a relatively natural state which can be actively enjoyed by man – in short, tourism could bring added wealth. Against the idea one must place the distastefulness of the culling, which is likely to be less humane than slaughtering domestic cattle, as is certainly the case at present with deer shot for venison in Britain. However, on some South African farms deer are kept under almost domesticated conditions, and methods of slaughter are similar to those used with cattle.

Third, and the top priority, is the conservation of as much of the natural vegetation that remains as possible, which only the firmest action by governments world-wide can achieve.

In the past national parks were mostly scenic, catering for man's recreation, for him to see and often be amazed at earth forms of all kinds, with their vegetation, and to find some measure of mental relaxation away from the pressures of life, and the hubbub and claustrophobic surroundings of most cities. Nature reserves were almost entirely set up for animals, either specially rare ones or more often natural assemblies of different ones as we have in African parks.

Extraordinarily enough, it was only in the 1980s that conservationists as a whole realized that you cannot protect animals without making sure that their environment remains unaltered; and the areas involved consist – surprise – basically of plants. The most complex of these assemblages of life are the rain forests, which contain two-thirds of the world's flowering plants. They have more variety of trees than any other forest; a hectare of Borneo forest, for instance, has been found to contain 400 kinds of tree, some of which may only appear once or twice. The trees are accompanied by epiphytes of many kinds, lianas, and a great range of lowly plants on the forest floor; a vast array of bacteria, fungi, mosses and algae accompany them; and these plants create the homes of incredible numbers of insects, worms, crustaceans, lizards, frogs, warm-blooded animals large and small, and birds. When the jungle is destroyed, all these tens of thousands of associated creatures disappear for ever. The rain forest is the most remarkable web of life on earth; it has taken millennia to develop and once it has gone it can never be recreated.

Forests of other kinds and the other vegetation complexes on the earth all have their own webs of life, though none so complex. The more intricate they are, the less easily can they recreate themselves.

The World Wildlife Fund was in 1984–5 making the first steps towards creating a new array of reserves to supplement those already in existence. Though a few may be designed to protect individual species, choice must be of areas with the greatest diversity of plants and hence animals, and of habitats as representative as possible of all the vegetation complexes that now exist on earth.

A major part of the problem facing planners of reserves is the need to educate and assist the often illiterate users of the land, whose immediate needs seem to them far more pressing than the decrees of governments, let alone conservationist scientists often from foreign countries. Managing reserves will often have to be done with human needs as a major goal. Getting the local people involved will be absolutely vital, as has been done here and there in India already. A whole new concept of conservation must arise and of managers to implement this.

Out of our human needs will probably come the most compelling arguments for preventing the continuing blind destruction of the plant world and its natural habitats. These will consist partly of the desire to preserve the animal life that occupies these areas, partly to make available the natural solace which the green world and unspoilt scenery provide. But mainly it will be to conserve representative examples of the most important habitat types that remain to us with their webs of life as intact as possible. They will be areas including the largest diversity of plants, especially those with relevance to man's future needs in food, medicine and other utilitarian purposes, from which we can extract seeds to cultivate and, with proper management, all manner of plant products. In the long term there seems no possible argument that without such reserves – most of them needing to be vast – the quality of human life will be seriously impaired. The time for action is short.

Of course many wild plants will be preserved in cultivation, in gardens, parks and especially botanic gardens. Most of the 600 important botanic gardens in the world are now involved with conservation, not just in growing endangered plants or preserving their seeds in seed banks, but in rescue operations, carrying out propagation of threatened species where the greed of collectors is apparent, and especially to educate. It is remarkable just how many exotic plants can be gathered together in a good park or garden. I was impressed not long ago by the enormous variety of plants from many different parts of the world brought together, however artificially, in the richer residential areas of Los Angeles, a city more often maligned for its smog and concrete than praised for the beauty of its gardens and the attractive planting of its freeways. A small plus, perhaps, in comparison with Brazil's forests reduced to ashes or Bougainville's open-cast mine; but it may be all we shall have.

Whatever is done, we are faced with the fact that man, unlike the plants, has evolved very rapidly indeed. Not for him the aeons of adjustment to changing circumstances, but an ever-accelerating development that seems unable to consider consequences; capable of fantastic technological achievement but apparently unable to look beyond the current year or even the day, to control his reproduction, nor his more violent emotions.

It is not, of course, beyond possibility that man will destroy himself by nuclear explosion and its aftermath, or by some more subtle biological weapon – perhaps deliberate, perhaps accidental. He seems equally likely to choke and starve himself through pollution and over-population, or perish in epidemics, revolution or war triggered off by these factors.

What are the possibilities for plants as we look at various alternative futures? If mankind settles its differences and agrees to control the environment and its exploitation fairly soon, and fairly effectively, the world of plants should go on more or less as we see it today. Species have no chance of altering in geographical distribution except as they are carried about the globe by man for his own purposes or by accident. Once in a while a plant will escape from its man-controlled habitat and spread itself around, but this will not happen to many, while already constricted natural habitats are likely to go on shrinking. If we succeed in reforesting large tracts of country, lesser plants will recolonize the forests and set up new communities of some sort, though the tropical ones are likely to be pale shadows of what exists now.

If mankind pursues its present policies and trends the outlook for wild plants seems bleak indeed. Cultivated plants will preponderate, whether as crops or for ornament. There will, one hopes, be tracts of wild plants preserved, especially if they can be combined with wild grazing animals or provide us with some special commodity, but there is a dreadful irreversibility about the present scale of their destruction. As a Russian diplomat called Golunsky said during the drawing up of the United Nations Charter, 'the human race is the equivalent of a highly noxious bacillus battening on, and thus destroying "nature". This bacillus, by some inscrutable decree of providence, has now got the upper hand and is no longer, as in the past, existing with "nature" in a state of symbiosis.' There is indeed every indication of the probability of a runaway crisis developing due to reduced crop yields, changing climate and the unstoppable increase of mankind.

If the 'human bacillus' continues to batten on nature, it will be the weeds that will tend to represent the wild plant, for they will find the niches on the borders of cultivation, on river banks and roadsides, in unremarked pockets of soil in cities, and adapt to them, however precarious such niches may be. It will be a plant world consisting largely of annual weeds capable of rapid and widepsread reproduction and of weed trees like the sycamore and acacia which grow so quickly.

It may be a world dominated by micro-organisms, some of which have begun to get the better of our technology: the bacteria, for instance, which feed on iron and steel and create corrosion, or attack concrete and stone with sulphuric acid as a by-product; the fungus that lives on hydrocarbon fuels, notably aviation fuel, reducing their efficiency, blocking filters and pipes, and causing short-circuits in electrical connections; or another fungus that lives on the linseed oil in the anti-corrosive red lead paint which is used within box girders, allowing damp and rust to penetrate. On the credit side there are micro-organisms which have been harnessed to useful ends, such as bacteria which will clean out residual oil in tankers, thus avoiding the dumping of oil in the sea, and those which destroy toxic chemicals like 2,4,5-T released into the environment; there are also fungi capable of breaking down chemicals like dioxin and DDT. Then there are useful fungi and yeasts like those which have been 'trained' to ferment ground-up rubber tyres, reducing the particles to about a tenth of their original size, and extracting over 50 per cent of the oil content within a week. The reduced particles can be used as soil conditioners and also to increase soil fertility. There are projects for using bacteria to feed on coal seams, breaking them down into coal dust and gas which can easily be brought to the surface, and for improving the secondary recovery of oil from exhausted oil fields.

Apart from such developments of weeds and lesser organisms, one cannot see the world of plants evolving much further on its own. Man will select and develop plants useful to him or others that he finds decorative; they will be transformed (apart from crops we have only to look at the truly amazing changes in garden flowers) but whether one ought to call this evolution is an open question.

If mankind ends up destroying itself as a result of a runaway crisis caused by reduced crop yields, changing climate, unstoppable reproduction and resulting epidemics, the probability is that most natural vegetable life will survive. Plants are essentially resilient. We have seen how they manage to exist in deserts, on mountains, in the harshest Arctic conditions. They recolonize volcanoes and new volcanic islands, they become adjusted to toxic mineral conditions. Morning glories germinated ten days after the Hiroshima atom bomb; the atoll of Bikini was soon covered once more with vegetation. The potential revenge of the vegetable kingdom would cover and disintegrate our cities as it has in the past choked the once-splendid towns and temples of Central America and Asia, forcing apart our fabrications with the inexorable penetration and swelling of their roots. After a cataclysm to man, most cultivated plants would undoubtedly perish.

But there is another scenario, for the other most likely end to the human race is world-wide nuclear war. Scientists evaluating the potential results of even a conservative 5,000 megaton exchange suggest that this would first result in an appalling firestorm, which would produce so much dust, smoke, soot and ash that the sun would literally be blotted out for months, probably over much of the world. At the same time as this nuclear darkness, temperatures would fall – the most pessimistic estimate being to −47°C.

In such circumstances no plant could continue to function; plankton would disappear from the seas; perhaps a few lichens could survive and those aeons-old micro-organisms that have reappeared so often in these pages.

If nuclear war is confined to the superpowers, the darkness effect will be mainly limited to the northern hemisphere, though drastic cooling will spread southwards across the equator, and tropical vegetation would probably be destroyed.

It is dismal indeed to have to end on such a note, and for the sake of the human race – one's *own* race – let us hope that nothing like this will come to pass. But in some ways, is not the prospect of a world with uncontrolled population expansion, ever-increasing desertification and consequent obliteration of natural vegetation even more horrific?

One can put it starkly but without exaggeration that several major countries can at this time be regarded as waging the equivalent of nuclear warfare on their own territories: the sight of devastated areas in the tropics is comparable to that of photographs of Hiroshima after the first atom bomb. And the after-effects are as awful in terms of human suffering. This has been called the Third World War – and we are losing it, destroying our green allies in the process.

Sense, money and concerted action are the alternatives to be deployed. The estimated cost of rehabilitating degraded land becoming desertified is $2·5 billion a year; only a quarter of this is being spent. Compare this sum with the astronomical amounts being expended on peaceful space research, let alone on 'star wars'. Is it not ironic that we know more about our planetary system than we do about the rain forests of Bolivia and Colombia? The planets will still be there in twenty years' time. Without an enormous, concerted swing in the thinking of governments – the Third World needing to take the action and the northern powers to back them up materially – the rain forests will not, and the physical state of the world as we know it will alter irrevocably. As President Houphouet-Boigny of the Ivory Coast has said, 'Man has gone to the moon but he does not know how to make a flame tree or a birdsong.'

Somehow, we *must* learn to curtail our activities in such a way as to strike a live-and-let-live bargain with the world of plants – the planet sharers with which, more cheerfully, I began this book, whose substance feeds us, which provide us with so many needs and luxuries, and whose diverse beauties help to keep us sane. They are too fascinating, ingenious and wonderful to be wantonly discarded.

A car sticker seen in the United States asked 'Have you thanked a green plant today?' If, effectively, we all thanked the green things of earth more actively, the future might well look brighter. In its modern way it is really the same sentiment as John Gerard's in the Introduction to his famous *Herball* of 1597: 'Who would therefore looke dangerously up at Planets, that might safely look downe at Plants?'

Select Bibliography

Many books and periodicals have been consulted in the preparation of this volume, some of them technical and specialized. The books listed below are those considered most likely to amplify specific areas for the general reader. Periodicals and learned journals have not been listed.

Andrews, H. N., Jr, *Studies in Palaeobotany*, Wiley (New York & London), 1961

Audus, L. J., *Plant Growth Substances*, Leonard Hill, 1959

Barton, Lela V., *Seed Preservation and Longevity*, Leonard Hill, 1961

Bell, P. and Woodcock, C., *The Diversity of Green Plants*, Arnold, 1968

Briggs, D. and Walters, S. M., *Plant Variation and Evolution*, World University Library, 1969

Carefoot, G. L. and Sprott, E. R., *Famine on the Wind – Plant Diseases and Human History*, Angus & Robertson, 1969

Caufield, Catherine, *In the Rainforest*, Heinemann, 1985

Chapman, V. J., *Seaweeds and their Uses*, Methuen, 1970

Cailborne, Robert, *Climate, Man and History*, Angus & Robertson, 1973

Cobley, L. S., *The Botany of Tropical Crops*, Longmans, 1956

Corner, E. J. S., *The Life of Plants*, Weidenfeld & Nicolson, 1964

Corner, E. J. S., *The Natural History of Palms*, Weidenfeld & Nicolson, 1966

Corner, E. J. S., *Wayside Trees of Malaya*, Government Printing Office, Singapore, 1951

Croizat, Leon, *Space, Time, Form: The Biological Synthesis* (Caracas), 1962

Davis, P. H., Harper, P. C. and Hedge, I. C. (editors), *Plant Life of South-West Asia*, Botanical Society of Edinburgh, 1971

De Beer, Gavin, *A Handbook on Evolution*, British Museum, Natural History, 1970
Deverall, Brian J., *Defence Mechanisms of Plants*, Cambridge University Press, 1977
Ehrlich, Paul and Anne, *Extinction*, Gollancz, 1982
Evenari, M., Shanan, L. and Tadmor, N., *The Negev – the Challenge of a Desert*, Harvard University Press, 1971
Fogg, G. E., *The Growth of Plants*, Pelican, 1963
Good, Ronald, *Features of Evolution in the Flowering Plants*, Longmans, 1956
Good, Ronald, *The Geography of the Flowering Plants*, Longmans, 1953
Hale, Mason E., Jr, *The Biology of Lichens*, Arnold, 1967
Harrison, Gordon, *Earthkeeping*, Hamish Hamilton, 1972
Hederg, Olav, *Features of Afroalpine Plant Ecology*, Almqvist & Wiksells (Uppsala), 1964
Hutchinson, J., *Evolution and Phylogeny of Flowering Plants*, Academic Press, 1969
Huxley, Anthony, *Green Inheritance*, Collins/HaRvill, 1984
Jones, David A. and Wilkins, Dennis A., *Variations and Adaptation in Plant Species*, Heinemann, 1971
Kerner von Marilaun, A., translated by Oliver, F. W., *The Natural History of Plants*, Blackie, 1894
Koopowitz, Harold, and Kaye, Hilary, *Plant Extinction: A Global Crisis*, Stone Wall Press (Washington), 1983
Kuijt, Job, *The Biology of Parasitic Flowering Plants*, University of California Press, 1969
Kullenberg, B., *Studies in Ophrys Pollination*, Almqvist & Wiksells (Uppsala), 1961
Lawrence, W. C. J., *Plant Breeding*, Arnold, 1968
Menninger, Edwin A., *Fantastic Trees*, Viking (New York), 1967
Merrill, E. D., *Plant Life of the Pacific World*, Macmillan (USA), 1946
Moore, David M. (editor), *Green Planet*, Cambridge University Press, 1982
Myers, Norman (editor), *The Gaia Atlas of Planet Management*, Pan, 1985
National Academy of Sciences (Washington, DC), *Underexploited Tropical Plants with Promising Economic Value*, 1975
Oakley, K. P. and Muir-Wood, H. M., *The Succession of Life through Geological Time*, British Museum, Natural History, 1967
Ordish, George, *Biological Methods in Pest Control*, Constable, 1967
Percival, Mary S., *Floral Biology*, Pergamon, 1965
Polunin, Nicholas, *Introduction to Plant Geography*, Longmans, 1960
Proctor, M. and Yeo, P., *The Pollination of Flowers*, Collins New Naturalist, 1973

Pyke, Magnus, *Man and Food*, World University Library, 1970

Ramsbottom, J., *Mushrooms and Toadstools*, Collins New Naturalist, 1953

Ridley, H. N., *The Dispersal of Plants throughout the World*, Reeve, 1930

Sauer, Carl O., *Agricultural Origins and Dispersals*, MIT Press (USA), 1969

Schery, Robert W., *Plants for Man*, Allen & Unwin, 1954

Scott, George D., *Plant Symbiosis*, Arnold, 1969

Seward, A. C., *Plant Life through the Ages*, Cambridge University Press, 1931

Simmonds, N. W. (editor), *Evolution of Crop Plants*, Longmans, 1976

Smith, Anthony, *Mato Grosso*, Michael Joseph, 1971

Sporne, K. R., *The Morphology of Gymnosperms – the Structure and Evolution of Primitive Seed-Plants*, Hutchinson, 1965

Stanley, Wendell M. and Valens, Evans G., *Viruses and the Nature of Life*, Methuen, 1962

Street, H. E. and Öpik, Helgi, *The Physiology of Flowering Plants*, Arnold, 1970

Struever, Stuart (editor), *Prehistoric Agriculture*, Natural History Press (New York), 1971

Sweeney, Beatrice M., *Rhythmic Phenomena in Plants*, Academic Press, 1969

Takhtajan, A., Translated by Jeffrey, C., *Flowering Plants – Origin and Dispersal*, Oliver & Boyd, 1969

Tivy, Joy, *Biogeography*, Oliver & Boyd, 1971

Ucko, P. J. and Dimbleby, G. W. (editors), *The Domestication and Exploitation of Plants and Animals*, Duckworth, 1969

Valentine, D. H., *Taxonomy, Phytogeography and Evolution*, Academic Press, 1972

Van der Pijl, L. and Dodson, S. H., *Orchid Flowers: their Pollination and Evolution*, University of Miami Press (USA), 1967

Van der Pijl, L., *Principles of Dispersal in Higher Plants*, Springer-Verlag (Berlin), 1969

Vavilov, N. I., *Cultivated Plants*, Chronica Botanica (USA), 1951

Walter, H., *Ecology of Tropical and Subtropical Vegetation*, Oliver & Boyd, 1973

Wickler, Wolfgang, *Mimicry in Plants and Animals*, World University Library, 1968

Williams, W., *Genetical Principles and Plant Breeding*, Blackwell, 1964

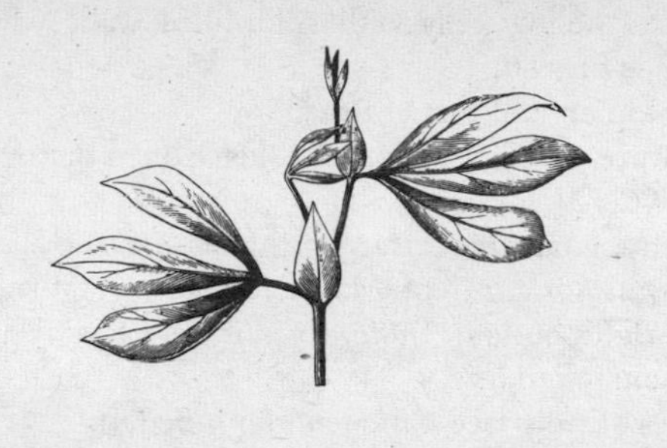

Index to Illustrations

All illustrations are from *The Natural History of Plants*, by Kerner & Oliver (1894), unless otherwise specified. Numbers refer to pages in the text.

Index